APPLICATIONS OF ARTIFICIAL INTELLIGENCE IN MEDICAL IMAGING

Artificial Intelligence Applications in Healthcare and Medicine

APPLICATIONS OF ARTIFICIAL INTELLIGENCE IN MEDICAL IMAGING

Edited by

ABDULHAMIT SUBASI
Department of Computer Science, College of Engineering, Effat University, Jeddah, Saudi Arabia;
Institute of Biomedicine, Faculty of Medicine, University of Turku, Turku, Finland

ELSEVIER
ACADEMIC PRESS
An imprint of Elsevier

Academic Press is an imprint of Elsevier
125 London Wall, London EC2Y 5AS, United Kingdom
525 B Street, Suite 1650, San Diego, CA 92101, United States
50 Hampshire Street, 5th Floor, Cambridge, MA 02139, United States
The Boulevard, Langford Lane, Kidlington, Oxford OX5 1GB, United Kingdom

ISBN: 978-0-443-18450-5

For Information on all Academic Press publications
visit our website at https://www.elsevier.com/books-and-journals

Publisher: Stacy Masucci
Acquisitions Editor: Rafael E. Teixeira
Editorial Project Manager: Pat Gonzalez
Production Project Manager: Fahmida Sultana
Cover Designer: Harris Greg

Typeset by MPS Limited, Chennai, India

Dedication

A huge thanks to my parents for every time expecting me to do my best, and telling me I could accomplish anything, no matter what it was.

To my wife, Rahime, for her patience and support.

To my wonderful children, Seyma Nur, Tuba Nur, and Muhammed Enes. You are always in my heart and the joys in my life.

To those who read this book, and appreciate the work that goes into them. If you have any feedback, please let me know.

Abdulhamit Subasi

Contents

4. Breast cancer detection from mammograms using artificial intelligence 109

ABDULHAMIT SUBASI, AAYUSH DINESH KANDPAL, KOLLA ANANT RAJ AND ULAS BAGCI

5. Breast tumor detection in ultrasound images using artificial intelligence 137

OMKAR MODI AND ABDULHAMIT SUBASI

6. Artificial intelligence-based skin cancer diagnosis 183

ABDULHAMIT SUBASI AND SAQIB AHMED QURESHI

7. Brain stroke detection from computed tomography images using deep learning algorithms 207

AYKUT DIKER, ABDULLAH ELEN AND ABDULHAMIT SUBASI

8. A deep learning approach for COVID-19 detection from computed tomography scans 223

ASHUTOSH VARSHNEY AND ABDULHAMIT SUBASI

List of contributors

Ulas Bagci
Northwestern University, Chicago, IL, United States

Abhijit Bhattacharyya
Department of Electronics and Communication Engineering, National Institute of Technology Hamirpur, Hamirpur, Himachal Pradesh, India

Aykut Diker
Department of Software Engineering, Faculty of Engineering and Natural Sciences, Bandirma Onyedi Eylul University, Bandirma, Balikesir, Turkey

Abdullah Elen
Department of Software Engineering, Faculty of Engineering and Natural Sciences, Bandirma Onyedi Eylul University, Bandirma, Balikesir, Turkey

Emrah Hancer
Department of Software Engineering, Mehmet Akif Ersoy University, Burdur, Turkey

Aayush Dinesh Kandpal
Department of Metallurgical and Materials Engineering, National Institute of Technology Rourkela, Rourkela, Odisha, India

Manav Nitin Kapadnis
Department of Electrical Engineering, Indian Institute of Technology Kharagpur, Kharagpur, West Bengal, India

Omkar Modi
Indian Institute of Technology, Kharagpur, West Bengal, India

Abhranta Panigrahi
Department of Biotechnology and Medical Engineering, National Institute of Technology Rourkela, Rourkela, Odisha, India

Sohan Patnaik
Department of Mechanical Engineering, Indian Institute of Technology, Kharagpur, West Bengal, India

Saqib Ahmed Qureshi
Indian Institute of Technology Kharagpur, Kharagpur, West Bengal, India

Kolla Anant Raj
Department of Electrical Engineering, Indian Institute of Technology Kharagpur, Kharagpur, West Bengal, India

Aayush Rajput
Indian Institute of Technology, Kharagpur, West Bengal, India

Siddhesh Shelke
Indian Institute of Technology, Indore, Madhya Pradesh, India

Abdulhamit Subasi
Institute of Biomedicine, Faculty of Medicine, University of Turku, Turku, Finland; Department of Computer Science, College of Engineering, Effat University, Jeddah, Saudi Arabia

Ashutosh Varshney
Department of Computer Science and Engineering, Indian Institute of Technology, Kharagpur, West Bengal, India

Series preface

Artificial intelligence (AI) is a concept, which allows for the optimization of performance criteria utilizing a set of data and some experience. The learning process is essentially the execution of the model parameter optimization with a training dataset or past experience. Models can be predictive, for predicting the future, descriptive, for extracting knowledge from input data, or both. Two fundamental activities are accomplished in machine learning: (1) processing massive amounts of data and optimizing the model and (2) testing the model and efficiently displaying the solution.

The process of applying AI methods to a large dataset is called data mining. The logic behind the data mining is that a large volume of raw data is processed, and efficient prediction model is constructed, with high predictive accuracy. AI applications are present in different areas: in finance, for credit scoring, fraud detection, or stock market prediction; in manufacturing, for optimization, control, and troubleshooting; in medicine, for efficient medical diagnosis; in telecommunication, for network and quality of service optimization. Furthermore, AI enables algorithms to learn and adapt to changes in a variety of situations. Artificial neural networks, fuzzy logic, support vector machines, decision tree algorithms, and deep learning algorithms are all examples of AI techniques.

AI tools have been employed in different areas for several years. New AI methods such as deep learning have assisted uncover information which entirely altered the approach in different areas. AI has reached certain maturity as an academic subject, and there are many useful books related to this subject. Since AI is an interdisciplinary subject, it should be implemented in different ways depending on the application field. Nowadays, there is a great interest in AI applications in several disciplines. This book series presents how AI and machine learning methods can be used in the analysis of different data. Different AI applications in different fields, including biomedical engineering, electrical engineering, computer science, information technology, medical science, healthcare, finance, and economy are the applications of engineering techniques to problems in these fields.

This book series will consist of numerous volumes, each of which will cover an application of AI techniques in a different field. This series will benefit a wide range of readers, including academicians, professionals, graduate students, and researchers from a variety of fields who are exploring AI applications. The book series will provide an in-depth account of recent research in this emerging topic, and the principles discussed here will spark additional research in this multidisciplinary field.

The target audience is widespread since this series will include several areas of applications of AI, machine learning, and deep learning which are given below. Hence, the audience can be computer scientists, biomedical engineers and healthcare scientists, financial engineers, economists, researchers and consultants in electrical, computer engineering and science, finance, economy, and security. Nowadays AI is one of the hot topics, which is used in several data analysis in the world. Hence, there are many applications areas and audience of this book series. This book series will include many new AI trends in the application of several fields. Among them the most important ones are healthcare, cyber security, and finance.

Abdulhamit Subasi

Preface

Artificial intelligence (AI) plays an important role in the field of medical image analysis, including computer-aided diagnosis, image-guided therapy, image registration, image segmentation, image annotation, image fusion, and retrieval of image databases. With advances in medical imaging, new imaging methods and techniques are needed in the field of medical imaging, such as cone-beam/multislice CT, MRI, positron emission tomography (PET)/CT, 3D ultrasound imaging, diffuse optical tomography, and electrical impedance tomography, as well as new AI algorithms/applications. To provide adequate results, single-sample evidence given by the patient's imaging data is often not appropriate. It is usually difficult to derive analytical solutions or simple equations to describe objects such as lesions and anatomy in medical images, due to wide variations and complexity. Tasks in medical image analysis therefore require learning from examples for correct image recognition (IR) and prior knowledge. This book offers advanced or up-to-date medical image analysis methods through the use of algorithms/techniques for AI, machine learning (ML), and IR. A picture or image is worth a thousand words, indicating that, for example, IR may play a critical role in medical imaging and diagnostics. Data/information can be learned through AI, IR, and ML in the form of an image, that is, a collection of pixels, as it is impossible to recruit experts for big data.

AI tools have been employed in different areas for several years. New AI methods such as deep learning have assisted to uncover information which entirely altered the approach in different areas. AI has reached certain maturity as an academic subject, and there are many useful books related to this subject. Since AI is an interdisciplinary subject, it must be implemented in different ways depending on the application field. Nowadays, there is a great interest in AI applications in several disciplines. This edited book presents how AI and ML methods can be used in the medical image analysis. Different AI applications in different fields, including biomedical engineering, electrical engineering, computer science, information technology, medical science, and healthcare, are the applications of AI to problems in these fields.

This book provides the description of various biomedical image analyses in several disease detection using AI and can therefore be used to incorporate knowledge obtained from different medical imaging devices such as CT, X-ray, PET, and ultrasound. In this way, a more integrated and, thus, more holistic research on biomedical image analysis may contribute significantly to the successful enhancement of a single patient's clinical knowledge.

This book includes several medical image analysis techniques using AI approaches including deep learning techniques. Deep learning algorithms such as convolutional neural networks and transfer learning techniques are widely used in medical imaging. Medical image analysis using AI is widely employed in the area of medical image classification, segmentation, and detection. The applications of AI in medical imaging are widely used as a decision support system for physicians. AI can be used in the diagnosis of different types of cancers including cervical cancer, ovarian cancer, breast cancer, prostate cancer, lung cancer, and liver cancer.

The author of this book has a lot of hands-on experience using Python and MATLAB® to solve real-world problems in the context of the ML ecosystem. *Applications of Artificial Intelligence in Medical Imaging* aims to provide the readers of various skill levels with the knowledge and experience necessary to develop useful AI solutions. Additionally, this book serves as a solution manual for creating sophisticated real-world systems. This provides a structured framework with guidelines, instructions, real-world examples, and code. Additionally, this book benefits from the crucial knowledge that its readers require to comprehend and resolve a variety of ML difficulties.

The book covers different subjects, involving cancer diagnosis, including lung cancer, prostate cancer, breast cancer, and skin cancer; COVID-19 detection; histopathological image classification; classification of diabetic retinopathy lesions using CT, MRI, X-ray, and ultrasound; and pathological medical imaging. This book consists of 14 chapters. Chapter 1 presents topics relevant to the numerous AI methodologies, such as supervised and unsupervised learning. Hence, the key AI algorithms are discussed briefly in this chapter. Relevant PYTHON programming codes and routines are provided in each section. Chapter 2 provides lung cancer detection from histopathological lung tissue images using AI. Chapter 3 provides MRI-based automated brain tumor detection by means of AI. Chapter 4 presents breast cancer detection from mammograms using AI. Chapter 5 includes AI-based breast tumor detection using ultrasound images. Chapter 6 includes AI-based skin cancer diagnosis. Chapter 7 presents brain stroke detection from CT images using deep learning algorithms. Chapter 8 provides a deep learning approach for COVID-19 detection from CT scans. Chapter 9 includes detection and classification of diabetic retinopathy lesions using deep learning. Chapter 10 presents automated detection of colon cancer using histopathological images. Chapter 11 includes brain hemorrhage detection in CT images utilizing deep learning. Chapter 12 presents AI-based retinal disease classification using OCT images. Chapter 13 presents diagnosis of breast cancer from histopathological images with deep learning architectures. Chapter 14 includes AI-based Alzheimer disease detection using MRI images.

Abdulhamit Subasi

Acknowledgments

First of all, I would like to thank my publisher Elsevier and its team of dedicated professionals who have made this book writing journey very simple and effortless and many who have worked in the background to make this book a success.

I would like to thank Rafael Teixeira, Linda Versteeg-Buschman, and Pat Gonzalez, who provided excellent support and did a lot of work for this book. Additionally, I would like to thank to Fahmida Sultana for being patient in getting everything necessary completed for this book.

Abdulhamit Subasi

CHAPTER

1

Introduction to artificial intelligence techniques for medical image analysis

Abdulhamit Subasi[1,2]

[1]Institute of Biomedicine, Faculty of Medicine, University of Turku, Turku, Finland [2]Department of Computer Science, College of Engineering, Effat University, Jeddah, Saudi Arabia

OUTLINE

1.1 Introduction

Artificial intelligence (AI) model is defined by a set of parameters that are optimized using training data or previous experience to generate a computer program. AI generates models using statistical analysis since the main purpose is to make inferences from a training sample. In some circumstances, the training algorithm's efficacy is just as important as its classification accuracy. AI techniques are employed as a decision assistance system in a variety of fields [1,2].

Learning is a multidisciplinary phenomenon that includes parts of statistics, mathematics,

DOI: https://doi.org/10.1016/B978-0-443-18450-5.00010-4

computer science, physics, economics, and biomedicine. Surprisingly, not all human tasks are linked to intelligence, therefore there are certain situations when a computer can do better. There are some intelligent activities which humans are incapable of performing, and which robots can perform better than humans. Classical machine learning (ML) techniques in complicated systems cannot provide the essential intelligent response since important activities and decision-making make it vital to comprehend the model response and components for effective decision-making. Every behavior, activity, or decision has a systemic understanding. In contrast, one activity might be the outcome of another event or set of events from a systematic standpoint. Those are convoluted and difficult to grasp connections. As system models and robots are expected to perform intelligently even in nonpredictive scenarios, learning ideas and models must be seen through the lens of new expectations. These expectations necessitate ongoing learning from many sources of knowledge. The analysis and customization of data for these techniques, as well as their effective use, are crucial [2,3].

Medical imaging is concerned with many sorts of images utilized in medical applications. Medical images include X-ray images, magnetic resonance imaging (MRI), computed tomography (CT) images, ultrasound (US) images, and others that radiologists utilize in the diagnostic process to detect and analyze anomalies in the human body [4]. The image data may be deteriorated owing to a variety of circumstances including natural events. Various devices will be utilized to capture the images. These equipment are not always flawless, and they might degrade the quality of images captured. Some issues arise throughout the acquisition process that might help reduce image quality. Medical imaging technologies are more precise and give high-quality medical images, but they might also be contaminated by environmental noises [5,6]. Medical images are impacted by numerous noise signals, such as Gaussian noise [7], speckle, and so on, and the image quality might be poor as a result. Therefore radiologists may make incorrect interpretations. As a result, before studying these images, noise signals must be suppressed. Denoising, which is a preprocessing activity in the discipline of image processing, is performed not only to reduce noise signals but also to retain image relevant information such as edges, texture details, fine structures, and so on [8]. Numerous approaches for image denoising have been developed, but none of them produce efficient results for various sorts of noise issues. As a result, a framework must be designed to eliminate noise signals while preserving image data [9,10].

Medical image processing is a powerful tool for disease prognosis and diagnosis. Nevertheless, as the volume of digital data increases, so does the requirement for accuracy and efficacy in medical image processing procedures. Researchers can construct a computer-aided diagnostic (CAD) system for disease categorization using medical imaging and current advances in AI. Physicians can now view hidden features in medical photos thanks to the advancement of the most modern imaging technology. As a result, computer assistance is not only essential but also indispensable in the physician's diagnosis procedure. Automated approaches can be used to help clinicians in the early detection of diseases, reducing the need on invasive procedures for diagnosis [11,12]. However, there are number of drawbacks for using MRI and CT. CT scans are ideal for bone fractures, chest disorders, and the identification of abdominal malignancies. On the other hand, MRI is appropriate for assessing brain tumors and soft tissues. A CT scan takes around 5 minutes, whereas an MRI can take up to 30 minutes. MRI does not utilize ionized radiation, but there is a potential risk of radiation exposure with CT. The MRI frequently creates claustrophobia, but CT scan does not. Furthermore, CT scans are less expensive than MRIs [13].

The use of recent developments in healthcare technologies has considerably enhanced human health. Various medical imaging technologies, such as computer-assisted image processing, help in the early detection of disorder. The rising volume of medical imaging data, such as CT, MRI, and US, places a significant diagnostic strain on radiologists. In this context, automated diagnostic systems will improve diagnostic accuracy while also lowering costs and increasing efficiency. Recently, digital imaging data has grown exponentially, exceeding the number of radiologists' availability. This increase in workload has a direct influence on the performance of radiologists. As a result, human analysis is out of sync with the volume of data to be processed. As a result, it is needed to create a computer-assisted image classification and analysis structure to help radiologists deal with such massive amounts of data. Furthermore, because the gray-level intensities in nature overlap, it is impossible to create the CAD utilizing texture information [14–19]. The researchers' main challenge is creating an effective, accurate, and robust CAD framework for the classification of tissues since representations of tissues are overlapping in nature. Furthermore, the accuracy is insufficient for commercial usage of these diagnostic devices. As a result, there is an urgent need to develop a diagnostic framework capable of properly and rapidly characterizing tissues such as tumors, cysts, stones, and normal tissues [13]. The building blocks of CAD framework are shown in Fig. 1.1.

1.2 Artificial intelligence for image classification

The classification of distinct objects in images is referred to as image classification. The various items or areas of image must be recognized and classified. The accuracy of the

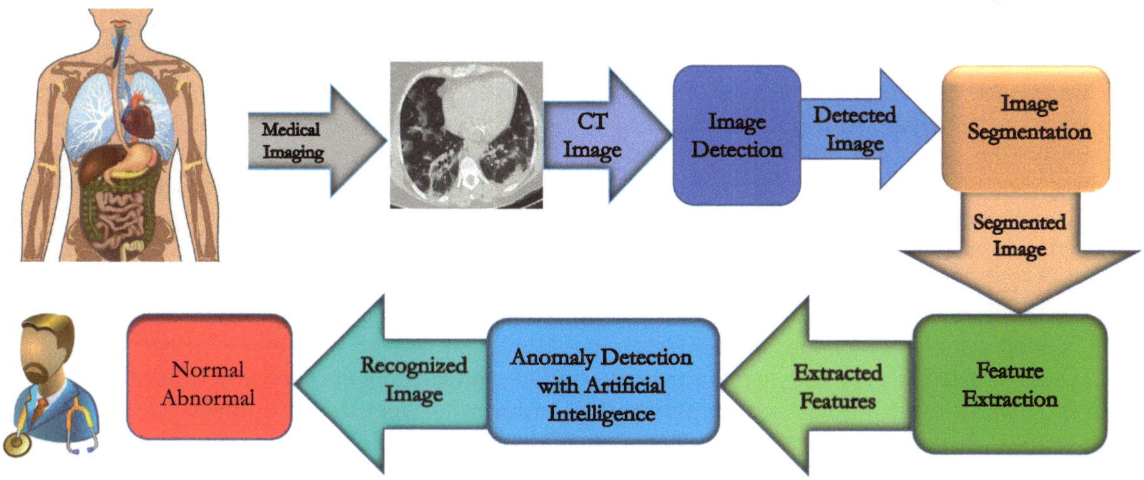

FIGURE 1.1 The building blocks of CAD framework. *CAD*, Computer-aided diagnostic.

outcome is determined by the classification algorithm. It is frequently based on a single image or a collection of images. When image sets are employed, the set will comprise many images of the same object from various angles and under various situations. When compared to classifying with single images, it will be more successful since the algorithm can adapt variable situations such as variations in backdrop, lighting, or appearances. It is also insensitive to image rotation and other transformations. The algorithm is fed image pixels, which has numeric characteristic as input. The result is a single value or a series of values representing the class. The algorithm is a mapping function, which converts pixel data into the proper class. The classification process might be either unsupervised or supervised. A set of training data containing class information is provided in supervised classification, and the number of classes is known. It is similar to learning from a trainer. On the other hand, the number of classes is unknown in unsupervised classification, and no training data are supplied. It is necessary to learn the link (or mapping) between the data to be categorized and the distinct classifications. It is similar to learning without a trainer. Unsupervised and supervised approaches can be coupled to generate semisupervised methods if some information about the mapping of data to classes is known. The most essential factors connected with input data that are used to classify the data are known as features. Defining specific qualities of an item called features is critical in classification. For classification, features extracted from visual objects are employed [20].

ML is a branch of AI, which allows computer systems to learn from input/training data. There are three types of learning: unsupervised, supervised, and semisupervised. Unsupervised learning occurs when learning occurs with unknown input data. In supervised learning, the algorithm is trained with a set of training data for a specific objective. Semisupervised learning falls somewhere in the middle of these two groups. The training inputs and intended outputs are provided in supervised learning, and the algorithm learns the relationship between input and output. The mapping between input and output is already established. The inputs are provided in unsupervised learning, and the algorithm learns to uncover patterns or characteristics to create the output. The method does not need to know the number of outputs ahead of time. The training inputs and intended outputs are only partially provided in semisupervised learning, and the algorithm learns to uncover the missing patterns and relations [20].

Image classification is one of the most difficult problems for an algorithm to learn and complete. When numerous images are provided, the human brain learns and classifies both existing and new images with near-perfect accuracy. AI algorithms are created to precisely imitate the activity of the human brain. When the images are taken in various circumstances, such as changing the lighting, rotating, or translating the items in the image, having hidden or incomplete objects, the task becomes more complex. Such circumstances result in hundreds of distinct images having the same item, further complicating the classification/recognition task. Image categorization can be done pixel-by-pixel or object-by-object. The properties of each pixel are retrieved in pixel-based classification to designate it as belonging to a certain class. In order to extract areas or objects in an image and assess their properties, segmentation is used in object-based classification. In order to do classification, features or properties must be retrieved. The algorithm's efficiency is determined by the amount of features employed in the process. This raises the issue of the "curse of dimensionality." Dimension reduction is required that equates to feature reduction and thus lower the computational complexity. More processing and data storage are required as the number of characteristics increases. This raises the algorithm's time complexity. More efficient algorithms identify things with the fewest characteristics and in the least amount of time [20].

Since the AI algorithms have the potential to learn, they are being employed for image classification. A collection of training data is provided, and the network must be fed correlations between training inputs and desired outputs. The network is trained using known data to detect and classify new input. AI algorithms were created to facilitate learning, with the logic being learned by the algorithm during training. Inputs include patterns or data, outputs are defined, and the algorithm learns to determine the relation between inputs and outputs. When the issue is difficult, such as image classification, additional hidden layers are needed, causing the neural network to become "deep." Hundreds of hidden layers enhance classification accuracy, and the learning becomes "deep learning" [20].

1.3 Unsupervised learning (clustering)

Clustering is one of the most frequently utilized approaches for analyzing experimental data. People strive to achieve an initial understanding of their results in various areas, from social sciences to computer science to biology, by creating expressive groups among the data points. Companies, for instance, cluster clients based on their customer profiles for focused marketing, astronomers cluster stars based on their unique proximity, and bioinformation cluster genes based on similarities in their expression. Clustering is intuitively defined as the act of grouping a collection of objects so that similar objects are grouped in the same class and dissimilar ones are divided into distinct classes. This term is certainly vague and maybe ambiguous. Furthermore, it is difficult to find a more precise description. There are various causes for this problem. One major issue is that the two aims expressed in the abovementioned sentence frequently contradict one another. Closeness (or similarity) is not a transitive connection, but cluster involvement is an equivalence relation, and specifically a transitive one. This input is clustered by splitting it horizontally on both lines by a clustering method that emphasizes not separating close-by points [2,21].

Another basic issue with clustering is a lack of "ground truth," which is a typical issue with unsupervised learning. So far, the book has mostly dealt with supervised learning. The goal of supervised learning is straightforward: we would like to train a classifier to predict the labels of future samples as accurately as possible. A supervised learner can also quantify the accomplishment or possibility of hypotheses by calculating the empirical loss using labeled training data. On the other hand, clustering is an unsupervised learning issue in which no labels are predicted. Rather, we would like to find a realistic approach to arrange the data. As a result, there is no simple approach for assessing clustering performance. Furthermore, even with complete understanding of the underlying data distribution, it is unclear what the "right" clustering is for that data or how to evaluate a proposed clustering [2,21].

Clustering is a method that groups together similar things. We might utilize one of different kinds of inputs. The input to the algorithm in similarity-based clustering is a dissimilarity matrix or distance matrix D. Similarity-based clustering provides the benefit of easily incorporating domain-specific similarity or kernel functions. The advantage of feature-based clustering is that it may use "raw" data that is hypothetically noisy. Aside from the two input kinds, there exist two possible output types: hierarchical clustering, in which a nested partition tree is produced, and flat clustering, also known as partition clustering, in which the objects are divided into disjoint sets. Some methods state that D is a true distance matrix, whereas others do not. If we have a similarity matrix S, we may transform it to a dissimilarity matrix by using any monotonically decreasing function. The most frequent technique to describe item dissimilarity is through the dissimilarity of their properties. Some typical attribute dissimilarity functions include the hamming distance, city block distance, square (Euclidean) distance, and correlation coefficient [2,22].

Clustering is one of the simple techniques employed by humans to accommodate the massive quantity of information they get every day. Handling each piece of information as a separate object would be tough. As a result, humans appear to group things into clusters. Each cluster then characterizes the precise qualities of the entities that form it. As with supervised learning, it is assumed that all patterns are described in terms of features that constitute one-dimensional feature vectors. In a number of circumstances, a stage known as the clustering inclination should be present. It covers a few tests that determine if there is a clustering pattern in the provided data or not. For example, if the dataset is completely random, attempting to untangle clusters is futile. Different feature options and proximity measurements are available. Clustering criteria and clustering methods may provide wildly differing clustering results [2,23].

1.3.1 Image segmentation with clustering

Images are widely recognized as one of the most important approaches of delivering information. An example would be the usage of images for robotic navigation. Other uses, such as removing cancerous tissues from body scans, are an important aspect of medical diagnostics. One of the initial steps in image recognition is to segment them and discover distinct things inside them. This may be accomplished through the use of features such as frequency-domain transformations and histogram plots [2,24].

Image segmentation is a critical preprocessing step in computer vision and image recognition. Image segmentation, which is the breakdown of an image into a number of nonoverlapping relevant sections with the same qualities, is a critical method in digital image processing, and segmentation accuracy has a direct impact on the efficacy of subsequent activities [25]. Because image segmentation is critical in many image processing applications, various image segmentation algorithms were built during the last few decades. However, these methods are always being sought since image segmentation is a difficult issue, which necessitates a better solution for the successive image processing stages. Although the clustering approach was not designed specifically for image processing, it is utilized for image segmentation by the computer vision community. The k-means clustering method, for example, requires previous information of the number of clusters (k) to be categorized into. Every pixel in the picture is iteratively and repeatedly assigned to the cluster, the centroid of which is closest to the pixel. The centroid of each cluster is identified based on the pixels assigned to that cluster. Both the selection of pixel membership in the clusters and the computation of the centroids are based on distance calculations. Because it is straightforward to compute, the Euclidean distance is the mostly utilized. The utilization of Euclidean distance produces error in the final image segmentation [2,26]. A simple k-means clustering Python code for image segmentation is given below.

```
import numpy as np # linear algebra
import pandas as pd # data processing, CSV file I/O (e.g. pd.read_csv)
from IPython.display import Image
import matplotlib.pyplot as plt
from mpl_toolkits.mplot3d import Axes3D
from sklearn.cluster import KMeans
import cv2

img = cv2.imread(filepath)
img = cv2.cvtColor(img, cv2.COLOR_BGR2RGB)
r, g, b = cv2.split(img)
r = r.flatten()
g = g.flatten()
b = b.flatten()

K=3
attempts=10
ret,label,center=cv2.kmeans(vectorized,K,None,criteria,attempts,cv2.KMEANS_RANDOM_CENTERS)
label = label.flatten()

center = np.uint8(center)
res = center[label.flatten()]
result_image = res.reshape((img.shape))

plt.imshow(result_image)
plt.show()
```

1.4 Supervised learning

1.4.1 K-nearest neighbor

When given big training sets, the k-nearest neighbor (k-NN) approach is labor demanding and did not obtain popularity until more computing power became available. It has now become extensively utilized in pattern recognition. k-NN classifiers are based on analogy learning, which involves comparing a particular test instance to training instances, which are similar to that. A k-NN classifier explores the pattern space for the k training instances, which are closest to the unknown instance. These k training instances are k "nearest neighbors" of the unknown instance. A distance metric, such as Euclidean distance, is employed to define "closeness." We usually normalize the values of each variable to prevent attributes with high start ranges from outweighing attributes with smaller initial ranges. The unknown instance is allocated to the most frequent class among its k-NNs in k-NN classification. When $k = 1$, the unknown instance is allocated to the class of the training instance in pattern space that is closest to it. Moreover k-NN classifiers can be utilized for numeric prediction, which is when a given unknown instance is given a real-valued prediction.

In this scenario, the classifier provides the average of the real-valued labels related to the unknown instance's k-NNs. A test set is employed to assess the error rate of the classifier starting with k = 1. This method is employed as many times as needed by increasing k to achieve one additional neighbor. The k value, which generates the lowest error rate, is selected. Distance-based assessments are utilized by k-NN classifiers, which assign equal weight to each attribute. As a result, when given noisy or irrelevant qualities, they may suffer from low accuracy. However, the approach has been tweaked to include attribute weighting as well as the pruning of noisy data instances. The distance measure you choose can have a big impact. Other distance metrics, such as the Manhattan (city block) distance, may also be used. When classifying test instances, nearest-neighbor classifiers can be incredibly slow. The use of partial distance computations and modifying the stored instances are two further methods for reducing classification time. The distance is computed using a subset of the n characteristics in the partial distance technique. If the distance surpasses a certain threshold, the process will stop working on the current stored instance and go on to the next one. The editing approach gets rid of any training instances that are not useful. Because it minimizes the overall amount of instances saved, this strategy is also known as pruning [27]. A simple Python code for k-NN is given below.

1.4.1.1 Support vector machine

Support vector machine (SVM) is a classification technique for both linear and nonlinear data. SVM transforms the original training data into a higher dimension via a nonlinear mapping. It seeks the linear optimum separation hyperplane inside this new dimension. The SVM employs support vectors and margins to get hyperplane. Despite the fact that even the quickest SVMs have a long training period, they are remarkably precise due to their ability to predict complicated nonlinear decision limits. They have a lower risk of overfitting than other approaches. The discovered support vectors also serve as a concise representation of the learnt model. SVMs may be used for both classification and numerical prediction. Medical imaging, object recognition, handwritten digit recognition, and speaker identification are just a few of the domains where they have been used [27].

Separating lines can be drawn in an infinite number of ways. We want to find the "best" one, which will (hopefully) have the least amount of classification error on previously unseen dataset. How are we going to find the best line? We want to find the optimal hyperplane when we generalize to n dimensions. Regardless of the amount of input features, we'll refer to the decision boundary we're looking for as a "hyperplane." To put it another way, how can we determine the optimum hyperplane? An SVM solves this problem by investigating for the maximum marginal hyperplane. Nevertheless, we anticipate the hyperplane with the bigger margin

```
#Import k-NN Model
from sklearn.neighbors import KNeighborsClassifier
#Create the Model
clf = KNeighborsClassifier(n_neighbors = 5)
#Train the model with Training Dataset
clf.fit(Xtrain,ytrain)
#Test the model with Testset
ypred = clf.predict(Xtest)
```

to be more accurate than the hyperplane with the lower margin in classifying future data instances. The related margin creates the most significant distinction between categories [27]. A simple Python code for SVM is given below.

```
#Import SVM Model
from sklearn import svm
C = 10.0 # SVM regularization parameter
#Create the Model
clf =svm.SVC(kernel = 'linear', C = C)
#Train the model with Training Dataset
clf.fit(Xtrain,ytrain)
#Test the model with Testset
ypred = clf.predict(Xtest)
```

1.4.2 Decision tree

Decision tree is a nonparametric technique and generates binary trees from data with both discrete and continuous attributes. The impurity reduction criteria are used to assess split characteristics, with impurity specified as the so-called Gini (diversity) index. It is also feasible to utilize entropy or any other measure of impurity instead of the Gini index. Breiman et al. [28] also suggested twoing, a strategy for dealing with multiclass issues using two-class criteria. Instead of analyzing the original classes, twoing entails dividing the classes into two superclasses and doing two-class analysis on the groupings. Naturally, if one tries to examine all potential groups when the number of classes is big, the number of alternative groupings might create combinatorial explosion. Breiman et al. [28] suggested an efficient approach for determining ideal superclasses for each potential split instead of examining all splits for all possible class groups. The approach works for impurity criteria with two classes (compatible with Gini index). There exists the option of taking misclassification costs into account while making judgments. Surrogate splits are used to deal with missing data values. When a data item lacks a value required for a tree node test, it is forwarded to another test, which uses a different characteristic to build a split that is as close to the original as feasible. For data with missing values, many surrogate splits can be discovered and utilized in the correct sequence [29]. A simple Python code for decision tree is given below.

```
#Import Decision Tree Model
from sklearn import tree
#Create the Model
clf = tree.DecisionTreeClassifier()
#Train the Model with Training dataset
clf.fit(Xtrain,ytrain)
#Test the Model with Testing dataset
ypred = clf.predict(Xtest)
```

1.4.3 Random forest

Random forest (RF) is a collection of classifiers in which each classifier in the ensemble is a decision tree classifier. To calculate the split, the individual decision trees are built employing a random selection of characteristics at each node. In more technical terms, each tree is determined by the values of a random vector sampled separately and with the same distribution across all trees in the forest and each tree votes during classification. Bagging in conjunction with random attribute selection may be used to construct RFs. A training set of d tuples, D, is provided. The following is the general process for producing k decision trees for the ensemble. A training set, D_i, of d tuples is sampled with replacement from D for every iteration, i (i = 1, 2,..., k). Because each D_i is a bootstrap sample of D, certain tuples may appear several times in D_i while others may be eliminated. Let F be the number of features that will be utilized to calculate the split at each node, where F is substantially less than the number of accessible features. To build a decision tree classifier, M_i randomly selects F characteristics as candidates to split each node. The trees are allowed to develop to their full potential without being trimmed. Forest-RI is a name given to RFs generated in this manner using random input selection. Forest-RC, another type of RF, employs random linear combinations of the input characteristics. Rather than picking a subset of the characteristics at random, it generates new features that are a linear combination of the current features. To put it another way, a feature is created by providing L, the number of original features to be merged. At each node, L characteristics are chosen at random and appended utilizing coefficients that are uniform random integers on [1,1]. F linear combinations are created, and the optimal split is found by searching through them. When there are only a few characteristics available, this type of RF is beneficial for reducing the correlation between individual classifiers. RFs have the same accuracy as AdaBoost but are more resistant to errors and outliers. As the number of trees in a forest grows vast, the generalization error converges. As a result, overfitting is not an issue. The accuracy of an RF is defined by the strength of the individual classifiers as well as a measure of their interdependence. The ideal situation is to keep individual classifiers strong while decreasing their correlation. The number of characteristics chosen for consideration at each split has no effect on RFs. Typically, values up to $\log 2d + 1$ are used. RFs are efficient on very big databases because they examine many fewer features for each split. They have the potential to be faster than either bagging or boosting. Internal estimations of variable significance are provided by RFs [27]. A simple Python code for RF is given below.

```
#Import Random Forest Ensemble Model
from sklearn.ensemble import RandomForestClassifier
#Create the Model
clf = RandomForestClassifier(n_estimators = 200)
#Train the model with Training set
clf.fit(Xtrain,ytrain)
#Test the model with Test set
ypred = clf.predict(Xtest)
```

1.4.4 Bagging

Combination of the decisions of many models entails combining the numerous outputs into a single forecast. In the case of classification, the simplest method is to have a vote; in the case of numerical prediction, the simplest method is to compute the average. Assume that three training datasets of the same size are chosen at random from the problem area to introduce bagging by employing a specific ML approach to create a decision tree for every dataset. You may anticipate these trees to be almost identical and to predict the same thing for each new test case. Remarkably, this assumption is frequently incorrect, especially when the training datasets are small. This is a really alarming truth that appears to put a shadow on the entire enterprise! Since decision tree induction is an unstable method, small changes in the training data can simply achieve a new feature being selected at a certain node, with substantial consequences for the subtree below that node. Consider the experts to be decision trees in their own right. By allowing them to vote on each test case, we can merge the trees. If one class obtains more votes than the others, that class is assumed to be accurate. The more the better, in general: As more votes are counted, the predictions generated by voting grow more reliable [30].

By recreating the process outlined earlier with a specific training set, bagging seeks to offset the volatility of learning methods. Rather than selecting a new, independent training dataset every time, the old training data is modified by removing certain occurrences and reproducing others. Examples from the original dataset are randomly picked and replaced to generate a new dataset of the same size. This sampling approach invariably duplicates some instances while eliminating others. When presenting the bootstrap technique for assessing the generalization error of a learning system, this concept may strike a chord; fact, the phrase bagging stands for bootstrap aggregating. Bagging uses a learning method to each of these artificially produced datasets and the classifiers that result vote for the class to be predicted. The way the training datasets are generated differs between bagging and the idealized technique outlined above. Bagging simply resamples the existing training data rather than generating independent datasets from the domain. The resampled datasets are distinct from one another, yet they are not independent because they are all based on the same dataset [30]. A simple Python code for Bagging is given below.

```
#Import Bagging Ensemble Model
from sklearn.ensemble import BaggingClassifier
#Create a Bagging Ensemble Classifier
bagging = BaggingClassifier(tree.DecisionTreeClassifier(),max_samples = 0.5,
                                                           max_features=0.5)
#Train the model using the training set
bagging.fit(Xtrain,ytrain)
#Predict the response for test set
ypred = bagging.predict(Xtest)
```

1.4.5 Boosting

The boosting approach for merging several models takes use of this knowledge by looking for models. Boosting, like bagging, combines the output of individual models by voting for classification or averaging for numeric prediction. It combines models of the same type, such as decision trees, in the same way as bagging does. Boosting, on the other hand, is an iterative process. Individual models are produced individually in bagging, but the performance of prior models influences the performance of subsequent models. By providing more weight to examples handled improperly by previous models, boosting promotes subsequent models to become experts for those situations. Finally, rather than assigning equal weight to all models, boosting weighs the contribution of a model by its confidence [31].

1.4.6 AdaBoost

The technique of boosting has several variations. The widely known is AdaBoost approach, which is a frequently used classification method. It may be used to any classification learning method, just like bagging. To make things easier, we will suppose that the learning algorithm can manage weighted instances, in which each instance's weight is a positive number. When instance weights are included, the error of a classifier is determined as the sum of the weights of the misclassified instances divided by the total weight of all instances, rather than the proportion of misclassified instances. The learning algorithm may be pushed to focus on a certain group of examples, notably those with a high weight, by weighting instances. Because there is a higher motivation to appropriately identify such cases, they become much more essential [31].

The boosting process starts by giving all samples in the training data the same weight.

The learning process is then used to create a classifier for this data, and every sample is reweighted based on the classifier's output. The weight of correctly classified samples is reduced, while the weight of incorrectly classified samples increases. This results in a collection of "easy" occurrences with low weight and a collection of "hard" samples with high weight. A classifier is developed for the reweighted data in the following iteration—and all future ones—which concentrates on successfully categorizing the difficult incidents. The weights of the instances are then increased or decreased based on the output of this new classifier. Hence, certain difficult situations may become even more difficult, and easier situations may become even easier; on the other hand, some difficult situations may turn out to be easier, and easier ones harder. The weights represent how often the samples have been misclassified by the classifiers developed thus far after each iteration. This approach gives a sophisticated approach by producing a sequence of experts that complement one another by retaining a measure of "hardness" with each occurrence. Boosting frequently results in classifiers, which are substantially more precise on new data than those produced by bagging. Unlike bagging, however, boosting may occasionally fail in practice [31]. A simple Python code for AdaBoost is given below.

1.4.7 XGBoost

XGBoost [32] is an efficient gradient tree boosting technique which generates decision trees sequentially. In all computer platforms, it has the capacity to do relevant computations pretty fast. As a result, XGBoost is extensively employed for its ability to model newer features and classify labels. With its implementations in tabular and structured datasets, the XGBoost algorithm has attracted a considerable amount of interest. The XGBoost method began with a decision tree-based technique, in which graphical representations of probable choice solutions are produced based on particular circumstances. Then, using a majority voting mechanism, an ensemble meta-algorithm called "bagging" was devised to aggregate forecasts from numerous decision trees. This bagging method extended further to create a forest or aggregation of decision trees by randomly choosing attributes. The performance of the models was improved by decreasing the errors associated with the construction of sequential models. As an additional enhancement, the gradient decent approach was used to eliminate errors in the sequential model. Finally, the XGBoost method was recognized as a useful way for optimizing the gradient boosting algorithm by deleting missing data and avoiding overfitting concerns utilizing

```
#Import Adaboost ensemble model
from sklearn.ensemble import AdaBoostClassifier
#Create an Adaboost Ensemble Classifier
clf = AdaBoostClassifier(tree.DecisionTreeClassifier(),n_estimators = 10,
                                  algorithm='SAMME',learning_rate = 0.5)
#Train the model using the training set
clf.fit(Xtrain,ytrain)
#Predict the response for test set
ypred = clf.predict(Xtest)
```

parallel processing [33]. A simple Python code for XGBoost is given below.

```
% pip install xgboost
#Import XGBoost ensemble model
from xgboost import XGBClassifier
# Create XGB model
model = XGBClassifier()
#Train the model using the training set
model.fit(Xtrain, ytrain)
# make predictions for test data
ypred = model.predict(Xtest)
```

1.4.8 Artificial neural networks

Artificial neural networks (ANNs) are computational models, which are modeled after the biological neural networks, which make up the human brain. An ANN is made up of artificial neurons, which are a collection of linked units. Each neuronal connection has the ability to send a signal from one to the other. The signal(s) can be processed by the receiving neuron, which can then transfer them to downstream neurons linked to it. Neurons have states that are normally characterized by real numbers, ranging from 0 to 1. Besides, neurons contain weights, which change as they learn, allowing them to change the intensity of the signal they transmit downstream to other neurons. They may also have a threshold, with the downstream signal only being transmitted if the combined output is below (or above) that level. Neurons are generally arranged in layers. On their inputs, different layers conduct various modifications [20].

They have become more popular in applications that are tough to represent using standard rule-based programming. The initial aim of the neural network approach was to resolve problems in the same manner as a human brain would. Then, interest shifted to selecting certain mental skills, resulting in biological abnormalities such as backpropagation, and modifying the network to indicate feedback information. Computer (machine) vision, image classification, medical diagnosis, pattern/object identification, machine translation, speech recognition, social network filtering, video game play, and many more disciplines have all used neural networks [20]. A simple Python code for ANN is given below.

```
#Import ANN model
from sklearn.neural_network import MLPClassifier
#Create the Model
mlp = MLPClassifier(hidden_layer_sizes = (100, ), learning_rate_init = 0.001,
                                          alpha = 1, momentum = 0.9,max_iter = 1000)
#Train the Model with Training set
mlp.fit(Xtrain,ytrain)
#Test the Model with Test set
ypred = mlp.predict(Xtest)
```

The selection of an activation function of the neural network design process is crucial. The necessity to forecast a binary class label drives the selection of the sign activation function in the case of the perceptron. Other scenarios in which other goal variables might be predicted are also possible. Nonlinear functions such as the hyperbolic tangents, sigmoid, or sign can be utilized in different layers. The identity or linear activation function is the simplest fundamental activation function Φ(.) represented by a linear function:

$$\Phi(v) = v$$

When the target is a real value, the linear activation function is frequently utilized at the output node. When a smoothed surrogate loss function is required for discrete outputs, it is also employed. The hyperbolic tangent, sigmoid, and sign functions were the classic activation functions utilized early in the construction of neural networks. The sign activation cannot be utilized to create the loss function at training time due to its nondifferentiability, since it may be utilized to map to binary outputs at prediction time. The sigmoid activation produces a result in the range of (0, 1), which is useful for calculations, which should be interpreted as probabilities. The tanh function is shaped similarly to the sigmoid function, with the exception that it is horizontally and vertically rescaled to $[-1, 1]$:

$$\tanh(v) = 2 \cdot \text{sigmoid}\,(2v) - 1$$

When the outputs of the calculations must be both positive and negative, the tanh function is preferred to the sigmoid. It is also easier to train owing of its mean-centering and bigger gradient (due to stretching) as compared to the sigmoid. The tanh and sigmoid functions have been widely used for introducing nonlinearity into neural networks. However, recently, a variety of piecewise linear activation functions have gained popularity [34]:

$$\Phi(v) = \max\{v, 0\}(\text{Rectified Linear Unit [ReLU]})$$

$$\Phi(v) = \max\{\min[v, 1], -1\}(\text{hard tanh})$$

1.4.9 Deep learning

It has been discovered that deep neural networks are best fitted for the image processing application. Conventional neural networks contain one input layer, two or three hidden layers, and one output layer. Deep neural networks contain one input layer, many hidden layers, and one output layer. The greater the number of hidden layers, the deeper the network. The layers are linked, with the previous layer's output becoming the next layer's input. The network's performance is determined by the weights of its inputs and outputs. Training the network includes specifying the right weights for numerous layers. Deep networks need more computational speed, processing capacity, a huge database, and the right parallel processing software [20].

Deep learning is an AI area that focuses on building huge neural network models capable of generating correct data-driven decisions. Deep learning is best suited to scenarios where the data is complicated and huge datasets are accessible. Deep learning is used in the health-care industry to interpret medical images (X-rays, MRI scans, and CT scans) to diagnose health issues [34]. Based on the given data, ML algorithms build their own logic. The algorithm learns on its own, thus no coding is required to tackle every problem. A large number of medical images must be given into the algorithm in order for it to learn to classify. It is supervised learning if the images have previously been categorized and fed. Otherwise, it is unsupervised learning. The most basic application is categorizing a pattern into two groups, such as determining whether a medical image belongs to tumor tissue or not [20]. A simple Python code for deep neural network is given below.

```
from keras.models import Sequential
from keras.layers import Dense
# define the keras model
model = Sequential()
model.add(Dense(10, input_shape=(4,), activation='relu', name='fc1'))
model.add(Dense(10, activation='relu', name='fc2'))
model.add(Dense(3, activation='softmax', name='output'))
# compile the keras model
model.compile(loss='binary_crossentropy', optimizer='adam',
              metrics=['accuracy'])

# Train the model
model.fit(Xtrain, ytrain, verbose=2, batch_size=10, epochs=200)
# Test on unseen data
results = model.evaluate(Xtest, ytest)
# Pront the results
print('Final test set loss: {:4f}'.format(results[0]))
print('Final test set accuracy: {:4f}'.format(results[1]))
```

1.4.10 The overfitting problem in neural network training

Despite neural networks' impressive reputation as universal function approximators, important obstacles such as overfitting remain in training neural networks. The overfitting problem implies the fact that fitting a model to a specific training dataset does not ensure that it will perform well on unknown test data. When the models are sophisticated and the dataset is small, there is always a performance difference between training and test data [34].

If we applied our solution to unknown test data, we would most likely get extremely poor results since the learnt parameters are erroneously implied and are not likely to generalize well to new situations. This form of erroneous inference is generated by a lack of training data, which allows random details to be encoded into the model. Hence, the solution fails to generalize effectively to previously unknown test data. Increasing the number of training examples enhances the model's generalization power but increasing the model's complexity decreases its generalization power. Simultaneously, when there is a large amount of training data available, a too simple model is not likely to catch complicated correlations between characteristics and targets. A decent rule of thumb is that the total amount of training data points should be at least 2 to 3 times greater than the number of neural network parameters. In ML, the concept of overfitting is frequently viewed in terms of the trade-off between bias and variance. In general, even when a significant quantity of data is available, neural networks need rational strategy to limit the detrimental consequences of overfitting [34].

1.4.10.1 Regularization

Because a higher number of parameters lead to overfitting, constraining the model to utilize fewer nonzero parameters is an obvious solution. When the amount of data accessible is small, regularization is very critical. Regularization, according to one biological explanation, relates to progressive forgetting, where "less important" examples are erased.

Generally, more complicated models with regularization are preferable to simpler models without regularization [34].

1.4.11 Convolutional neural networks

Convolutional neural network (CNN) is a deep learning network, which has grown in popularity for image categorization. Fig. 1.2 depicts the CNN architecture. It is made up of an input layer as well as hundreds of feature detection layers. Feature detection layers conduct one of the three actions. Convolution, pooling, and Rectified Linear Unit (ReLU) are all terms used to describe the process of convolution [20]. CNNs are biologically inspired networks utilized in computer vision to classify images and recognize objects. A convolution process is specified for the convolution layers, where a filter is employed to transfer the activations from one layer to the next. A convolution operation employs a three-dimensional weighted filter with the same depth as the current layer just a smaller spatial area. The dot product of all the weights in the filter and any choice of spatial region in a layer describes the value of the hidden state in the subsequent layer. The interaction between the filter and the spatial area in a layer is performed at every available location to create the next layer, in which the activations keep their spatial connections from the prior layer. Since every activation in a specific layer is a function of just a small spatial region in the preceding layer, the connections in a CNN are relatively sparse. Except for the final pair of two of three levels, all layers retain their spatial structure. As a result, it is feasible to physically visualize which elements of a picture impact which portions of activations in a layer. Lower level layer features capture lines or other primitive forms, but higher level layer features catch more complex shapes. As a result, subsequent layers can generate numbers by assembling the shapes in these intuitive characteristics. Furthermore, a subsampling layer simply averages the data in the local areas of size 2×2 to reduce the geographical footprints of the layers by a factor of 2. CNNs were the most effective of all forms of neural networks in the past. They are commonly employed in image identification, object detection and location, and even language processing [35].

Convolution processes an image by passing it through convolution filters, which trigger

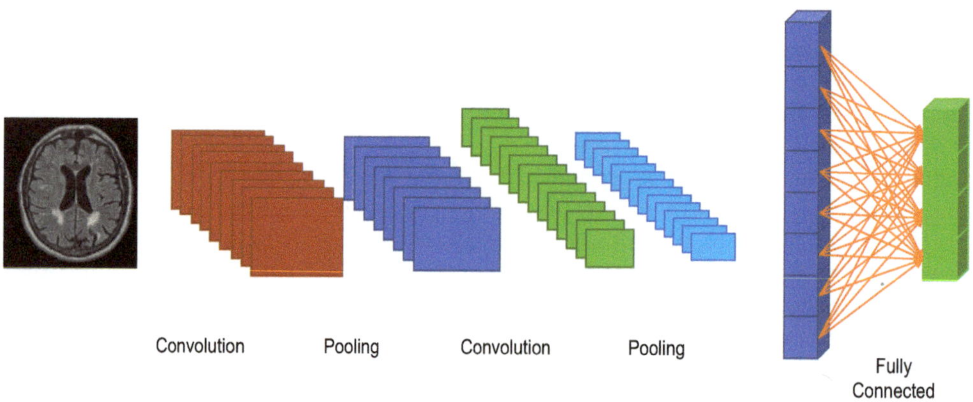

FIGURE 1.2 CNN architecture. *CNN*, Convolutional neural network.

particular characteristics in the image. Pooling uses nonlinear downsampling to decrease the amount of data that must be processed. ReLU keeps positive values while negative values are set to zero. The classification layer is located before to the output layer. It is a fully linked layer with an N-dimensional output, where N is the number of categories to be classified. This layer produces an N-dimensional vector, with each member representing the likelihood that the input image belongs to one of the N classes. The categorized output is provided by the final output layer, which employs a softmax function. Each layer of the network processes data, which is then passed on to the next layer. CNNs are based on the biological anatomy of the visual cortex. The simple and complicated cells of the visual cortex activate receptive field subregions of a visual field. Instead of being totally linked, the subregions of the preceding layer are linked to the neurons of a layer in CNN. Other subregions have no effect on the neurons. Unlike traditional neural nets, the subregions are permitted to overlap and thus provide spatially related results. This is the primary distinction between CNN and other neural networks [20].

1.4.11.1 *Functioning of convolutional neural network*

In the operation of a CNN, an image containing one or more items to be categorized will be sent into the CNN. The amount of input values will be determined by the image size and pixel depth. These are merely integers that must be interpreted or recognized as objects of a particular type. CNN attempts to mimic the human visual cortex, which possesses sensitivity to specific areas of the visual field in tiny clusters of cells. Some neurons in the brain are susceptible to specific information in visual images, such as curves, edges, and so on. The convolution layer is the first layer of CNN, and it conducts spatial convolution of a predefined mask with pixel values. This is analogous to a linear filtering procedure, and the convolution result is determined by the predefined mask. Another feature map at a higher level than the first is created, when the first layer's activation output (feature map) is fed into the second hidden layer and the convolution procedure is repetitive. This approach generates distinct activation maps for various image characteristics, including complex features. At the network's end, there is a completely linked layer that accepts input from the preceding layer and creates N-dimensional output. For an N-class issue, it may return N alternative probability values, each expressing the likelihood that the item belongs to that class. For the classification to be accurate, the CNN must be trained on millions of pictures using the backpropagation technique. The procedure is nonlinear since each convolution layer is followed by an activation ReLU layer. The negative activation values are set to zero, while the positive activation values are kept. Following the ReLU levels are pooling layers, the most well-known of which is the max-pooling layer, which uses the activation inputs and creates a downsampled output. During the network's training phase, there are additional dropout layers that remove specific activation outputs. Overfitting is avoided as a result of this [20].

The conditions in every layer of CNNs are organized according to a spatial grid pattern. The spatial associations are passed down from one layer to the next, since each feature value is dependent on a tiny local geographic location in the preceding layer. It is critical to retain these spatial correlations among the grid cells, as the convolution operation and the transition to the next layer are heavily based on these linkages. A layer's depth in a CNN must not be confused with the network's overall depth. The CNN works similar to a standard feed-forward neural network, with the exception that the procedures in its levels are spatially structured and the links between layers are sparse (and carefully planned). Convolution, pooling, and ReLU are the three types of layers that are typically seen in a CNN. The activation of the ReLU is similar to that of a standard neural network. Furthermore, a final set of layers is frequently completely linked and translates to a set of output nodes in an application-specific manner. The CNN's input data is structured into a two-dimensional grid structure, with pixels representing the values of individual grid points. As a result, each pixel in the picture correlates to a certain spatial position. However, a multidimensional array values at each grid place is required to represent the specific hue of the pixel. We have an intensity of the three primary hues in the RGB color scheme, which are red, green, and blue, respectively [35].

1.4.11.2 *Padding*

One thing to keep in mind is that the convolution procedure shrinks the $(q + 1)$th layer in contrast to the qth layer. In general, this style of image reduction is undesirable since it has a tendency to lose some information near the image's edges. Padding can be used to remedy this problem. To preserve the spatial footprint, padding is used to add (Fq 1)/2 "pixels" all around the edges of the feature map. In the case of padding hidden layers, these pixels are really feature values. Regardless of whether the input or hidden layers are padded, the value of each of these padded feature values is set to 0. As a result, the input volume's spatial height and breadth will both grow by (Fq −1), which is exactly what they would decrease after the convolution. Because their values are set to 0, the padded parts have no effect on the final dot product. Padding allows the convolution operation to be performed with a piece of the filter "sticking out" from the layer's boundaries, allowing the dot product to be performed just over the area of the layer wherever the values are defined [35].

1.4.11.3 *Strides*

Convolution can also be used to minimize the image's spatial footprint in various ways. The method described above executes convolution at every point in the feature map's spatial location. However, the convolution does not have to be performed at every spatial place in the layer. The concept of strides may be used to decrease the granularity of the convolution. A stride of one is the most usual, but a stride of two is also rarely utilized. In normal conditions, strides of more than two are uncommon. Larger strides can aid with memory constraints or decrease overfitting if the spatial resolution is excessively high. A bigger receptive field is beneficial to detect a complex feature in a greater spatial region of the image. The hierarchical feature engineering method of a CNN captures increasingly complicated forms in later layers. Historically, another process known as max-pooling has been used to improve the receptive fields [35].

1.4.11.4 *The Rectified Linear Unit layer*

The pooling and ReLU operations are combined with the convolution operation. The activation of the ReLU is similar to how it is done in a standard neural network. Because the ReLU is a basic one-to-one mapping of activation values, it has no effect on the layer's dimensions. The activation function is paired with a linear transformation with a matrix of weights to produce the next layer of activations in classic neural networks. An ReLU layer is frequently not clearly displayed in graphical representations of convolution neural network designs, and it often follows a convolution operation. It's worth noting that the ReLU activation function is a very new addition to neural network architecture. Saturating activation functions like tanh and sigmoid were employed in the past. In terms of speed and accuracy, the ReLU has a significant advantage over these activation functions. Improved speed is also linked to precision since it permits for the adoption of more complex models and the training of them for longer periods of time. The ReLU activation function has progressively substituted the other activation functions in CNN architecture in recent years, to the point that the ReLU will be used as the default activation function [35].

1.4.11.5 Pooling

The pooling process, on the other hand, is quite different. Unlike filters, the pooling process applies on tiny grid patches of size $Pq \times Pq$ in each layer and creates a new layer of the same depth. The maximum of these values is provided for each square region with the size $Pq \times Pq$ in each of the dq activation maps. This method is stated as max-pooling. The spatial dimensions of each activation map are dramatically reduced when they are pooled. Pooling takes place at the level of each activation map, unlike convolution processes. As a result, the operation of pooling has no effect on the amount of feature maps. In other words, the depth of the pooled layer is the same as the depth of the layer on which the pooling operation was conducted. The usual size Pq of the region across which pooling is performed is 2×2. At a stride of 2, there would be no overlap between the several zones being pooled, and this kind of arrangement is extremely frequent. However, it has been argued that having at least some overlap among the geographical units at which the pooling is conducted is advantageous since it reduces the likelihood of overfitting [35].

1.4.11.6 Fully connected layers

Every hidden state in the first fully connected layer is coupled to every feature in the final spatial layer. This layer performs the same duties as a standard feed-forward network. In most circumstances, more than one fully connected layer may be used to enhance the computing capacity toward the end. The connections between these levels are arranged in the same way that a traditional feed-forward network is. The fully connected layers include many parameters, as the fully connected layers are densely connected. The fully connected layers generally contain more connections than the convolutional layers, despite the fact that the convolutional layers have more activations. Based on the characteristics of the application, one may utilize logistic, softmax, or linear activation. To construct a single value, one option for employing completely linked layers is to apply average pooling throughout the whole spatial region of the final set of activation maps [35].

1.4.11.7 Training a convolutional network

The backpropagation technique is used to train a CNN. The convolution, ReLU, and max-pooling layers are the three main types of layers. Because the ReLU is similar to a standard neural network, it is quite simple to backpropagate via it. To max-pool with no overlap across pools, all that is required is to figure out which unit is the pool's maximum value. With regard to the pooled state, the partial derivative of the loss flows back to the unit with the highest value. Except for the largest entry in the grid, all other entries are given a value of 0. ReLU procedures and backpropagation through maximizing are similar to those used in conventional neural networks [35].

1.4.11.8 Dropout

Dropout is a method for creating a neural network ensemble that employs node sampling rather than edge sampling. When a node is removed, all incoming and outgoing connections to and from that node must be removed as well. For the nodes, just the network's input and hidden layers are sampled. Dropout is a technique of combining node sampling with weight sharing. Backpropagation is then used by the training procedure to update the weights of the sampled network utilizing a single sampled example. Dropout has the major consequence of incorporating regularization into the learning process. Dropout efficiently introduces noise into both the input data and the hidden interpretations by dropping both input and hidden units. Regularization is a type of noise addition. Dropout prevents hidden units from adapting to one other's features, a process known as feature coadaptation. Because the impact of dropout is a masking noise that eliminates part of the hidden units, this method imposes some redundancy between the characteristics learnt at the various hidden units. Increased resilience is the result of this form of redundancy. Since every one of the sampled subnetworks is trained with a limited number of sampled examples, dropout is efficient. As a result, just the additional labor of sampling the hidden units is required. However, because dropout is a regularization approach, it limits the network's expressive ability. Thus, in order to fully benefit from dropout, one must employ larger models and more units. As a result, there is a hidden computational overhead. Moreover, if the initial training dataset is big enough to limit the chance of overfitting, the computational benefits of dropout may be minor but still noticeable [35].

1.4.11.9 *Early stopping*

Early stopping is a popular type of regularization, where the gradient descent is stopped after just a few iterations. Keeping a portion of the training data and then assessing the model's error on the hold-out set is one technique to determine the stopping point. When the error on the hold-out set starts to grow, the gradient-descent strategy is stopped. The size of the parameter space is effectively reduced to a smaller neighborhood inside the starting values of the parameters when early stopping is used. Early stopping functions as a regularizer in this case since it effectively limits the parameter space [34].

1.4.11.10 *Batch normalization*

Batch normalization is a relatively new technique for dealing with vanishing and exploding gradient issues that leads to activation gradients in consecutive layers to either decrease or increase in magnitude. Internal covariate shift is another significant issue in deep network training. The issue is that throughout training, the parameters are adjusted, and hence the hidden variable activations are adjusted as well. The goal of batch normalization is to create features with identical variance by adding extra "normalization layers" between hidden layers, which withstand this sort of actions. This extra node must be taken into account by the backpropagation algorithm in order to guarantee the loss derivative of layers before the batch normalization layer compensates for the transformation entailed with these new nodes. Batch normalization has the unique characteristic of acting as a regularizer. It is worth noting that the same data point might result in slightly distinct adjustments based on which batch it is in. This impact might be viewed as a form of noise that is introduced to the updating process. Adding a tiny bit of noise to the training data is a common way to accomplish regularization. Although there is no perfect agreement on this topic, it has been empirically noted that regularization approaches such as dropout do not enhance the performance once batch normalization is applied [35]. A simple Python code for CNN is given below.

```
from tensorflow.keras.preprocessing.image import load_img ,img_to_array
import matplotlib.pyplot as plt
from tensorflow.keras.models import Sequential
from tensorflow.keras.layers import Conv2D
from tensorflow.keras.layers import MaxPooling2D,AveragePooling2D
from tensorflow.keras.layers import BatchNormalization
from tensorflow.keras.layers import Dense,Activation
from tensorflow.keras.layers import Flatten,Dropout,SpatialDropout2D
from tensorflow.keras.optimizers import SGD
from tensorflow.keras.optimizers import Adam
from tensorflow.keras.callbacks import EarlyStopping, ModelCheckpoint
early_stopping = EarlyStopping(monitor='val_loss', mode='min', patience=10,
                               restore_best_weights=True, verbose=1)

#Create the CNN Model
model = Sequential()
#1st Convolutional Layer
model.add(Conv2D(32, (3, 3), padding='valid', strides=(1, 1),input_shape=img_shape))
model.add(BatchNormalization())
model.add(Activation('relu'))
model.add(MaxPooling2D(pool_size=(2,2), strides=(1,1), padding='same'))

#2nd Convolutional Layer
model.add(Conv2D(32, (3, 3), padding='same', strides=(1, 1)))
model.add(BatchNormalization())
model.add(Activation('relu'))
model.add(MaxPooling2D(pool_size=(2, 2),strides=(2,2), padding='same'))
model.add(Dropout(0.2))

#3rd Convolutional Layer
model.add(Conv2D(64, (3, 3), padding='same', strides=(1, 1)))
model.add(BatchNormalization())
model.add(Activation('relu'))
model.add(MaxPooling2D(pool_size=(2, 2),strides=2))

#4th Convolutional Layer
model.add(Conv2D(64, (3, 3), padding='same', strides=(1, 1)))
model.add(BatchNormalization())
model.add(Activation('relu'))
model.add(MaxPooling2D(pool_size=(2, 2),strides=2))
```

```
#Passing it to a Fully Connected layer
model.add(Flatten())
# 1st Fully Connected Layer
model.add(Dense(64, input_shape=(32,32,3,)))
model.add(BatchNormalization())
model.add(Activation('relu'))
# Add Dropout to prevent overfitting
model.add(Dropout(0.4))

#2nd Fully Connected Layer
model.add(Dense(32))
model.add(BatchNormalization())
model.add(Activation('relu'))
# Add Dropout to prevent overfitting
model.add(Dropout(0.4))

#Output Layer
model.add(Dense(3))
model.add(BatchNormalization())
model.add(Activation('softmax'))

model.compile(optimizer='adam',loss='categorical_crossentropy',metrics=['acc'])

# Train the model
history = model.fit(
    Xtrain,
    ytrain,
    epochs=200,
    validation_split=0.25,
    batch_size=10,
    verbose=2,
    callbacks=[early_stopping])
# Test on unseen data
results = model.evaluate(Xtest, ytest)
# Pront the results
print('Final test set loss: {:4f}'.format(results[0]))
print('Final test set accuracy: {:4f}'.format(results[1]))

# plot training history
from matplotlib import pyplot
pyplot.plot(history.history['loss'], label='train')
pyplot.plot(history.history['val_loss'], label='validation')
pyplot.legend()
pyplot.show()
```

1.4.12 Recurrent neural networks

All neural networks are built to handle multidimensional data with features that are mainly independent to each other. Particular data types, such as biological data, text, and time series, do, nevertheless, have sequential relationships between the properties. The following are some examples of such dependencies:

1. The values on successive time stamps in a time-series data collection are tightly tied to one another. If the values of these time stamps are treated as separate characteristics, important information about the connections between them is lost. While the values at various time stamps are processed separately; however, this causes information lost.
2. Even though text is frequently processed as a bag of words, the sequencing of the words might provide superior semantic insights. In such instances, it is critical to build models that account for the sequencing information. The most typical use of recurrent neural networks (RNNs) is text data.
3. Sequences in biological data are often seen, and the symbols may relate to one of the nucleobases or amino acids, which comprise DNA's building blocks.

Particular values in a sequence might be actual or symbolic in nature. Time series is another term for real-valued sequences. For any form of data, RNNs may be employed. Symbolic values are more commonly used in practical applications. The vanishing and exploding gradient problems are one of the most crucial problems in this field. This problem is especially prominent in deep networks, such as RNNs. Hence, a variety of RNN types have been developed, including the gated recurrent unit and long short-term memory (LSTM). Image captioning, sequence-to-sequence learning, sentiment analysis, and machine translation are just a few of the areas where RNNs and their derivatives have been applied [35]. A simple Python code for RNN is given below.

```
from keras.layers import Dense
from keras.models import Sequential
from keras.layers import SimpleRNN
#Create Model
model = Sequential()
model.add(SimpleRNN(32))
model.add(Dense(1, activation = 'sigmoid'))
# Compile model
model.compile(optimizer = 'rmsprop', loss = 'binary_crossentropy', metrics
= ['acc'])
# Fit model
history = model.fit(Xtrain, ytrain,epochs = 100, batch_size = 20, validation_split=0.2)
# Evaluate the model
_, train_acc = model.evaluate(Xtrain, ytrain, verbose = 0)
_, test_acc = model.evaluate(Xtest, ytest, verbose = 0)
print('Training Accuracy: %.3f, Testing Accuracy: %.3f' % (train_acc,test_acc))
```

1.4.13 Long short-term memory

Problems with vanishing and exploding gradients affect RNNs. This is a typical dilemma in neural network updates, in which repeated multiplication by the matrix is intrinsically unstable; the gradient either vanishes during backpropagation or blows up to huge values in an unstable manner. At various time stamps, consecutive multiplication with the weight matrix causes this form of instability. One way to look at this issue is that a neural network, which only employs multiplicative updates, is only excellent at learning short sequences and, hence, has good short-term memory but poor long-term memory by default. One approach to solving this challenge is to use the LSTM and long-term memory to change the recurrence equation for the hidden vector. The operations of LSTM are designed to give the user fine-grained control over the data kept in this long-term memory [35]. A simple Python code for LSTM is given below.

1.4.14 Data augmentation

Data augmentation is a popular approach to reduce overfitting in CNN. New training patterns are created utilizing transformations on the existing samples in data augmentation. Data augmentation works particularly effectively in the field of image processing. Because numerous transformations, such as rotation, patch extraction, reflection, and translation, do not mainly alter the attributes of an image's object. Moreover, they improve the dataset's generalization ability when trained with the supplemented dataset. Because many of these types of data augmentation involve relatively little work, the augmented images do not require to be produced clearly beforehand. They can instead be generated during the training process when an image is being examined. In many circumstances, employing picture patches to create the training dataset might be beneficial [35]. A simple

```
from keras.layers import Dense
from keras.models import Sequential
from keras.layers import LSTM
#Create Model
model = Sequential()
model.add(LSTM(32))
model.add(Dense(1, activation = 'sigmoid'))
# Compile model
model.compile(optimizer = 'rmsprop', loss = 'binary_crossentropy', metrics= ['acc'])
# Fit model
history = model.fit(Xtrain, ytrain,epochs = 100,batch_size = 10, validation_split = 0.2)
# Evaluate the model
_, train_acc = model.evaluate(Xtrain, ytrain, verbose = 0)
_, test_acc = model.evaluate(Xtest, ytest, verbose = 0)
print('Training Accuracy: %.3f, Testing Accuracy: %.3f' % (train_acc, test_acc))
```

Python code for data augmentation is given below. For detailed examples, check the machinelearningmastery.com website[1].

```
#Example of horizontal shift image augmentation from machinelearningmastery.com
from numpy import expand_dims
from keras.preprocessing.image import load_img
from keras.preprocessing.image import img_to_array
from keras.preprocessing.image import ImageDataGenerator
from matplotlib import pyplot
# load the image
img = load_img('img.jpg')
# convert to numpy array
data = img_to_array(img)
# expand dimension to one sample
samples = expand_dims(data, 0)
# create image data augmentation generator
datagen = ImageDataGenerator(width_shift_range=[-200,200])
# prepare iterator
it = datagen.flow(samples, batch_size=1)
# generate samples and plot
for i in range(9):
  # define subplot
  pyplot.subplot(330 + 1 + i)
  # generate batch of images
  batch = it.next()
  # convert to unsigned integers for viewing
  image = batch[0].astype('uint8')
  # plot raw pixel data
  pyplot.imshow(image)
# show the figure
pyplot.show()
```

1.4.15 Generative adversarial networks

In generative adversarial networks (GANs), two neural network models are used at the same time. First, a generative model generates synthetic instances of images, which are comparable to those found in a real repository. Moreover, the objective is to produce synthetic images, which are so genuine that a trained observer will be unable to tell whether an item is produced synthetically or belongs to the original dataset. The created objects are frequently used to generate vast volumes of synthetic data for AI algorithms, and they may also be used to augment data. Furthermore, by providing context, this

[1] https://machinelearningmastery.com/how-to-configure-image-data-augmentation-when-training-deep-learning-neural-networks/.

technique may be used to generate objects with different aspects. The parameters of the generator and discriminator are simultaneously updated throughout the training phase of a GAN. The discriminator and the generator are both neural networks. The generator can be compared to the decoder part of a variational autoencoder. However, the training procedure differs significantly from that of a variational autoencoder. GAN is frequently used to generate image with a variety of contexts. The image setup is, without a doubt, the most prevalent use of GANs. The image setting generator is known as a deconvolutional network. As a result, the matching GAN is also known as a deep convolutional generative adversarial network (DCGAN) [35]. A simple Python code for GAN is given below. For details, check the related KERAS web page[2].

```
import matplotlib.pyplot as plt
import tensorflow as tf
import tensorflow_datasets as tfds
from tensorflow import keras
from tensorflow.keras import layers

# DCGAN generator
def get_generator():
    noise_input = keras.Input(shape=(noise_size,))
    x = layers.Dense(4 * 4 * width, use_bias=False)(noise_input)
    x = layers.BatchNormalization(scale=False)(x)
    x = layers.ReLU()(x)
    x = layers.Reshape(target_shape=(4, 4, width))(x)
    for _ in range(depth - 1):
        x = layers.Conv2DTranspose(
            width, kernel_size=4, strides=2, padding="same", use_bias=False,
        )(x)
        x = layers.BatchNormalization(scale=False)(x)
        x = layers.ReLU()(x)
    image_output = layers.Conv2DTranspose(
        3, kernel_size=4, strides=2, padding="same", activation="sigmoid",
    )(x)

    return keras.Model(noise_input, image_output, name="generator")
```

[2] https://keras.io/examples/generative/gan_ada/.

```
# DCGAN discriminator
def get_discriminator():
    image_input = keras.Input(shape=(image_size, image_size, 3))
    x = image_input
    for _ in range(depth):
        x = layers.Conv2D(
            width, kernel_size=4, strides=2, padding="same", use_bias=False,
        )(x)
        x = layers.BatchNormalization(scale=False)(x)
        x = layers.LeakyReLU(alpha=leaky_relu_slope)(x)
    x = layers.Flatten()(x)
    x = layers.Dropout(dropout_rate)(x)
    output_score = layers.Dense(1)(x)

    return keras.Model(image_input, output_score, name="discriminator")

class GAN_ADA(keras.Model):
    def __init__(self):
        super().__init__()

        self.augmenter = AdaptiveAugmenter()
        self.generator = get_generator()
        self.ema_generator = keras.models.clone_model(self.generator)
        self.discriminator = get_discriminator()

        self.generator.summary()
        self.discriminator.summary()

    def compile(self, generator_optimizer, discriminator_optimizer, **kwargs):
        super().compile(**kwargs)

        # separate optimizers for the two networks
        self.generator_optimizer = generator_optimizer
        self.discriminator_optimizer = discriminator_optimizer

        self.generator_loss_tracker = keras.metrics.Mean(name="g_loss")
        self.discriminator_loss_tracker = keras.metrics.Mean(name="d_loss")
        self.real_accuracy = keras.metrics.BinaryAccuracy(name="real_acc")
        self.generated_accuracy = keras.metrics.BinaryAccuracy(name="gen_acc")
        self.augmentation_probability_tracker = keras.metrics.Mean(name="aug_p")
        self.kid = KID()
```

```
@property
def metrics(self):
    return [
         self.generator_loss_tracker,
         self.discriminator_loss_tracker,
         self.real_accuracy,
         self.generated_accuracy,
         self.augmentation_probability_tracker,
         self.kid,
    ]

def generate(self, batch_size, training):
    latent_samples = tf.random.normal(shape=(batch_size, noise_size))
    # use ema_generator during inference
    if training:
        generated_images = self.generator(latent_samples, training)
    else:
        generated_images = self.ema_generator(latent_samples, training)
    return generated_images

def adversarial_loss(self, real_logits, generated_logits):
    # this is usually called the non-saturating GAN loss

    real_labels = tf.ones(shape=(batch_size, 1))
    generated_labels = tf.zeros(shape=(batch_size, 1))

    # the generator tries to produce images that the discriminator considers as real
    generator_loss = keras.losses.binary_crossentropy(
        real_labels, generated_logits, from_logits=True
    )
    # the discriminator tries to determine if images are real or generated

    discriminator_loss = keras.losses.binary_crossentropy(
        tf.concat([real_labels, generated_labels], axis=0),
        tf.concat([real_logits, generated_logits], axis=0),
        from_logits=True,
    )

    return tf.reduce_mean(generator_loss), tf.reduce_mean(discriminator_loss)

def train_step(self, real_images):
    real_images = self.augmenter(real_images, training=True)
```

```
# use persistent gradient tape because gradients will be calculated twice
with tf.GradientTape(persistent=True) as tape:
     generated_images = self.generate(batch_size, training=True)
     # gradient is calculated through the image augmentation
     generated_images = self.augmenter(generated_images, training=True)

     # separate forward passes for the real and generated images, meaning
     # that batch normalization is applied separately
     real_logits = self.discriminator(real_images, training=True)
     generated_logits = self.discriminator(generated_images, training=True)


     generator_loss, discriminator_loss = self.adversarial_loss(
         real_logits, generated_logits
     )

# calculate gradients and update weights
generator_gradients = tape.gradient(
    generator_loss, self.generator.trainable_weights
)
discriminator_gradients = tape.gradient(
    discriminator_loss, self.discriminator.trainable_weights
)
self.generator_optimizer.apply_gradients(
    zip(generator_gradients, self.generator.trainable_weights)
)
self.discriminator_optimizer.apply_gradients(
    zip(discriminator_gradients, self.discriminator.trainable_weights)
)

# update the augmentation probability based on the discriminator's performance

self.augmenter.update(real_logits)
self.generator_loss_tracker.update_state(generator_loss)
self.discriminator_loss_tracker.update_state(discriminator_loss)
self.real_accuracy.update_state(1.0, step(real_logits))
self.generated_accuracy.update_state(0.0, step(generated_logits))
self.augmentation_probability_tracker.update_state(self.augmenter.probability)
```

```
        # track the exponential moving average of the generator's weights to decrease
        # variance in the generation quality
        for weight, ema_weight in zip(
            self.generator.weights, self.ema_generator.weights
        ):
            ema_weight.assign(ema * ema_weight + (1 - ema) * weight)

        # KID is not measured during the training phase for computational efficiency
        return {m.name: m.result() for m in self.metrics[:-1]}

    def test_step(self, real_images):
        generated_images = self.generate(batch_size, training=False)

        self.kid.update_state(real_images, generated_images)

        # only KID is measured during the evaluation phase for computational efficiency
        return {self.kid.name: self.kid.result()}

    def plot_images(self, epoch=None, logs=None, num_rows=3, num_cols=6, interval=5):
        # plot random generated images for visual evaluation of generation quality
        if epoch is None or (epoch + 1) % interval == 0:
            num_images = num_rows * num_cols
            generated_images = self.generate(num_images, training=False)

            plt.figure(figsize=(num_cols * 2.0, num_rows * 2.0))
            for row in range(num_rows):
                for col in range(num_cols):
                    index = row * num_cols + col
                    plt.subplot(num_rows, num_cols, index + 1)
                    plt.imshow(generated_images[index])
                    plt.axis("off")
            plt.tight_layout()
            plt.show()
            plt.close()
```

```
# create and compile the model
model = GAN_ADA()
model.compile(
    generator_optimizer=keras.optimizers.Adam(learning_rate, beta_1),
    discriminator_optimizer=keras.optimizers.Adam(learning_rate, beta_1),
)

# save the best model based on the validation KID metric
checkpoint_path = "gan_model"
checkpoint_callback = tf.keras.callbacks.ModelCheckpoint(
    filepath=checkpoint_path,
    save_weights_only=True,
    monitor="val_kid",
    mode="min",
    save_best_only=True,
)

# run training and plot generated images periodically
model.fit(
    train_dataset,
    epochs=num_epochs,
    validation_data=val_dataset,
    callbacks=[
        keras.callbacks.LambdaCallback(on_epoch_end=model.plot_images),
        checkpoint_callback,
    ],
)

# load the best model and generate images
model.load_weights(checkpoint_path)
model.plot_images()
```

1.4.16 Transfer learning

One of the obstacles of the image analysis is that for a specific use, labeled training data might not still be accessible. Taking into account the condition where one has a set of images, which require to be employed for image retrieval. It is crucial for the features to be semantically consistent since the labels do not exist in retrieval applications. The ImageNet dataset includes more than a million images taken from 1000 classes experienced in daily life. The idea behind the selection of 1000 classes is that they can be used to derive image features for general-purpose settings. For instance, the features extracted from the ImageNet data can be employed to define an entirely separate image dataset by passing it through a pretrained CNN and extracting the multidimensional features from the fully connected layers. This new form can also be used for a completely other purpose, such as clustering or retrieval. This technology is so widely used that CNNs are seldom trained from scratch. Because a shared resource such as ImageNet could be employed to extract features in situations when appropriate training data is not available to address various issues, this kind of off-the-shelf feature extraction technique can be employed as a type of transfer learning. If any extra training data is accessible and can only be utilized to fine-tune the deeper layers that are closer to the output layer. The weights (closer to the input) of the early layers are fixed. While holding the early layers fixed, the purpose for training only the deeper layers is that the earlier layers catch only simple features such as edges, whereas the deeper layers capture more complex features. For the application at hand, the simple features do not alter too much, while the deeper features might be vulnerable to the desired application [35].

1.4.16.1 AlexNet

It is worth noting that the original AlexNet architecture had two parallel processing pipelines, which are controlled by two GPUs cooperating to create the training model at a faster speed and with memory sharing. After each convolutional layer, the ReLU activation function was used, followed by normalization and max-pooling. The second convolutional layer filters the first convolutional layer's response-normalized and pooled output. The third, fourth, and fifth convolutional layers have no intervening pooling or normalizing layers. The fully connected layers contain 4096 neurons. To accomplish classification, AlexNet's final layer employs a 1000-way softmax. It is worth noting that the last layer of 4096 activations is frequently utilized to generate a flat 4096-dimensional representation of an image for purposes other than classification. These characteristics may be extracted from any out-of-sample image by simply feeding it through the trained neural network. These characteristics frequently transfer well to various tasks and datasets. In most CNNs today, the activation function is virtually entirely centered on the ReLU, which was not the case prior to AlexNet. In order to enhance generalization, dropout with L2-weight decay was utilized [35]. A simple Python code for AlexNet is given below.

```
from tensorflow.keras.models import Sequential
from tensorflow.keras.layers import Conv2D
from tensorflow.keras.layers import MaxPooling2D,AveragePooling2D,BatchNormalization
from tensorflow.keras.layers import Dense,Activation
from tensorflow.keras.layers import Flatten,Dropout,SpatialDropout2D
from tensorflow.keras.optimizers import SGD
from tensorflow.keras.optimizers import Adam
from tensorflow.keras.callbacks import EarlyStopping, ModelCheckpoint
early_stopping = EarlyStopping(monitor='val_loss', mode='min', patience=10,
                               restore_best_weights=True,verbose=1)

AlexNet = Sequential()

#1st Convolutional Layer
AlexNet.add(Conv2D(filters=96, input_shape=img_shape, kernel_size=(11,11),
                                   strides=(4,4),padding='same'))
AlexNet.add(BatchNormalization())
AlexNet.add(Activation('relu'))
AlexNet.add(MaxPooling2D(pool_size=(2,2), strides=(2,2), padding='same'))

#2nd Convolutional Layer
AlexNet.add(Conv2D(filters=256, kernel_size=(5, 5), strides=(1,1), padding='same'))
AlexNet.add(BatchNormalization())
AlexNet.add(Activation('relu'))
AlexNet.add(MaxPooling2D(pool_size=(2,2), strides=(2,2), padding='same'))
```

```
#3rd Convolutional Layer
AlexNet.add(Conv2D(filters=384, kernel_size=(3,3), strides=(1,1), padding='same'))
AlexNet.add(BatchNormalization())
AlexNet.add(Activation('relu'))

#4th Convolutional Layer
AlexNet.add(Conv2D(filters=384, kernel_size=(3,3), strides=(1,1), padding='same'))
AlexNet.add(BatchNormalization())
AlexNet.add(Activation('relu'))

#5th Convolutional Layer
AlexNet.add(Conv2D(filters=256, kernel_size=(3,3), strides=(1,1), padding='same'))
AlexNet.add(BatchNormalization())
AlexNet.add(Activation('relu'))
AlexNet.add(MaxPooling2D(pool_size=(2,2), strides=(2,2), padding='same'))

#Passing it to a Fully Connected layer
AlexNet.add(Flatten())
# 1st Fully Connected Layer
AlexNet.add(Dense(4096, input_shape=(32,32,3,)))
AlexNet.add(BatchNormalization())
AlexNet.add(Activation('relu'))
# Add Dropout to prevent overfitting
 AlexNet.add(Dropout(0.2))

#2nd Fully Connected Layer
AlexNet.add(Dense(4096))
AlexNet.add(BatchNormalization())
AlexNet.add(Activation('relu'))
#Add Dropout
AlexNet.add(Dropout(0.2))

#3rd Fully Connected Layer
AlexNet.add(Dense(1000))
AlexNet.add(BatchNormalization())
AlexNet.add(Activation('relu'))
#Add Dropout
AlexNet.add(Dropout(0.2))
```

```
#Output Layer
AlexNet.add(Dense(3))
AlexNet.add(BatchNormalization())
AlexNet.add(Activation('softmax'))

AlexNet.compile(optimizer='adam',loss='categorical_crossentropy',metrics=['acc'])

# Train the model
history = AlexNet.fit(
    Xtrain,
    ytrain,
    epochs=200,
    validation_split=0.25,
    batch_size=10,
    verbose=2,
    callbacks=[early_stopping])
# Test on unseen data
results = AlexNet.evaluate(Xtest, ytest)
# Pront the results
print('Final test set loss: {:4f}'.format(results[0]))
print('Final test set accuracy: {:4f}'.format(results[1]))

# plot training history
from matplotlib import pyplot
pyplot.plot(history.history['loss'], label='train')
pyplot.plot(history.history['val_loss'], label='test')
pyplot.legend()
pyplot.show()
```

1.4.16.2 Visual geometry group

Visual geometry group (VGG) [36] also noted the rising trend of increased network depth. The studied networks were built in a variety of topologies with layer sizes ranging from 11 to 19, with 16 or more layers being the most effective. VGG's significant breakthrough was that it lowered filter sizes while increasing depth. It is critical to recognize this type of a smaller filter size since it needs a greater depth. A tiny filter can only capture a limited part of the image if the network is not deep. VGG always employs filters with a spatial footprint of 3×3 and a pooling size of 2×2. The convolution is performed using stride 1 and padding of 1. Stride 2 is applied for pooling. Another intriguing feature of VGG's architecture is that the number of filters is frequently doubled after each max-pooling. The aim is to constantly double the depth when the spatial footprint decreased by a factor of two. This design idea leads in some degree of balance in processing attempt between layers that is utilized by certain subsequent topologies such as ResNet. The usage of deep configurations has the disadvantage of increasing sensitivity with startup that is known to create instability. Pretraining was used to tackle this challenge, where a shallow architecture was initially trained, and then more layers were included [35]. A simple Python code for VGG16 is given below.

```
from tensorflow.keras.models import Sequential
from tensorflow.keras.layers import BatchNormalization
from tensorflow.keras.layers import Dense,Activation
from tensorflow.keras.layers import Flatten,Dropout
from tensorflow.keras.optimizers import Adam
from tensorflow.keras.callbacks import EarlyStopping, ModelCheckpoint
early_stopping = EarlyStopping(monitor='val_loss', mode='min', patience=10,
                                     restore_best_weights=True, verbose=1)

base_model = tf.keras.applications.VGG16(
    include_top=False,
    weights="imagenet",
    input_tensor=None,
    input_shape=img_shape,
    pooling=None
)

for l in base_model.layers:
    l.trainable = False

model = Sequential()
model.add(base_model)

model.add(Flatten())
model.add(BatchNormalization())
model.add(Dense(512,activation='relu'))
model.add(Dropout(0.5))
model.add(Dense(256,activation='relu'))
model.add(Dropout(0.5))
model.add(Dense(128,activation='relu'))
model.add(Dropout(0.5))
model.add(Dense(64,activation='relu'))
model.add(Dropout(0.5))
model.add(Dense(3,activation='softmax'))

model.compile(optimizer='adam',loss='categorical_crossentropy',
              metrics=['acc'])
```

```
# Train the model
history = model.fit(
    Xtrain,
    ytrain,
    epochs=200,
    validation_split=0.25,
    batch_size=10,
    verbose=2,
    callbacks=[early_stopping])
# Test on unseen data
results = model.evaluate(Xtest, ytest)
# Pront the results
print('Final test set loss: {:4f}'.format(results[0]))
print('Final test set accuracy: {:4f}'.format(results[1]))

# plot training history
from matplotlib import pyplot
pyplot.plot(history.history['loss'], label='train')
pyplot.plot(history.history['val_loss'], label='test')
pyplot.legend()
pyplot.show()
```

1.4.16.3 ResNet

ResNet [37] included 152 layers that is nearly an order of size more than existing architectures. Training a 152-layered architecture is often not achievable unless some significant advances are used. The fundamental problem with training such deep networks is to hinder the gradient flow between layers by the enormous number of operations in deep layers, which might change the magnitude of the gradients. Increased depth causes issues such as vanishing and exploding gradients. Even though some deep networks have substantial differences between training and test error, many deep networks have high error on both training and test data. This means that the optimization process has not progressed sufficiently. In neural network learning, layer-wise implementations need all ideas in the image to be at the same level of abstraction, while hierarchical feature engineering is the golden standard. Some concepts need fine-grained connections, whereas others can be learned employing shallow networks [35]. A simple Python code for ResNet101 is given below.

```
from tensorflow.keras import Model
from tensorflow.keras.models import Sequential
from tensorflow.keras.layers import Conv2D
from tensorflow.keras.layers import MaxPooling2D,AveragePooling2D,BatchNormalization
from tensorflow.keras.layers import Dense,Activation
from tensorflow.keras.layers import Flatten,Dropout,SpatialDropout2D,AveragePooling2D,
                                                    GlobalAveragePooling2D
from tensorflow.keras.optimizers import SGD
from tensorflow.keras.optimizers import Adam
import tensorflow as tf
from tensorflow.keras.callbacks import EarlyStopping, ModelCheckpoint

early_stopping = EarlyStopping(monitor='val_loss', mode='min', patience=10,
                               restore_best_weights=True, verbose=1)

base_model = tf.keras.applications.ResNet101(
                include_top=False,
                weights="imagenet",
                input_tensor=None,
                input_shape=img_shape,
                pooling=None,
            )
for layer in base_model.layers:
    layer.trainable = False

model = Sequential()
model.add(base_model)

model.add(Flatten())
model.add(BatchNormalization())
model.add(Dense(128,activation='relu'))
model.add(Dropout(0.2))
model.add(Dense(64,activation='relu'))
model.add(Dropout(0.2))
model.add(Dense(32,activation='relu'))
model.add(Dropout(0.2))
model.add(Dense(3,activation='softmax'))

model.compile(optimizer='adam',loss='categorical_crossentropy',
              metrics=['acc'])
```

```
# Train the model
history = model.fit(
    Xtrain,
    ytrain,
    epochs=200,
    validation_split=0.25,
    batch_size=10,
    verbose=2,
    callbacks=[early_stopping])
# Test on unseen data
results = model.evaluate(Xtest, ytest)
# Pront the results
print('Final test set loss: {:4f}'.format(results[0]))
print('Final test set accuracy: {:4f}'.format(results[1]))

# plot training history

from matplotlib import pyplot
pyplot.plot(history.history[ 'loss'], label='train')
pyplot.plot(history.history[ 'val_loss'], label='test')
pyplot.legend()
pyplot.show()
```

1.4.16.4 MobileNet architecture

The MobileNet model is developed on depthwise separable convolutions, a type of factorized convolution that divides a regular convolution into a depthwise convolution and a 1×1 convolution known as a pointwise convolution. In MobileNets, a single filter is added to each input channel with depthwise convolution. In order to integrate the depthwise convolution outputs, the pointwise convolution is then followed by a 1×1 convolution. In a single step, in order to generate a new set of outputs, a standard convolution filters and mixes inputs. The depthwise separable convolution divides this into two layers, one for filtering and another for combining. In order to break the interaction between the kernel size and the number of output channels, MobileNet employs depthwise separable convolutions. The ordinary convolution includes the impact of filtering and merging data based on the convolutional kernels to build a new representation. For a substantial reduction in computational cost, the filtering and combining stages can be split into two parts by means of factorized convolutions called depthwise separable convolutions. Depthwise separable convolutions are composed of two layers: depthwise and pointwise convolutions. The output of the depthwise layer is then linearly combined by means of pointwise convolution, a basic 1×1 convolution. MobileNets utilize batchnorm and ReLU nonlinearities for both layers. Depthwise convolution is far more efficient than ordinary convolution. It does, however, simply filter input channels and does not merge them to generate new features. An extra layer, which computes a linear combination of the output of depthwise convolution with 1×1 convolution is essential in order to generate these new features. Downsampling is employed in the depthwise convolutions as well as the first layer using strided convolution. A last average pooling decreases the spatial resolution to 1, before the completely linked layer. MobileNet includes 28 layers if depthwise and pointwise convolutions are included as independent layers [38]. A simple Python code for MobileNet is given below.

```
from tensorflow.keras.models import Sequential
from tensorflow.keras.layers import  BatchNormalization
from tensorflow.keras.layers import Dense,Activation
from tensorflow.keras.layers import Flatten,Dropout,SpatialDropout2D
from tensorflow.keras.optimizers import SGD
from tensorflow.keras.optimizers import Adam
import tensorflow as tf
from tensorflow.keras.callbacks import EarlyStopping, ModelCheckpoint
early_stopping = EarlyStopping(monitor='val_loss', mode='min', patience=10,
                              restore_best_weights=True, verbose=1)

base_model = tf.keras.applications.MobileNet(
                alpha = 0.75,
                include_top=False,
                weights="imagenet",

                 input_tensor=None,
                 input_shape=img_shape,
                 pooling=None,
            )
 for layer in base_model.layers:
     layer.trainable = False

 model = Sequential()
 model.add(base_model)

 model.add(Flatten())
 model.add(BatchNormalization())
 model.add(Dense(256,activation='relu'))
 model.add(Dropout(0.2))
 model.add(Dense(128,activation='relu'))
 model.add(Dropout(0.2))
 model.add(Dense(64,activation='relu'))
 model.add(Dropout(0.2))
 model.add(Dense(32,activation='relu'))
 model.add(Dropout(0.2))
 model.add(Dense(3,activation='softmax'))

 model.compile(optimizer='adam',loss='categorical_crossentropy',
              metrics=['acc'])
```

```
# Train the model
history = model.fit(
    Xtrain,
    ytrain,
    epochs=200,
    validation_split=0.25,
    batch_size=10,
    verbose=2,
    callbacks=[early_stopping])
# Test on unseen data
results = model.evaluate(Xtest, ytest)
# Pront the results
print('Final test set loss: {:4f}'.format(results[0]))
print('Final test set accuracy: {:4f}'.format(results[1]))

# plot training history
from matplotlib import pyplot
pyplot.plot(history.history[ 'loss'], label='train')
pyplot.plot(history.history[ 'val_loss'], label='test')
pyplot.legend()
pyplot.show()
```

1.4.16.5 Inception-v4 and Inception-ResNet

The Inception architecture is highly adjustable, which means that the amount of filters in the various layers can be changed without affecting the quality of the fully trained network. To balance the calculation among the numerous model subnetworks and improve training speed, the layer sizes are adjusted carefully. For Inception-v4, identical selections are made for the Inception blocks across all grid sizes in order to get rid of unnecessary baggage. Cheaper Inception blocks are employed for the residual versions of the Inception networks than the original Inception. Each Inception block is followed by a filter-expansion layer that is used to increase the dimensionality of the filter bank before adding it to meet the depth of the input criteria. This is required to compensate for the reduction in dimensionality caused by the Inception block. The residual version of Inception is tested with many variations. Only two of them are discussed in detail here. The first, "Inception-ResNet-v1," essentially corresponds to the computational cost of Inception-v3, while "Inception-ResNet-v2" corresponds to the raw cost of the recently proposed Inception-v4 network. Another minor technical difference between residual and nonresidual Inception variations is that batch normalization is utilized only on top of the standard layers, not on top of the summations, in the case of Inception-ResNet. Although it is rational to predict that extensive usage of batch normalization would be beneficial, it is aimed to make each model replica trainable on a single GPU. The overall number of Inception blocks is significantly increased by eliminating the batch normalization on top of those layers. The more usage of computer resources will eliminate the need for this trade-off [39]. A simple Python code for Inception-ResNet-V2 is given below.

```
from tensorflow.keras.models import Sequential
from tensorflow.keras.layers import BatchNormalization
from tensorflow.keras.layers import Dense,Activation
from tensorflow.keras.layers import Flatten,Dropout,SpatialDropout2D
from tensorflow.keras.optimizers import Adam
import tensorflow as tf
from tensorflow.keras.callbacks import EarlyStopping, ModelCheckpoint
early_stopping = EarlyStopping(monitor='val_loss', mode='min', patience=10,
                               restore_best_weights=True, verbose=1)

base_model = tf.keras.applications.InceptionResNetV2(
                include_top=False,
                weights="imagenet",
                input_tensor=None,
                input_shape=img_shape,
                pooling=None,
            )
for layer in base_model.layers:
    layer.trainable = False

model = Sequential()
model.add(base_model)


model.add(Flatten())
model.add(BatchNormalization())
model.add(Dense(256,activation='relu'))
model.add(Dropout(0.2))
model.add(Dense(128,activation='relu'))
model.add(Dropout(02))
model.add(Dense(64,activation='relu'))
model.add(Dropout(0.2))
model.add(Dense(32,activation='relu'))
model.add(Dropout(0.2))
model.add(Dense(3,activation='softmax'))

model.compile(optimizer='adam',loss='categorical_crossentropy'
              metrics=['acc'])
```

```
# Train the model
history = model.fit(
    Xtrain,
    ytrain,
    epochs=200,
    validation_split=0.25,
    batch_size=10,
    verbose=2,
    callbacks=[early_stopping])
# Test on unseen data
results = model.evaluate(Xtest, ytest)
# Pront the results
print('Final test set loss: {:4f}'.format(results[0]))
print('Final test set accuracy: {:4f}'.format(results[1]))

# plot training history
from matplotlib import pyplot
pyplot.plot(history.history[ 'loss'], label='train')
pyplot.plot(history.history['val_loss'], label='test')
pyplot.legend()
pyplot.show()
```

1.4.16.6 Xception

Chollet [40] suggested a depthwise separable CNN model based purely on convolution layers. In fact, the mapping of cross-channel and spatial correlations in CNN feature maps may be completely dissociated. As this theory is a more powerful version of the hypothesis that underpins the Inception design, and the suggested architecture is termed Xception (short for "Extreme Inception"). Based on this stronger premise, an "extreme" version of an Inception module would first utilize a 1×1 convolution to map cross-channel correlations and then individually map the spatial correlations of each output channel. This extreme version of an Inception module is almost equivalent to a depthwise separable convolution, which is a neural network building procedure that has gained popularity as its adoption in the TensorFlow framework. In deep learning frameworks such as TensorFlow and Keras, a depthwise separable convolution consists of a depthwise convolution. The Xception architecture contains 36 convolutional layers that serve as the network's feature extraction foundation. Because image classification is the only focus of the experimental assessment, the suggested convolutional base will be followed by a logistic regression layer. Before the logistic regression layer, fully connected layers can be added if desired. The 36 convolutional layers are divided into 14 modules, each of which is surrounded by linear residual connections, except for the first and last modules. In a nutshell, the Xception architecture is a linear stack of depthwise detachable convolution layers with residual links. This allows the architecture to characterize and reconfigure very simply; it only takes 30–40 lines of code to define and modify using a high-level library such as Keras or TensorFlow, similar to architectures such as VGG-16, but dissimilar architectures such as Inception V2 or V3 that are far more complicated to describe [40]. A simple Python code for Xception is given below.

```
from tensorflow.keras.models import Sequential
from tensorflow.keras.layers import BatchNormalization
from tensorflow.keras.layers import Dense,Activation
from tensorflow.keras.layers import Flatten,Dropout,SpatialDropout2D
from tensorflow.keras.optimizers import Adam
import tensorflow as tf
from tensorflow.keras.callbacks import EarlyStopping, ModelCheckpoint
early_stopping = EarlyStopping(monitor='val_loss', mode='min', patience=10,
                               restore_best_weights=True, verbose=1)

base_model = tf.keras.applications.Xception(
                include_top=False,
                weights="imagenet",
                input_tensor=None,
                input_shape=img_shape,
                pooling=None,
            )
for layer in base_model.layers:
    layer.trainable = False

model = Sequential()
model.add(base_model)

model.add(Flatten())
model.add(BatchNormalization())
model.add(Dense(256,activation='relu'))
model.add(Dropout(0.2))
model.add(Dense(128,activation='relu'))

model.add(Dropout(02))
model.add(Dense(64,activation='relu'))
model.add(Dropout(0.2))
model.add(Dense(32,activation='relu'))
model.add(Dropout(0.2))
model.add(Dense(3,activation='softmax'))

model.compile(optimizer='adam',loss='categorical_crossentropy',
              metrics=['acc'])
```

```
# Train the model
history = model.fit(
    Xtrain,
    ytrain,
    epochs=200,
    validation_split=0.25,
    batch_size=10,
    verbose=2,
    callbacks=[early_stopping])
# Test on unseen data
results = model.evaluate(Xtest, ytest)
# Pront the results
print('Final test set loss: {:4f}'.format(results[0]))
print('Final test set accuracy: {:4f}'.format(results[1]))

# plot training history
from matplotlib import pyplot
pyplot.plot(history.history['loss'], label='train')
pyplot.plot(history.history['val_loss'], label='test')
pyplot.legend()
pyplot.show()
```

1.4.16.7 Densely connected convolutional networks

One benefit of ResNets is that the gradient can flow straight via the identity function from later layers to previous levels. Yet, the identity function and the output of H′ are merged through summing, that might restrict information flow in the network. A novel connectivity pattern is suggested with direct connections from any layer to all following layers, in order to increase information flow between layers even further. This network design is known as a dense convolutional network (DenseNet) due to its dense connectivity. Downsampling layers, which vary the size of feature maps, are an important component of convolutional networks. In this architecture, the network is partitioned into numerous densely linked dense blocks to assist downsampling. Layers between blocks are referred to as transition layers because they perform convolution and pooling. In testing, a batch normalization layer is employed with a 1×1 convolutional layer, and a 2×2 average pooling layer as transition layers. DenseNet can have very small layers, which distinguishes it from conventional network topologies [41]. A simple Python code for DenseNet121 is given below.

```
from tensorflow.keras.models import Sequential
from tensorflow.keras.layers import BatchNormalization
from tensorflow.keras.layers import Dense,Activation
from tensorflow.keras.layers import Flatten,Dropout,SpatialDropout2D
from tensorflow.keras.optimizers import Adam
import tensorflow as tf
from tensorflow.keras.callbacks import EarlyStopping, ModelCheckpoint
early_stopping = EarlyStopping(monitor='val_loss', mode='min', patience=10,
                               restore_best_weights=True, verbose=1)

base_model = tf.keras.applications.DenseNet121(
                include_top=False,
                weights="imagenet",
                input_tensor=None,
                input_shape=img_shape,
                pooling=None,
            )
for layer in base_model.layers:
    layer.trainable = False

model = Sequential()
model.add(base_model)

model.add(Flatten())
model.add(BatchNormalization())
model.add(Dense(256,activation='relu'))
model.add(Dropout(0.2))
model.add(Dense(128,activation='relu'))
model.add(Dropout(02))
model.add(Dense(64,activation='relu'))
model.add(Dropout(0.2))
model.add(Dense(32,activation='relu'))
model.add(Dropout(0.2))
model.add(Dense(3,activation='softmax'))

model.compile(optimizer='adam',loss='categorical_crossentropy',
              metrics=['acc'])
```

```
# Train the model
history = model.fit(
    Xtrain,
    ytrain,
    epochs=200,
    validation_split=0.25,
    batch_size=10,

    verbose=2,
    callbacks=[early_stopping])
# Test on unseen data
results = model.evaluate(Xtest, ytest)
# Pront the results
print('Final test set loss: {:4f}'.format(results[0]))
print('Final test set accuracy: {:4f}'.format(results[1]))

# plot training history
from matplotlib import pyplot
pyplot.plot(history.history['loss'], label='train')
pyplot.plot(history.history['val_loss'], label='test')
pyplot.legend()
pyplot.show()
```

1.4.16.8 Feature extraction with pretrained models

Pretrained CNNs from publicly available resources such as ImageNet are frequently available for usage in different applications and datasets. This is accomplished by retaining the majority of the pretrained weights in the neural network with the exception of the final classification layer. The final classification layer's weights are determined by the dataset at hand. The last layer must be trained since the class labels in a given situation may differ from those in ImageNet. Nonetheless, the weights in the early layers are still valuable since they learn different sorts of shapes in images which can be used for nearly any type of classification application. In addition, feature activations in the penultimate layer can be employed for unsupervised applications. It is worth noting that the usage of pretrained convolutional networks is so widespread that training is almost never initiated from scratch [35].

Many deeper architectures with feed-forward topologies include numerous layers where consecutive transformations of the preceding layer's inputs lead to progressively complex data representations. In the output layer, properly transformed feature interpretations are more susceptible to basic sorts of predictions. The nonlinear activations in intermediate layers are responsible for this level of sophistication. Tanh and sigmoid activation functions were traditionally the most common options in the hidden layers, but the ReLU activation has gained popularity in recent years due to the attractive attribute, which is better at avoiding the disappearing and bursting gradient difficulties. One way to look at the division of work between the hidden layers and the final prediction layer is that the early layers provide a feature representation, which is more appropriate for the task at hand. This learnt feature representation is subsequently used by the final layer. The characteristics learned in the hidden layers are mostly generalizable to different datasets and problem settings in the same domain. This characteristic may be used in a variety of ways by simply replacing a pretrained network's output node(s) with a new application-specific output layer for problem at hand and the dataset. Each hidden layer produces a transformed feature representation of the data, the dimensionality of which is specified by the number of units in that layer. This approach may be viewed as a type of hierarchical feature engineering, with features in earlier layers representing rudimentary data qualities and those in later layers representing complicated data attributes having semantic meaning to the class labels. Transfer learning is another term for the method of employing pretrained models [35]. A DenseNet121 deep feature extraction example is given below. For the details check the machinelearningmastery.com web page[3]

```
# Example of using the DenseNet121 model as a feature extraction model
                                  (Adapted from machinelearningmastery.com)
from keras.preprocessing.image import load_img
from keras.preprocessing.image import img_to_array
from keras.applications.densenet import preprocess_input
from keras.applications.densenet import decode_predictions
from keras.applications.densenet import DenseNet121
from keras.models import Model
from pickle import dump
# load an image from file
image = load_img('BrainMRI.jpg', target_size=(224, 224))
# convert the image pixels to a numpy array
image = img_to_array(image)
# reshape data for the model
image = image.reshape((1, image.shape[0], image.shape[1], image.shape[2]))
# prepare the image for the DenseNet121 model
image = preprocess_input(image)
# load model
model = DenseNet121()
# remove the output layer
model = Model(inputs=model.inputs, outputs=model.layers[-2].output)
# get extracted features
features = model.predict(image)
print(features.shape)
```

[3] https://machinelearningmastery.com/how-to-use-transfer-learning-when-developing-convolutional-neural-network-models/.

References

[1] E. Alpaydin, Introduction to Machine Learning, MIT press, 2014.

[2] A. Subasi, Practical Machine Learning for Data Analysis Using Python, Academic Press, 2020.

[3] P. Kulkarni, Reinforcement and Systemic Machine Learning for Decision Making, 1, John Wiley & Sons, 2012.

[4] B.C. Patel, G. Sinha, Abnormality detection and classification in computer-aided diagnosis (CAD) of breast cancer images, J. Med. Imaging Health Inform. 4 (6) (2014) 881–885.

[5] V. N. P. Raj, T. Venkateswarlu, Denoising of medical images using undecimated wavelet transform, 2011, pp. 483–488.

[6] D.-H. Trinh, M. Luong, F. Dibos, J.-M. Rocchisani, C.-D. Pham, T.Q. Nguyen, Novel example-based method for super-resolution and denoising of medical images, IEEE Trans. Image Process. 23 (4) (2014) 1882–1895.

[7] J.E. Fowler, The redundant discrete wavelet transform and additive noise, IEEE Signal. Process. Lett. 12 (9) (2005) 629–632.

[8] A. Choubey, G. Sinha, S. Choubey, A hybrid filtering technique in medical image denoising: Blending of neural network and fuzzy inference, in: IEEE 3rd International Conference on Electronics Computer Technology. Kanyakumari, India, 2011, vol. 1, pp. 170–177.

[9] H.R. Sheikh, A.C. Bovik, Image information and visual quality, IEEE Trans. Image Process 15 (2) (2006) 430–444.

[10] D. Bhonsle, G. Sinha, V.K. Chandra, Medical image de-noising using combined Bayes shrink and total variation techniques, Artificial Intelligence and Machine Learning in 2D/3D Medical Image Processing, CRC Press, 2020, pp. 31–52.

[11] B.I. Choi, The current status of imaging diagnosis of hepatocellular carcinoma, Liver Transpl. 10 (S2) (2004) S20–S25.

[12] R.B. Freeman, et al., Optimizing staging for hepatocellular carcinoma before liver transplantation: a retrospective analysis of the UNOS/OPTN database, Liver Transpl. 12 (10) (2006) 1504–1511.

[13] K. Rawal, G. Sethi, D. Ghai, Medical imaging in healthcare applications, Artificial Intelligence and Machine Learning in 2D/3D Medical Image Processing, CRC Press, 2020, pp. 97–106.

[14] G. Sethi, B.S. Saini, D. Singh, Segmentation of cancerous regions in liver using an edge-based and phase congruent region enhancement method, Comput. Electr. Eng. 53 (2016) 244–262. Available from: https://doi.org/10.1016/j.compeleceng.2015.06.025. Jul.

[15] G. Sethi, B. Saini, Segmentation of abdomen diseases using active contour models in CT images, Biomed. Eng. Appl. Basis Commun. 27 (05) (2015) 1550047.

[16] A. Quatrehomme, I. Millet, D. Hoa, G. Subsol, W. Puech, Assessing the classification of liver focal lesions by using multi-phase computer tomography scans, Medical Content-Based Retrieval for Clinical Decision Support, Springer, 2012, pp. 80–91.

[17] S.G. Mougiakakou, I.K. Valavanis, N.A. Mouravliansky, A. Nikita, K.S. Nikita, DIAGNOSIS: a telematics-enabled system for medical image archiving, management, and diagnosis assistance, IEEE Trans. Instrum. Meas. 58 (7) (2009) 2113–2120.

[18] B. Gopinath, N. Shanthi, Computer-aided diagnosis system for classifying benign and malignant thyroid nodules in multi-stained FNAB cytological images, Australas. Phys. Eng. Sci. Med. 36 (2) (2013) 219–230.

[19] S.S. Kumar, R.S. Moni, J. Rajeesh, An automatic computer-aided diagnosis system for liver tumours on computed tomography images, Comput. Electr. Eng. 39 (5) (2013) 1516–1526. Available from: https://doi.org/10.1016/j.compeleceng.2013.02.008. Jul.

[20] A. Vasuki, S. Govindaraju, Deep neural networks for image classification, Deep Learning for Image Processing Applications, 31, IOS Press, 2017, p. 27.

[21] S. Shalev-Shwartz, S. Ben-David, Understanding Machine Learning: From Theory to Algorithms, Cambridge University Press, 2014.

[22] K.P. Murphy, Machine Learning: A Probabilistic Perspective, MIT press, 2012.

[23] S. Theodoridis, A. Pikrakis, K. Koutroumbas, D. Cavouras, Introduction to Pattern Recognition: A MATLAB Approach, Academic Press, 2010.

[24] S. Tatiraju, A. Mehta, Image segmentation using k-means clustering, EM and normalized cuts, Dep. EECS 1 (2008) 1–7.

[25] X. Zheng, Q. Lei, R. Yao, Y. Gong, Q. Yin, Image segmentation based on adaptive K-means algorithm, EURASIP J. Image Video Process 2018 (1) (2018) 68.

[26] J. Gaura, E. Sojka, M. Krumnikl, Image segmentation based on k-means clustering and energy-transfer proximity, in: ISVC'11: Proc. 7th International Conference on Advances in Visual Computing - Volume Part II, 2011, pp. 567–577.

[27] J. Han, J. Pei, M. Kamber, Data Mining: Concepts and Techniques, Elsevier, 2011.

[28] L. Breiman, J.H. Friedman, R.A. Olshen, C.J. Stone, Classification and Regression Trees, Hall/CRC, Boca Raton, 1984.

[29] K. Grąbczewski, , 1 Meta-Learning in Decision Tree Induction, Springer, 2014.

[30] I.H. Witten, E. Frank, M.A. Hall, C.J. Pal, Data Mining: Practical Machine Learning Tools and Techniques, Morgan Kaufmann, 2016.

[31] M. Hall, I. Witten, E. Frank, Data Mining: Practical Machine Learning Tools and Techniques, Kaufmann. Burlingt. (2011).
[32] T. Chen, C. Guestrin, Xgboost: a scalable tree boosting system, 2016, pp. 785–794.
[33] S. Bhattacharya, et al., A novel PCA-firefly based XGBoost classification model for intrusion detection in networks using GPU, Electronics 9 (2) (2020) 219.
[34] J.D. Kelleher, Deep Learning, Mit Press, 2019.
[35] C.C. Aggarwal, Neural Networks and Deep Learning, Springer, 2018.
[36] K. Simonyan, A. Zisserman, Very deep convolutional networks for large-scale image recognition, ArXiv Prepr. ArXiv14091556 (2014).
[37] K. He, X. Zhang, S. Ren, J. Sun, Deep residual learning for image recognition, 2016 IEEE Conference on Computer Vision and Pattern Recognition (CVPR), 2016, pp. 770–778.
[38] A.G. Howard, et al., Mobilenets: efficient convolutional neural networks for mobile vision applications, ArXiv Prepr. ArXiv170404861 (2017).
[39] C. Szegedy, S. Ioffe, V. Vanhoucke, A. Alemi, Inception-v4, inception-ResNet and the impact of residual connections on learning, 2017, vol. 31, no. 1.
[40] F. Chollet, Xception: deep learning with depthwise separable convolutions, 2017, pp. 1251–1258.
[41] G. Huang, Z. Liu, L. Van Der Maaten, K.Q. Weinberger, Densely connected convolutional networks, 2017, pp. 4700–4708.

CHAPTER

2

Lung cancer detection from histopathological lung tissue images using deep learning

Aayush Rajput[1] *and Abdulhamit Subasi*[2,3]

[1]Indian Institute of Technology, Kharagpur, India [2]Institute of Biomedicine, Faculty of Medicine, University of Turku, Turku, Finland [3]Department of Computer Science, College of Engineering, Effat University, Jeddah, Saudi Arabia

OUTLINE

2.1 Introduction

Cancer is the term used for a collection of similar types of diseases in which the tissues of a body organ start growing abnormally. When we inhale air, it goes through the windpipe (trachea), then air is divided through smaller pipes known as bronchi, further, it is divided into finer tubes called bronchioles. Alveoli are the tiny sacs present at the ending of bronchioles. Lung cancer starts mainly from the bronchioles, alveoli, and bronchi. Lung cancer

Applications of Artificial Intelligence in Medical Imaging
DOI: https://doi.org/10.1016/B978-0-443-18450-5.00008-6

is a deadly disease with a 5-year survival rate of (18.6%) lower than other cancers such as breast and prostate [1]. It is the most common type of cancer, specifically in the United States, responsible for 154,050 deaths there alone, around a quarter of all the deaths are due to cancer in the country [2]. Lung cancer can be broadly classified into nonsmall-cell lung cancer (NSCLC) and small-cell lung cancer (SCLC). The NSCLC is the most common type of lung cancer, which is further classified into three main categories: adenocarcinoma, squamous-cell carcinoma, and large-cell carcinoma. The adenocarcinoma begins from the mucus-producing cell. Adenocarcinoma is found in the outer part of the lungs and can be detected before its spread. It is a common type in other types of cancer such as breast, colorectal, and prostate. The second one, squamous-cell carcinoma, begins its growth from the squamous cells. It is primarily found in the people who smoke. In the lungs, it affects the central part near the bronchus. The third type, large-cell carcinoma cancer, does not involve a particular area in the lungs. It can grow in any region and is difficult to diagnose with a faster growth, which makes it more complex to treat. NSCLC has some other types, but they are not common compared to these three types, for example, adenosquamous carcinoma and sarcomatoid carcinoma. SCLC covers only 10%–15% of all lung cancers [3].

There are cases in which cancer starts from one part of the body and, after spreading, affects another part of the body, but the cancer type will be based on the organ that is affected the first. For example, breast cancer can sometimes affect the lungs, but it will not be called lung cancer and the treatment will be to cure breast cancer. Till now, no one knows how to prevent the danger of lung cancer completely. But there are some ways through which the risk factor of cancer can be reduced. The first and foremost way is to avoid smoking [4]. In the early stage of lung cancer, the lung tries to repair the tissue, but the lung will not improve if the person smokes during that phase. Thus by quitting smoking, person can reduce the danger of lung cancer to a large extent. Keeping a minimum distance from the people diagnosed with cancer is also a way to protect ourselves from the cancer-causing agents. Taking a balanced and healthy diet also helps in reducing the risk factor of lung cancer. But the reduction in risk factors is way less than the increase in it due to smoking. So, first of all, quitting smoking should be the first step. However, the risk from lung cancer cannot be zero for anyone as there are instances of people having lung cancer who do not have any high-risk factors. The early detection of lung cancer is the most common factor among all the successful lung cancer treatments. Some types of lung cancer start showing the symptoms in the early stage of its development, through which it can be detected. The symptoms of lung cancer are most likely to be caused by other reasons, but it is essential to see the doctor so that if lung cancer is causing the symptoms, it can be detected and treated in its early phases. The most common symptoms of lung cancer include a last-longing cough, rust color phlegm, blood coughing, loss of appetite, breath shortness, the feeling of being tired, chest pain, persistent infections such as pneumonia, and wheezing. Widespread lung cancer can also affect the other parts of the body, which can also cause some symptoms such as bone pain, headache, balance problems or seizures, and yellowing of skin and eyes [5]. These all result from lung cancer spreading to the other parts such as the liver and nervous system.

There are several tests a doctor can use for testing lung cancer. Tests include imaging tests, sputum cytology, and biopsy [6]. An X-ray image is used in an imaging test to detect the abnormal mass present in the lung tissues. CT scan can be used to see the more minor details which X-ray might not notice. In sputum cytology, the sputum is used and observed under the microscope. The sputum of a person with lung cancer contains lung cancer cells that can be detected under the microscope. A sample of abnormal cells is taken from the potentially infected areas for diagnosis in a tissue sample or biopsy.

The treatment of lung cancer also has some possible side effects. Different persons react in different ways to the treatment process. The person should be fully prepared for these effects before the treatment. Some of the side effects of the treatment process are anemia as the chemotherapy lowers blood counts, nausea, vomiting, diarrhea, constipation, hair loss, changes in sexual functioning, and effects on fertility [7]. The most severe side effects are lowering white blood cell counts, increasing the infection, and heart disfunctioning. Patients who have a current heart problem or a history of heart problems should consult the doctor about it to avoid further complications. Palliative or supportive care is a way to ease the path of treatment for the patient and family. The care is not for treating specifically lung cancer; instead, it can be given to anyone of any age and stage of cancer. It is provided along with the curative treatment. This care improves the quality of life of the patient and provides an extra shield of protection.

The process of biopsy is often time-consuming and required very skilled practitioners or pathologists for the detection of the type of lung cancer. The human detection of the kind of lung cancer sometimes leads to a wrong prediction which can cause a massive cost to the patient in the form of side effects and financial burden, and sometimes even life. People suffered from it have a possibility of cancer coming back. Doctors advise regular check-ups and therapies such as radiotherapy and chemotherapy to control and minimize the likelihood of cancer attacking back. The side effects of the treatment can last very long, and it is not easy for the patient to learn to live with them. Any symptoms or difficulty should be immediately consulted with the doctor as soon as possible to avoid further complications. Sometimes the patient can feel very depressed facing all the complex situations. The family members and friends of the patient must help the patient emotionally. Deep learning (DL) is a subfield of artificial intelligence (AI) that enables computers to make decisions based on the data. DL can speed up the process of lung cancer detection through images to a vast extent. Transfer learning is a technique through which the prediction accuracy can be improved to a greater extent. The pretrained DL models trained on a huge dataset for weeks can be directly used through transfer learning without wasting time in its training.

In this chapter, we have used different DL techniques and combined transfer learning with classical machine learning models such as support vector machine (SVM), Random Forest, k-nearest neighbors (kNN), XGBoost, AdaBoost, and others to make it possible model to give accurate results. Different metrics are used so that the actual performance of the model can be known. The images are passed through the pretrained models, and the essential features are extracted and trained the classical machine learning models on these features.

2.2 Literature review

Atsushi et al. [8] used the deep convolutional neural network model for automating the process of lung cancer detection using the patch-based segmentation of malignant regions in images. The images were taken from the camera attached with the microscope of a $40\times$ objective lens. Original images were of 224×224-pixel resolution. A total of 306 benign and 315 malignant images were used to generate 60,000 patch images using data augmentation using techniques such as flipping, rotation, filtering, and color adjustment. The model used for the prediction was the deep convolutional neural network model based on fine-tuned VGG16 model with three layers having 1024, 256, and 2 units replacing the fully connected layers of the original VGG16 model. This method achieved an area under the ROC curve (AUC) score of 0.872 for the patch-based classification using augmented images.

Ausawalaithong et al. [9] used DL and transfer learning to classify lung cancer using chest X-ray images. The proposed model provides a heat map for identifying the location of the lung nodule. The dataset used in the study was taken from more than one source. They used JSRT (Japanese Society of Radiology Technology) datasets consisting of 247 frontal chest X-ray images with 154 images with lung nodules and 93 images without lung nodules. All images have a resolution of 2048 × 2048 pixels. The data was also taken from the Chest X-ray14 Dataset containing 112,120 frontal chest X-ray images. Every image has a resolution of 1024 × 1024 pixels. However, the dataset does not contain any lung cancer images. This dataset was used to compensate the lung cancer data that has only 100 cases by first training to recognize nodules, using nodule cases as positive and all remaining cases as negative. The 121-layered densely connected convolution network (DenseNet121) was used to replace the last fully connected layer with a single sigmoid node to get the output probability. The transfer learning was applied twice in the study, first for classification as "with nodule" or "without nodule" and second for classification as "with malignant nodule" and "without malignant nodule." The model has given the accuracy score, specificity, and sensitivity score of 74.43% ± 6.01%, 74.96% ± 9.85%, and 74.68% ± 15.33%.

Hatuwal et al. [10] used the images from LC25000 lung histopathological images. Images were resized to 180 × 180-pixel resolution, and the pixels were normalized. They trained a convolutional neural network (CNN) model having a neural network with three hidden layers, an input, and one fully connected layer. Their model achieved a training accuracy score and validation accuracy score of 96.11% and 97.2%, respectively.

Chakravarthy et al. [11] used CT lung Dicom images from Lung Image Database Consortium (LIDC) for the study. The MATLAB 2013a software was used for the preprocessing. The images were filtered by using the fast type of filter [12] for reducing the noise present in the images. First, the grayscale lung image was transformed to binary image type by replacing the more significant than a threshold value to 1 and the rest to 0. Then after the gray level co-occurrence matrices (GLCMs) feature extraction was done. Then chaotic crow search algorithm (CCSA) was used to select the GLCM features. A probabilistic neural network was used for training on the selected features by CCSA. The study results achieved the score of 95%, 85%, 90%, 86.36%, and 94.44% for sensitivity, specificity, accuracy, precision, and negative predictive value, respectively.

Sasikala et al. [13] used the CNN for the classification of tumors as malignant or benign. The images used were chest CT images taken from the LIDC and Image Database Resource Initiative (IDRI) [14], consisting of 1000 scans of malignant and benign lung tumors. The lung region was first extracted from the images, and then slices were segmented to get the tumor region. In the preprocessing, the median filter was used. Backpropagation algorithms train the model with a Rectified Linear Unit (ReLU) activation function and a softmax layer as the output layer. The data normalization step was not performed on the images while giving input to the model for training. The whole study was done using MATLAB software. The model gave specificity, sensitivity, and accuracy scores of 1, 0.875, and 0.96, respectively.

Serj et al. [15] also used CNN for lung cancer diagnosis. They proposed a new DL model for better accuracy and low variance in binary classification tasks. The dataset used was taken from the Kaggle Data Science Bowl 2017 (KDSB17) [16]. The proposed model has two max-pooling layers, a fully body convolution layer, and one fully connected layer with two softmax units. The input size of the image used was 120×120 pixels. The activation function used was the ReLU to produce output. The loss function used was cross-entropy which aims to maximize the multinomial logistic regression objective. Sensitivity, specificity, and F1 metrics were used to evaluate the performance of the model. The complete process was completed using MATLAB software, and the result obtained sensitivity, specificity, and F1 scores of 0.87, 0.991, and 0.95, respectively.

Masood et al. [17] proposed a computer-assisted diagnostic system for lung cancer detection using a novel DL-based model. They used the DFCNet [18] model based on the deep, fully CNN and CNN to classify detected nodules into four lung cancer stages. The comparison was made with other existing CNN techniques. They proposed an IoT-based approach in which sensors are used to collect the information from the patient body. The data from sensors is then collected through a computer and then fed to the model to classify the nodules. The images were resized to 100×100 to input into the DFCNet with the Gabor filter for preprocessing and enhanced pixels for segmentation of region of interest. The CT scan images of 18 patients from the Shanghai Hospital were taken out of which 11 were used for training and rest for testing. The LIDC-IDRI, RIDER, SPIE challenge dataset, LUNA16, and Lung CT-Diagnosis dataset were also used for evaluating the model. The images were analyzed patch wise. The CNN architecture includes 7 convolution layers with Parametric ReLU activation function, 7 max-pooling layers, 7 batch normalization layers, and 2 dense layers having Leaky Rectified Linear Units. All layers used 3×3-sized filters except the last three convolutional layers, 7×7-sized filter was used for them. The main model performance was evaluated by the dice score. The result of achieved median dice score was 91.34% and accuracy score of CNN and DFCNet score was 77.6% and 84.58%.

2.3 Artificial intelligence models

2.3.1 Artificial neural networks

An artificial neural network (ANN) is a simple deep network closely representing the arrangement of neurons inside the human brain. In the linear model, the input layer is directly mapped to the output layers through a simple affine transformation. Still, the linear model has the problem of underfitting, and often their performance is not very good with the complex data. So, to avoid the limitations of the linear model, extra stacks of hidden layers are introduced into the input and output layer of a linear model. This architecture is called ANN or multilayer perceptron [41]. In ANN, the initial layer is the input layer. Each unit or neuron is connected to every unit of the next layer, and every layer except the initial layer is also connected to each unit of the previous layer. Every neuron of one layer influences every neuron in the next layer in the fully connected layer. The cost of parameterization of ANN is very high.

The overall structure of an ANN consists of three parts: input layer, hidden layers, and output layer. Every layer also has a bias neuron and is connected to every neuron of next layer. ANN having more than or equal to 2 layers is called a deep neural network (DNN). The first method to train an ANN is described by Rumelhart et al. [19] in the article proposing the backpropagation algorithm. The complete training data is passed to the model, and the result is calculated for each instance, then error between predicted and true result is calculated. This initializes the backpropagation step. The contribution to error due to each neuron is calculated, and the sum of error for every instance of data is minimized. For the proper working of the backpropagation algorithm, activation functions are used. Popular activation functions include hyperbolic tangent function tanh, logistic function, and ReLU function.

The output range in tanh function is from −1 to 1, that's why the output from it is normalized [42]. In the logistic function, the output is from 0 to 1, while in the ReLU function, the output is continuous. Practically ReLU is faster to compute than the other two mentioned activation functions.

The ANN is often used for the classification task where the n neurons in the output layer represent n classes, and the value of each neuron represents the probability class. The activation used for the output layer will be softmax. The ANN model can detect the complex features present in the data; hence, it is more helpful than other models. It has been observed that an ANN with only one hidden layer with more neurons can model the most complex functions. One of the essential features of using DNN is that a pretrained model can be used as the initial layers of another model with the same weights to extract the input features and thus make the model's training faster and perform better. In DL, the last part of the convolution neural network consists of the ANN layer [43].

2.3.2 Deep learning

DL is a subset of machine learning. Nowadays, computers are fast enough to train these vast networks. A large amount of data is used to train DL models and increase the performance of the model. Generally, with the increase in the quantity of data, the performance of the model increases. DL is used in many fields such as face recognition, speech recognition, etc. In the medical field, the use of DL is increasing very rapidly [44]. Practitioners use the DL model for the detection of tumors from X-rays or CT scan images. DL models can automatically extract features from the data, which can train another classical machine learning model for better results. One of the most significant advantages of DL is that the DL model can also be trained using analog data such as images with pixel and audio files. The techniques used in DL are achieving great success. Tasks that take a lot of time to complete through human efforts can be done in little time using DL techniques. DL is making a significant amount of contribution toward the betterment of the quality of human life. Many companies use DL to show personalized recommendations to the user, making the user experience better and making huge profits. DL makes the training and prediction of the model with significantly less manual effort. After training the whole process becomes completely automatic.

2.3.3 Convolutional neural networks

CNNs have emerged from the study of the brain's visual cortex. They are used for image recognition since the 1980s. In 1989 LeCun et al. [20] proposed a method to recognize the handwritten digits using CNN. With the increase in computational power and data, CNN can perform very well on image recognition tasks. Even for complex tasks, CNNs can be trained to get fast and accurate results. CNN is powering image search services, self-driving cars, and more.

The most basic block of CNN is the convolution layer. Unlike ANN in CNN, each neuron of a layer is not connected to every neuron of the next layer. They are connected to only the neurons of their respective fields. Through this architecture, CNN can concentrate on low-level features in the first hidden layer, and then the last layers are concentrated on high-level features. This is the reason why CNN performs the task of image recognition so well. A neuron located in the ith row and jth column of a layer is connected to the previous layer neuron present in rows i to $i+f_h-1$ and columns j to $j+f_w-1$ where f_h and f_w are the height and width of the respective field. For a layer to have the same height and width as previous layer, zeros are added around the layer, known as zero padding. To connect a large convolution layer to a small convolution layer, the distance between the respective fields is increased, called strides. The weights in CNN are represented using the matrix known as filters. During the training of CNNs, the weights of these filters are changed so that the prediction error is reduced. The filters detect the feature from the image and transmit it to the subsequent layers. A good CNN model can detect even the tiny features present in the image. So basically, the training of CNN aims to make the weights of a filter such that they can detect all the features present in an image. CNN, which can detect most features, gives accurate results [45]. The general framework for a CNN model is given in Fig. 2.1.

Many convolution layers are stacked in to make the CNN able to detect the features present in the image. The final layer of the CNN layer is flattened (not in a fully connected neural network) so that the ANN can be attached to the model and the last layer is the softmax layer in the classification task. However, limitation of using the convolutional neural network is that it requires a huge memory in RAM during training. During the backpropagation step, all the intermediate values calculated during the forwarding pass are required. To shirk, the output of a convolutional layer pooling layer is used. The pooling is done to reduce the computational load, several parameters, and memory usage. Reducing also makes the neural network tolerate image shift. In pooling layer, a neuron is connected to fixed limited neurons of the previous layer. It does not have any parameters. To define a CNN, its size, stride, padding type, and pooling layer are defined manually. A typical CNN architecture has a few convolution layers with any activation function such as ReLU followed by a pooling layer, then repeating the same. As a CNN gets deeper and deeper, the size of the CNN layer decreases, and the number of channels of the convolutional layer increases [46]. Some pretrained CNN architectures are trained on the high-speed processor for weeks and are very good at extracting features from images. They can be directly used without training for any image classification task. The process is called transfer learning [47]. Alex Net (2012), Google Net (2014), and ResNet (2015) are some examples of pretrained CNN models and winners of the ILSVRC ImageNet challenge. A simple Python for CNN model is given below.

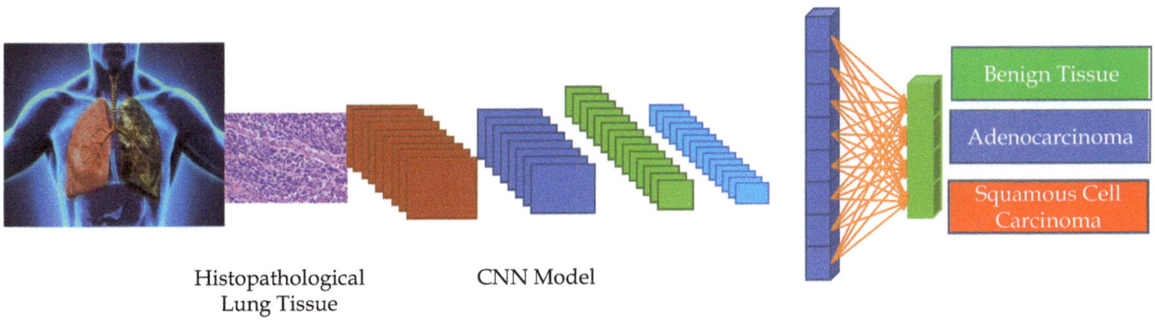

FIGURE 2.1 The general framework for a CNN model. *CNN*, Convolutional neural network.

```
from keras.layers import Dense, Flatten, Conv2D, MaxPooling2D, Dropout, Input
from keras.layers.normalization import BatchNormalization
from keras.models import Model, Sequential
inp = Input(shape = (128, 128, 3))
model = BatchNormalization()(inp)
model = Conv2D(filters = 64, kernel_size = (3, 3), padding = 'same', activation='relu')(model)
model = BatchNormalization()(model)
model = Conv2D(filters = 128, kernel_size = (3, 3), padding = 'same', activation='relu')(model)
model = MaxPooling2D()(model)
model = Dropout(0.2)(model)
model = Conv2D(filters = 128, kernel_size = (3, 3), padding = 'same', activation='relu')(model)
model = MaxPooling2D()(model)
model = Dropout(0.2)(model)
model = Conv2D(filters = 64, kernel_size = (3, 3), padding = 'same', activation='relu')(model)
model = MaxPooling2D()(model)
model = Dropout(0.2)(model)
model = Flatten()(model)
output = Dense(units = len(train_it.class_indices), activation = 'softmax')(model)
model = Model(inputs=inp, outputs=output)
```

2.4 Lung cancer detection using artificial intelligence

2.4.1 Feature extraction using deep learning

As described in the previous section, pretrained CNN models can be used to extract the feature on any dataset to improve the performance of our CNN model. Using feature extraction through transfer learning makes classification tasks for even a model with few parameters to achieve a very accurate result. This is the technique used to train models in our experiments. Each filter has the weights which are used to extract a particular feature from the image. After extracting the feature using pretrained models, the data for training reduces to a significantly smaller size. Now the training of the model can be done in significantly less time. So feature extraction is a technique used for dimensionality reduction. In fields other than image processing the feature extraction is also used. Features can also be extracted from the image without using the pretrained model. Each image is stored in the computer's memory in the form of pixels. Each image is a three-dimensional array with each value in the range of 0–255, generally three channels. The first value shows the concentration of red, the second of green, and the last of blue. To reduce the dimension of the image, the average values of the image's channel can be taken. Image can be loaded in grayscale also to reduce the dimension of data without losing any critical information. Preweighted filters can be applied to an image to extract information about the edges and sudden color changes. The simple Python code for deep feature extraction is given below.

```
def get_features(base_model, train, validate):
        X_train = base_model.predict(train)
        y_train = train.classes

        X_val = base_model.predict(validate)
        y_val = validate.classes

        X_val, X_test, y_val, y_test = train_test_split(X_val, y_val, test_size = 0.5,
                                                                       shuffle=True)
return (X_train, X_val, X_test, y_train, y_val, y_test)
```

2.4.2 Dimension reduction

Sometimes the data used for training the machine learning models have a lot of features. It is cumbersome to deal with the data with millions of features, and not everyone has enough resources such as RAM to training a model on such a large dataset. The training time also increases very drastically while using such a large dataset for training. To deal with this problem, various methods are used for dimensionality reduction, including principal component analysis (PCA) [21], locally linear embedding (LLE) [22], and generalized discriminant analysis [23]. The most commonly used method is PCA. The basic principle of PCA is that the hyperplane which lies closest to the data point is selected, and data is then projected on the hyperplane. After projecting the data on the selected hyperplane, the resulting data should have the variance as significant as possible. The data with large variance is less likely to lose the information [48].

Another way to choose the best hyperplane is to calculate the mean square error for every data point, and the plane having the lowest mean square error is selected. Scikit-Learn provides PCA [24,25] class to reduce data dimension quickly. The user has to give the number of dimensions to which dimension of data should be reduced. After reduction, the dimension data takes significantly less memory. After applying PCA on the data, data can be restored to the previous form. Still, data will not be identified as data is lost while doing dimension reduction using PCA. There are also some disadvantages to using dimension reduction like PCA to find a linear correlation between variables, which may not be the case every time. PCA is not an excellent choice to reduce the dimensionality of data when mean and covariance cannot define the dataset. To decide the number of features for accurate prediction process has to be repeated multiple times. Another technique used for dimensionality reduction is LLE. It does not use the projection of data on a hyperplane like the previous algorithm; instead, it computes how each training instance is linearly related to the closest neighbors. It reduces the dimensions where the relation with its closest neighbors is maintained. This technique can also be implemented using Scikit-Learn's class *LocallyLinearEmbedding* [26]. A simple Python code for PCA is given below.

```
from sklearn.decomposition import PCA
pca = PCA(n_components=7000)
X_train = pca.fit_transform(X_train)
X_val = pca.transform(X_val)
model.fit(X_train, y_train)
val_pred = model.predict(X_val)
```

2.4.3 Prediction and classification

Machine learning classification is the task of predicting the class in which any instance of data belongs. Classification can have binary or multiple classes. The classification task is a type of supervised learning in which the original class or label is given, and various models are used to find the best mapping to data to the classes. For the classification task using ANNs, softmax layers are added as the output layer. The number of units is equal to the number of classes in the input data. The value of each unit shows the possibility of the data instance being in that class—the sum of every unit value equal to one. There are two classifications: multiclass classification and multilabel classification; in this case, the problem is multiclass classification. In multiclass classification, a particular instance can only belong to one class, while in multilabel classification, one instance can belong to more than one class. Example of multiclass classification includes plant species classification and tumor detection. Algorithms such as KNN [27], decision tree [28], Naïve Bayes [29], and gradient boosting [30] can be used for multiclass classification. The general framework is given in Fig. 2.2.

In multilabel classification, the algorithms used for multiclass classification cannot be used directly. The modified version of the algorithm, such as multilabel decision trees, multilabel random forests, and multilabel gradient boosting, is used for this task. Another approach to solving the multilabel classification problem is to use different models for each class to predict whether the instance belongs to that class. In this way, the multilabel classification becomes like a binary classification, and every model can be used for the task. A simple Python code for the classification is given below.

```
from keras.models import Sequential
from sklearn.neighbors import KNeighborsClassifier
from sklearn.svm import SVC
from sklearn.ensemble import RandomForestClassifier, AdaBoostClassifier
from xgboost import XGBClassifier

def get_models():
       ANN = Sequential()
       ANN.add(Dense(128, input_dim = X_train.shape[1], activation = 'relu'))
       ANN.add(BatchNormalization())
       ANN.add(Dropout(0.2))
       ANN.add(Dense(64, activation='relu'))
       ANN.add(Dense(32, activation='relu'))
       ANN.add(Dense(16, activation='relu'))
       ANN.add(Dense(8, activation='relu'))
       ANN.add(Dense(len(train_it.class_indices), activation='softmax'))
       ANN.compile(loss='sparse_categorical_crossentropy', optimizer='adam', metrics=['accuracy'])

       KNN = KNeighborsClassifier()

       SVM = SVC(kernel = 'linear')

       RF = RandomForestClassifier(n_estimators = 50)

       ADB = AdaBoostClassifier()

       XGB = XGBClassifier(n_estimators = 50, use_label_encoder=False)

       return (ANN, KNN, SVM, RF, ADB, XGB)

def fit_model(model, X_train, y_train):
       model.fit(X_train, y_train)
       return model
```

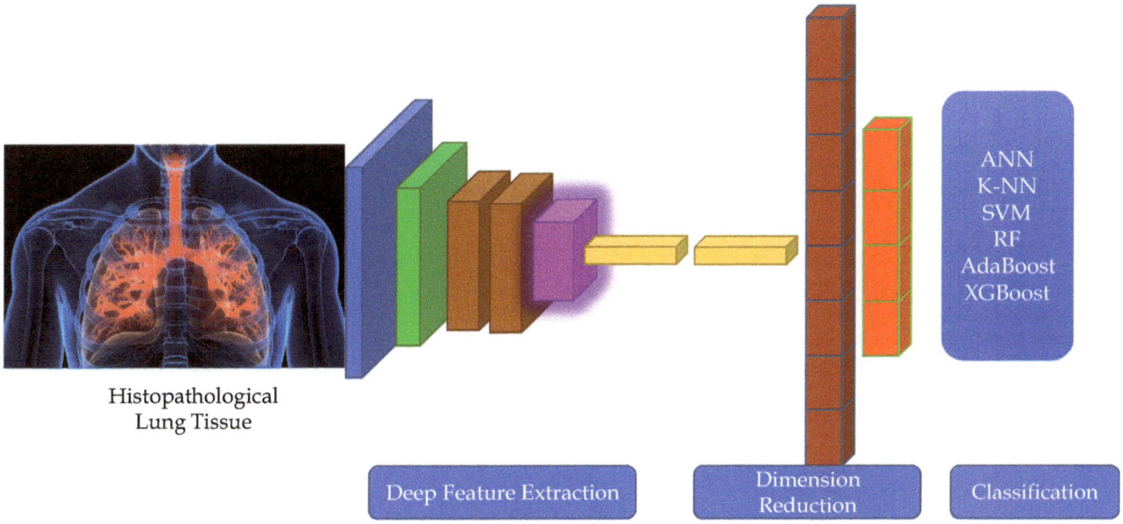

FIGURE 2.2 The general framework for histopathological image classification using deep feature extraction.

2.4.4 Experimental data

The experimental data used in this study was generated from an original sample of HIPAA compliant and validated sources, consisting of 750 total images of lung tissue (250 benign lung tissue, 250 lung adenocarcinomas, and 250 lung squamous-cell carcinomas) augmented to 15,000 using augmenter package. There are three classes in the dataset, each with 5000 images, (1) lung benign tissue, (2) lung adenocarcinoma, and (3) lung squamous-cell carcinoma. The dataset is available on the Kaggle as lung and colon cancer histopathological images [31,32]. Each image is in jpeg file format and has a size of 768×768 pixels. For reducing the size of data, images are decreased to a size of 128×128 pixels.

2.4.5 Performance evaluation measures

Various evaluation metrics are available to track how well a model can fit the data and give accurate results. Accuracy is not always the best metrics to evaluate the performance of a machine learning model. Sometimes, the model can predict every sample of data to one class and if the number of samples of that class is high, then its accuracy score will also be high. To avoid such a situation, evaluation metrics such as precision, recall, and F1 score are used. In this study, evaluation metrics calculate for each method are training accuracy, validation accuracy, test accuracy, F1 score, Cohen Kappa score, recall, precision, and AUC score. F1 score and AUC score are generally better evaluation metrics than accuracy score.

```
from sklearn.metrics import f1_score, roc_auc_score, cohen_kappa_score, precision_score,
                                        recall_score, accuracy_score, confusion_matrix
def get_accuracy_metrics(model, X_train, y_train, X_val, y_val, X_test, y_test):
   print("Train accuracy Score------------>")
   print ("{0:.3f}".format(accuracy_score(y_train, model.predict(X_train))*100), "%")
   print("Val accuracy Score--------->")
    val_pred = model.predict(X_val)
   print("{0:.3f}".format(accuracy_score(y_val, val_pred)*100), "%")

   predicted =  model.predict(X_test)
   print("Test accuracy Score--------->")
   print("{0:.3f}".format(accuracy_score(y_test, predicted)*100), "%")

   print("F1 Score--------------->")
   print("{0:.3f}".format(f1_score(y_test, predicted, average = 'weighted')*100), "%")

   print("Cohen Kappa Score------------->")
   print("{0:.3f}".format(cohen_kappa_score(y_test, predicted)*100), "%")

   print("Recall-------------->")
   print("{0:.3f}".format(recall_score(y_test, predicted, average = 'weighted')*100), "%")

   print("Precision-------------->")
   print("{0:.3f}".format(precision_score(y_test, predicted, average = 'weighted')*100), "%")

   cf_matrix_test = confusion_matrix(y_test, predicted)
   cf_matrix_val = confusion_matrix(y_val, val_pred)

    plt.figure(figsize = (12, 6))
    plt.subplot(121)
    sns.heatmap(cf_matrix_val, annot=True, cmap='Blues')
    plt.title("Val Confusion matrix")

    plt.subplot(122)
    sns.heatmap(cf_matrix_test, annot=True, cmap='Blues')
    plt.title("Test Confusion matrix")

    plt.show()
```

2.4.6 Experimental results

Different DL models are trained on the dataset, including the pretrained models. CNNs are also used for training. Pretrained models are also used for feature extraction and those features are used to train classical machine learning models. Pretrained models used in the study are ResNet50 [33], VGG16 [34], VGG19[34], InceptionV3 [35], MobileNet [36], DenseNet169 [37], DenseNet121 [37], InceptionResNetV2 [38], MobileNetV2 [39], and ResNet101 [33]. Classical machine learning models used are ANN, KNN, SVM [40], Random Forest, AdaBoost, and XGBoost).

From the Table 2.1, it can be seen that pretrained models are performing very well and giving very accurate results as compared to the CNN models, the accuracy of all the self-defined CNN model is always approximately around 33% which could be due to the model that is always predicting every image of the same class.

Among the pretrained models, VGG16, VGG19, and ResNet50 are giving the best results. The input layer has the shape of 128 × 128 × 3 in all of the CNN models, and at last, the output through models is flattened and then connected to 3-unit softmax output layer, one for each class. The best performing model is ResNet50 in the above table with a test accuracy of 98.711%, F1 score of 98.709%, Cohen Kappa score of 98.067%, and AUC Score 99.821%.

Pretrained models are used for extraction features and then trained using classical machine learning models. The results are shown in Tables 2.2–2.11.

VGG16 is used as feature extractor for the lung tissue images and the model is trained using clas-

TABLE 2.1 Performance of different convolutional neural network (CNN) models.

Classifier	Training accuracy	Validation accuracy	Test accuracy	F1 measure	KAPPA	AUC
CNN 2 Layer	0.30171	0.29467	0.28978	0.21283	-0.06434	0.4
CNN 3 Layer	0.35962	0.35156	0.36222	0.27079	0.03874	0.46776
CNN 4 Layer	0.33333	0.33511	0.33156	0.16511	0	0.5
CNN 5 Layer	0.42429	0.41867	0.43689	0.39193	0.1478	0.72973
CNN 6 Layer	0.372	0.37733	0.37333	0.3033	0.05282	0.70512
CNN 7 Layer	0.33	0.35	0.31956	0.15477	0	0.50139
CNN 8 Layer	0.31771	0.31689	0.32178	0.1692	-0.01475	0.5071
VGG16	0.99676	0.97289	0.97467	0.97467	0.962	0.99439
ResNet50	0.99914	0.98578	0.98711	0.98709	0.98067	0.99821
VGG19	0.99486	0.972	0.97556	0.97553	0.96332	0.99133
Inception_v3	0.84695	0.82311	0.82844	0.83234	0.7432	0.90396
MobileNet	0.85105	0.83378	0.84489	0.83623	0.76673	0.98124
DenseNet169	0.94543	0.93556	0.93644	0.93629	0.90466	0.9674
DenseNet121	0.91971	0.90533	0.90978	0.90953	0.86504	0.9679
InceptionResNetV2	0.42733	0.43778	0.43778	0.40157	0.16098	0.59807
MobileNetV2	0.9281	0.88889	0.89822	0.8974	0.84733	0.97831
ResNet101	0.98943	0.97022	0.97333	0.9733	0.96	0.99661

TABLE 2.2 Performance of VGG16 deep feature extraction with different machine learning models.

Model	Train accuracy	Val accuracy	Test accuracy	F1_score	Kappa	Recall	Precision
ANN	0.9981	0.98311	0.97822	0.97821	0.96733	0.97822	0.97833
KNN	0.915	0.88578	0.86889	0.86517	0.80361	0.86889	0.88749
SVM	1	0.97867	0.97422	0.97422	0.96133	0.97422	0.97425
Random Forest	1	0.95822	0.95378	0.9537	0.93065	0.95378	0.95421
AdaBoost	0.90476	0.90667	0.91156	0.91137	0.86739	0.91156	0.91451
XGBoost	1	0.97111	0.964	0.96404	0.94599	0.964	0.96425

TABLE 2.3 Performance of VGG19 deep feature extraction with different machine learning models.

Model	Train accuracy	Val accuracy	Test accuracy	F1_score	Kappa	Recall	Precision
ANN	0.99848	0.97778	0.97378	0.9738	0.96066	0.97378	0.97396
KNN	0.913	0.83378	0.83467	0.83424	0.7522	0.83467	0.84989
SVM	1	0.97733	0.96667	0.96667	0.95	0.96667	0.96668
Random Forest	1	0.95111	0.94	0.93979	0.90998	0.94	0.9403
AdaBoost	0.87305	0.86844	0.86578	0.86452	0.79883	0.86578	0.87284
XGBoost	1	0.97689	0.96356	0.96348	0.94533	0.96356	0.96344

TABLE 2.4 Performance of ResNet50 deep feature extraction with different machine learning models.

Model	Train accuracy	Val accuracy	Test accuracy	F1_score	Kappa	Recall	Precision
ANN	0.99714	0.98489	0.98622	0.98623	0.97933	0.98622	0.98624
KNN	0.936	0.86667	0.87956	0.88036	0.81938	0.87956	0.8895
SVM	1	0.988	0.988	0.988	0.982	0.988	0.988
Random Forest	1	0.95778	0.95956	0.95953	0.93933	0.95956	0.95952
AdaBoost	0.88181	0.87022	0.88844	0.8875	0.83282	0.88844	0.89912
XGBoost	1	0.97733	0.97644	0.97643	0.96466	0.97644	0.97642

TABLE 2.5 Performance of ResNet101 deep feature extraction with different machine learning models.

Model	Train accuracy	Val accuracy	Test accuracy	F1_score	Kappa	Recall	Precision
ANN	0.99714	0.97733	0.97156	0.97151	0.95734	0.97156	0.97194
KNN	0.92	0.85467	0.84978	0.84846	0.775	0.84978	0.87629
SVM	1	0.984	0.98578	0.98578	0.97866	0.98578	0.98578
Random Forest	1	0.95289	0.95511	0.95503	0.93266	0.95511	0.95497
AdaBoost	0.85124	0.83511	0.84356	0.83991	0.76568	0.84356	0.86475
XGBoost	1	0.97333	0.96622	0.96619	0.94933	0.96622	0.96618

TABLE 2.6 Performance of MobileNetV2 deep feature extraction with different machine learning models.

Model	Train accuracy	Val accuracy	Test accuracy	F1_score	Kappa	Recall	Precision
ANN	0.84019	0.81422	0.82533	0.81459	0.73737	0.82533	0.87165
KNN	0.888	0.85156	0.84889	0.84842	0.77344	0.84889	0.8514
SVM	1	0.88444	0.87333	0.87337	0.80998	0.87333	0.87342
Random Forest	0.9999	0.88578	0.89067	0.89013	0.83593	0.89067	0.89121
AdaBoost	0.76648	0.752	0.75156	0.72932	0.62836	0.75156	0.79502
XGBoost	1	0.91111	0.90089	0.90067	0.85131	0.90089	0.90053

TABLE 2.7 Performance of MobileNet deep feature extraction with different machine learning models.

Model	Train accuracy	Val accuracy	Test accuracy	F1_score	Kappa	Recall	Precision
ANN	0.94619	0.91689	0.912	0.91149	0.86788	0.912	0.92001
KNN	0.913	0.87467	0.85378	0.8516	0.78092	0.85378	0.86667
SVM	1	0.91022	0.90933	0.90938	0.864	0.90933	0.90947
Random Forest	1	0.90356	0.89244	0.89312	0.8386	0.89244	0.89495
AdaBoost	0.80324	0.79822	0.80133	0.78819	0.70266	0.80133	0.83618
XGBoost	1	0.92622	0.90978	0.91008	0.86464	0.90978	0.91059

TABLE 2.8 Performance of InceptionV3 deep feature extraction with different machine learning models.

Model	Train accuracy	Val accuracy	Test accuracy	F1_score	Kappa	Recall	Precision
ANN	0.87114	0.85022	0.85511	0.85808	0.78237	0.85511	0.87945
KNN	0.899	0.82756	0.83289	0.83133	0.74954	0.83289	0.84141
SVM	1	0.92089	0.91733	0.91741	0.87598	0.91733	0.91752
Random Forest	1	0.89511	0.88533	0.88512	0.82797	0.88533	0.8854
AdaBoost	0.81848	0.80667	0.804	0.79994	0.70617	0.804	0.80063
XGBoost	1	0.92089	0.92	0.91982	0.88	0.92	0.91971

TABLE 2.9 Performance of InceptionResNetV2 deep feature extraction with different machine learning models.

Model	Train accuracy	Val accuracy	Test accuracy	F1_score	Kappa	Recall	Precision
ANN	0.45714	0.46044	0.45822	0.39109	0.18746	0.45822	0.54051
KNN	0.82	0.70578	0.69422	0.69687	0.54147	0.69422	0.70166
SVM	0.7979	0.72711	0.72356	0.72421	0.58538	0.72356	0.7253
Random Forest	0.9999	0.73422	0.72844	0.72997	0.59249	0.72844	0.73276
AdaBoost	0.69552	0.66444	0.684	0.68581	0.52603	0.684	0.68812
XGBoost	0.984	0.74756	0.73333	0.73544	0.60001	0.73333	0.73812

TABLE 2.10 Performance of DenseNet169 deep feature extraction with different machine learning models.

Model	Train accuracy	Val accuracy	Test accuracy	F1_score	Kappa	Recall	Precision
ANN	0.96629	0.95778	0.95156	0.95152	0.92736	0.95156	0.9533
KNN	0.97	0.94756	0.94489	0.94477	0.91737	0.94489	0.94856
SVM	1	0.97644	0.97378	0.97379	0.96067	0.97378	0.97381
Random Forest	0.9999	0.94578	0.952	0.95208	0.92799	0.952	0.95233
AdaBoost	0.80629	0.80533	0.81067	0.80332	0.71653	0.81067	0.83895
XGBoost	1	0.97244	0.968	0.96801	0.952	0.968	0.96803

TABLE 2.11 Performance of DenseNet121 deep feature extraction with different machine learning models.

Model	Train accuracy	Val accuracy	Test accuracy	F1_score	Kappa	Recall	Precision
ANN	0.9641	0.94711	0.94444	0.94394	0.9167	0.94444	0.94627
KNN	0.977	0.94844	0.94356	0.9435	0.91535	0.94356	0.94428
SVM	1	0.97156	0.96356	0.96356	0.94533	0.96356	0.96356
Random Forest	1	0.94844	0.93822	0.93826	0.90732	0.93822	0.93847
AdaBoost	0.8039	0.79689	0.804	0.79248	0.70248	0.804	0.84563
XGBoost	1	0.96311	0.96222	0.96223	0.94332	0.96222	0.96248

sical machine learning (ML) models. The results are improved to a large extent as compared to the CNN models in Table 2.1. The highest accuracy on test data is achieved using ANN with a value of 97.82% and the VGG16 is performing most poorly with KNN with test accuracy of 86.88%.

```
def get_models():
    ANN = Sequential()
    ANN.add(Dense(128, input_dim = X_train.shape[1], activation = 'relu'))
    ANN.add(BatchNormalization())
    ANN.add(Dropout(0.2))
    ANN.add(Dense(64, activation='relu'))
    ANN.add(Dense(32, activation='relu'))
    ANN.add(Dense(16, activation='relu'))
    ANN.add(Dense(8, activation='relu'))
    ANN.add(Dense(len(train_it.class_indices), activation='softmax'))
    ANN.compile(loss='sparse_categorical_crossentropy', optimizer='adam', metrics=['accuracy'])

    KNN = KNeighborsClassifier()

    SVM = SVC(kernel = 'linear')

    RF = RandomForestClassifier(n_estimators = 50)

    ADB = AdaBoostClassifier()

    XGB = XGBClassifier(n_estimators = 50, use_label_encoder=False)

    return (ANN, KNN, SVM, RF, ADB, XGB)

def reshape_data(X_train, X_val, X_test):
    X_train = X_train.reshape(X_train.shape[0], -1)
    X_val = X_val.reshape(X_val.shape[0], -1)
    X_test = X_test.reshape(X_test.shape[0], -1)
    return (X_train, X_val, X_test)

def fit_ANN(model, X_train, y_train, X_val, y_test):
    es = EarlyStopping(monitor='val_loss', mode='min', verbose=1, patience=5)
    history = model.fit(X_train, y_train, validation_data=(X_val, y_test), epochs=10, verbose=1,
                                                                                    callbacks=[es])
    return model

def fit_model(model, X_train, y_train):
    model.fit(X_train, y_train)
    return model

base_model = VGG16(include_top=False, input_shape=(SIZE_X, SIZE_Y, 3), weights='imagenet')

for layer in base_model.layers:
    layer.trainable = False

model = Model(inputs=base_model.input, outputs=base_model.layers[-1].output)
model.summary()

X_train, X_val, X_test, y_train, y_val, y_test = get_features(model, train_it, validate_it)
X_train, X_val, X_test = reshape_data(X_train, X_val, X_test)
ANN, KNN, SVM, RF, ADB, XGB = get_models()

ANN = fit_ANN(ANN, X_train, y_train, X_val, y_val)
KNN = fit_model(KNN, X_train_scaled, y_train)
```

VGG19 and VGG16 are similar models, VGG19 has 19 layers, whereas VGG16 has 16. So results are also similar to what obtained using VGG16. Here, best results are obtained using ANN with test accuracy of 97.38% and worst are with KNN with value of 83.42%.

```
base_model = VGG19(include_top=False, input_shape=(SIZE_X, SIZE_Y, 3), weights='imagenet')

for layer in base_model.layers:
    layer.trainable = False

model = Model(inputs=base_model.input, outputs=base_model.layers[-1].output)
model.summary()

X_train, X_val, X_test, y_train, y_val, y_test = get_features(model, train_it, validate_it)
X_train, X_val, X_test = reshape_data(X_train, X_val, X_test)
ANN, KNN, SVM, RF, ADB, XGB = get_models()

ANN = fit_ANN(ANN, X_train, y_train, X_val, y_val)
KNN = fit_model(KNN, X_train_scaled, y_train)
```

ResNet50 is trained on over a million images on the ImageNet database and it has 50 layers, so it has very good weights which are able to extract many features of an image. Combining it with classical ML models is giving very good result. The SVM models is giving highest accuracy score of 98.8% and worst score is with KNN with test accuracy score of 87.95%.

```
base_model = ResNet50(include_top=False, input_shape=(SIZE_X, SIZE_Y, 3), weights='imagenet')

for layer in base_model.layers:
    layer.trainable = False

model = Model(inputs=base_model.input, outputs=base_model.layers[-1].output)
X_train, X_val, X_test, y_train, y_val, y_test = get_features(model, train_it, validate_it)
X_train, X_val, X_test = reshape_data(X_train, X_val, X_test)
ANN, KNN, SVM, RF, ADB, XGB = get_models()

KNN = fit_ANN(KNN, X_train, y_train, X_val, y_val)
SVM = fit_model(SVM, X_train_scaled, y_train)
```

ResNet101 is the upgraded version of ResNet50. Like ResNet50, ResNet101 is also giving good results. Here, the maximum test accuracy score achieved is 98.57% with SVM and worst test accuracy is given by KNN with value of 84.9%.

```
base_model = ResNet101(include_top=False, input_shape=(SIZE_X, SIZE_Y, 3), weights='imagenet')

for layer in base_model.layers:
    layer.trainable = False

model = Model(inputs=base_model.input, outputs=base_model.layers[-1].output)
X_train, X_val, X_test, y_train, y_val, y_test = get_features(model, train_it, validate_it)
X_train, X_val, X_test = reshape_data(X_train, X_val, X_test)
ANN, KNN, SVM, RF, ADB, XGB = get_models()

KNN = fit_model(KNN, X_train_scaled, y_train)
SVM = fit_model(SVM, X_train_scaled, y_train)
```

MobileNetV2 is 53-layer deep and is also trained on the ImageNet database. The resultant score obtained using MobileNetV2 are lower than the scores of VGG and ResNet models. Here, the best score is given by the XGBoost with a value of 90.08% and worst score is given by AdaBoost with value of 75.15%.

```
base_model = MobileNetV2(include_top=False, input_shape=(SIZE_X, SIZE_Y, 3), weights='imagenet')

for layer in base_model.layers:
    layer.trainable = False

model = Model(inputs=base_model.input, outputs=base_model.layers[-1].output)
X_train, X_val, X_test, y_train, y_val, y_test = get_features(model, train_it, validate_it)
X_train, X_val, X_test = reshape_data(X_train, X_val, X_test)
ANN, KNN, SVM, RF, ADB, XGB = get_models()

ADB = fit_model(ADB, X_train, y_train)
XGB = fit_model(XGB, X_train, y_train)
```

MobileNet is the previous version of MobileNetV2 model and are quite similar to one another. The results are also similar to the MobileNetV2 model. Here, the best result is obtained by training the ANN model on the features extracted by MobileNet with a test accuracy value of 90.97% and worst score is obtained using AdaBoost with a value of 80.13%.

```
base_model = MobileNet(include_top=False, input_shape=(SIZE_X, SIZE_Y, 3), weights='imagenet')

for layer in base_model.layers:
    layer.trainable = False

model = Model(inputs=base_model.input, outputs=base_model.layers[-1].output)
X_train, X_val, X_test, y_train, y_val, y_test = get_features(model, train_it, validate_it)
X_train, X_val, X_test = reshape_data(X_train, X_val, X_test)
ANN, KNN, SVM, RF, ADB, XGB = get_models()

ANN = fit_ANN(ANN, X_train, y_train, X_val, y_val)
ADB = fit_model(ADB, X_train, y_train)
```

InceptionV3 is a model of the family of Inception models. It consists of 48 layers. The best test accuracy score obtained is 92% with XGBoost model and worst score is 80.4% with AdaBoost model.

```
base_model = InceptionV3(include_top=False, input_shape=(SIZE_X, SIZE_Y, 3), weights='imagenet')

for layer in base_model.layers:
    layer.trainable = False

model = Model(inputs=base_model.input, outputs=base_model.layers[-1].output)
X_train, X_val, X_test, y_train, y_val, y_test = get_features(model, train_it, validate_it)
X_train, X_val, X_test = reshape_data(X_train, X_val, X_test)
ANN, KNN, SVM, RF, ADB, XGB = get_models()
ADB = fit_model(ADB, X_train, y_train)
XGB = fit_model(XGB, X_train, y_train)
```

InceptionResNetV2 is a network which is 164-layer deep. But the weights of this model are not good enough to extract important features required for the correct classification of this dataset. The best accuracy is given with XGBoost with the value of 73.33% and the worst accuracy is given with ANN with a value of 45.82%.

```
base_model = InceptionResNetV2(include_top=False, input_shape=(SIZE_X, SIZE_Y, 3),
                                weights='imagenet')

for layer in base_model.layers:
    layer.trainable = False

model = Model(inputs=base_model.input, outputs=base_model.layers[-1].output)
X_train, X_val, X_test, y_train, y_val, y_test = get_features(model, train_it, validate_it)
X_train, X_val, X_test = reshape_data(X_train, X_val, X_test)
ANN, KNN, SVM, RF, ADB, XGB = get_models()

ANN = fit_ANN(ANN, X_train, y_train, X_val, y_val)
XGB = fit_model(XGB, X_train, y_train)
```

DenseNet169 is also trained on ImageNet database and is a member of DenseNet family. The performance of this model is good on the dataset. The best score achieved is with SVM with the value of 97.37% and the worst accuracy score is given with AdaBoost with the value of 81.06%.

```
base_model = DenseNet169(include_top=False, input_shape=(SIZE_X, SIZE_Y, 3), weights='imagenet')

for layer in base_model.layers:
    layer.trainable = False

model = Model(inputs=base_model.input, outputs=base_model.layers[-1].output)
X_train, X_val, X_test, y_train, y_val, y_test = get_features(model, train_it, validate_it)
X_train, X_val, X_test = reshape_data(X_train, X_val, X_test)
ANN, KNN, SVM, RF, ADB, XGB = get_models()

SVM = fit_model(SVM, X_train, y_train)
ADB = fit_model(ADB, X_train, y_train)
```

DenseNet121 is another member of DenseNet family. The results obtained using this are similar to the results of DenseNet169. The best test accuracy score is given with SVM with a value of 96.35% and the worst score is given by using AdaBoost with the value of 80.4%.

```
base_model = DenseNet121(include_top=False, input_shape=(SIZE_X, SIZE_Y, 3), weights='imagenet')

for layer in base_model.layers:
    layer.trainable = False

model = Model(inputs=base_model.input, outputs=base_model.layers[-1].output)
X_train, X_val, X_test, y_train, y_val, y_test = get_features(model, train_it, validate_it)
X_train, X_val, X_test = reshape_data(X_train, X_val, X_test)
ANN, KNN, SVM, RF, ADB, XGB = get_models()
SVM = fit_model(SVM, X_train, y_train)
ADB = fit_model(ADB, X_train, y_train)
```

From the above tables, it can be seen that VGG16, VGG19, ResNet50, and ResNet101 can extract essential features from the images, and ANN gives the best result among all other models. The ANN used here has the shape of $128 \times 64 \times 32 \times 16 \times 8$ and is trained for 10 epochs. For training the KNN model, PCA is used to reduce data features as the training time grows exponentially with the number of data features. For the XGBoost and Random Forest model number of estimators is taken as 50. For all other models, default parameters are used.

The best model in Table 2.2 is ResNet50 with SVM giving test accuracy, F1 score, Cohen Kappa score, Recall, and Precision of 98.8%, 98.8%, 98.2%, 98.8%, and 98.8%, respectively.

For Tables 2.2–2.11, pretrained networks are used. As training, these huge networks required very huge data, computation power, and time. These models are developed by a team of researchers. These models are able to extract every small feature of an image. So these models are used to extract import features and then classical ML models ANN, KNN, SVM, Random

Forest, AdaBoost, and XGBoost are trained on the extracted features to give the output. The output given by pretrained models is flattened to give as input to the ML models. The best test accuracy score achieved using this method is 98.8% which is given by ResNet50 and SVM.

2.5 Discussion

The abovementioned results show that all transfer learning models achieved good performance. All the pretrained models are better than the self-designed CNN. Pretrained models are also reducing the time to train a model on such a vast dataset. Some models such as VGG16, VGG19, ResNet50, and ResNet101 are performing better than other models. The best results have the F1 score of 98.8% given by ResNet50 and SVM Classifier. Models such as Random Forest and XGBoost can fit very well on training data. But they are not giving good results on test and validation data. SVM can give more accurate results after training on the features extracted through a pretrained model than other classical machine learning models. KNN, AdaBoost, and Random Forest are not be able to give accurate results in most cases. Pretrained models are extracting important features with their weights which is enabling simple classical ML models to give very good results on the complex dataset. For making own designed CNN model able to make accurate predictions, a lot of computation power and time is required so it is always easy and better to just load the pretrained models.

2.6 Conclusion

Lung cancer is a very deadly disease, and everyone must take precautions to avoid every possibility of having lung cancer. Smoking should be avoided, and a healthy and balanced diet should be taken.

AI is helping the medical field to improve and give faster and better results. As we can see in the result, the accuracy of the machine learning model is almost equal to 1. Today, we have the data and resources to use machine learning to automate lung cancer detection. Through AI, it will be possible for each and every one to detect cancer in almost no time. Early diagnosis of cancer can be lifesaving in the initial stages, and the lung is more curable.

References

[1] U.S. National Institute of Health, National Cancer Institute. SEER Cancer Statistics Review, 1975–2015.

[2] Centres for Disease Control and Prevention, National Centre for Health Statistics. CDC WONDER On-Line Database, Compiled from Compressed Mortality File 1999–2016 Series 20 No. 2V, 2017.

[3] What is lung cancer? | Types of lung cancer. https://www.cancer.org/cancer/lung-cancer/about/what-is.html (accessed 14.05.21).

[4] Facebook and Facebook, 10 Tips for Preventing Lung Cancer, Verywell Health. <https://www.verywellhealth.com/tips-for-lung-cancer-prevention-2249286> (accessed 14.05.21).

[5] Healthline, Effects of lung cancer on the body, <https://www.healthline.com/health/lung-cancer/effects-on-body>, May 09, 2017 (accessed 14.05.21).

[6] How to detect lung cancer | Lung cancer tests. <https://www.cancer.org/cancer/lung-cancer/detection-diagnosis-staging/how-diagnosed.html> (accessed 14.05.21).

[7] Cancer Support Community, Coping with side effects of lung cancer treatment, <https://www.cancersupportcommunity.org/article/coping-side-effects-lung-cancer-treatment> (accessed 14.05.21).

[8] T. Atsushi, T. Tetsuya, K. Yuka, F. Hiroshi, Automated classification of lung cancer types from cytological images using deep convolutional neural networks, BioMed. Res. Int. 2017 (2017) 1–6. Available from: https://doi.org/10.1155/2017/4067832.

[9] W. Ausawalaithong, A. Thirach, S. Marukatat, T. Wilaiprasitporn, Automatic lung cancer prediction from chest X-ray images using the deep learning approach, in: 2018 11th Biomedical Engineering International Conference (BMEiCON), Nov. 2018, pp. 1–5. Available from: https://doi.org/10.1109/BMEiCON.2018.8609997.

[10] B. Hatuwal, H. Thapa, Lung cancer detection using convolutional neural network on histopathological images, Int. J. Comput. Trends Technol. 68 (2020)

21–24. Available from: https://doi.org/10.14445/22312803/IJCTT-V68I10P104.

[11] S.C. S R, H. Rajaguru, Lung cancer detection using probabilistic neural network with modified crow-search algorithm, Asian Pac. J. Cancer Prev. APJCP 20 (7) (2019) 2159–2166. Available from: https://doi.org/10.31557/APJCP.2019.20.7.2159.

[12] S.S. Chakravarthy, S.A. Subhasakthe, Adaptive median filtering with modified BDND algorithm for the removal of high-density impulse and random noise, Int. J. Comput. Sci. Mob. Comput. 4 (2015).

[13] S. Sasikala, M. Bharathi, B.R. Sowmiya, Lung cancer detection and classification using deep CNN, Int. J. Eng. Innov. Technol. Explo. Eng. 8 (25) (2018) 259–262.

[14] Lung Image Database Consortium, The Cancer Imaging Archive (TCIA) Public Access - Cancer Imaging Archive Wiki. <https://wiki.cancerimagingarchive.net/display/Public/Lung + Image + Database + Consortium> (accessed 14.05.21).

[15] M.F. Serj, B. Lavi, G. Hoff, D.P. Valls, A deep convolutional neural network for lung cancer diagnostic, (2018). *arXiv preprint arXiv:1804.08170.*

[16] Kaggle, Data Science Bowl 2017. <https://kaggle.com/c/data-science-bowl-2017>, 2017 (accessed 14.05.21).

[17] A. Masood, et al., Computer-assisted decision support system in pulmonary cancer detection and stage classification on C.T. images, J. Biomed. Inform. 79 (2018). Available from: https://doi.org/10.1016/j.jbi.2018.01.005.

[18] Q. Qiu, X. Cheng, R. Calderbank, G. Sapiro, DCFNet: deep neural network with decomposed convolutional filters, ArXiv180204145 Cs Stat, <http://arxiv.org/abs/1802.04145>, Jul. 2018 (accessed14.05.21).

[19] D.E. Rumelhart, G.E. Hinton, R.J. Williams, Learning representations by back-propagating errors, Nature 323 (6088) (1986). Available from: https://doi.org/10.1038/323533a0. Art. no. 6088, Oct.

[20] Y. LeCun, et al., Backpropagation applied to handwritten zip code recognition, Neural Comput. 1 (4) (1989) 541–551.

[21] S. Mishra, et al., Principal component analysis, Int. J. Livest. Res. 1 (2017). Available from: https://doi.org/10.5455/ijlr.20170415115235. Jan.

[22] W. Chojnacki, M. Brooks, A note on the locally linear embedding algorithm, IJPRAI 23 (2009) 1739–1752. Available from: https://doi.org/10.1142/S0218001409007752. Dec.

[23] G. Baudat, F. Anouar, Generalized discriminant analysis using a kernel approach, Neural Comput. 12 (2000) 2385–2404. Available from: https://doi.org/10.1162/089976600300014980.

[24] P.-G. Martinsson, V. Rokhlin, M. Tygert, A randomized algorithm for the decomposition of matrices, Appl. Comput. Harmon. Anal. 30 (1) (2011) 47–68. Available from: https://doi.org/10.1016/j.acha.2010.02.003.

[25] N. Halko, P.G. Martinsson, J.A. Tropp, Finding structure with randomness: probabilistic algorithms for constructing approximate matrix decompositions, SIAM Rev. 53 (2) (2011) 217–288. Available from: https://doi.org/10.1137/090771806.

[26] sklearn.manifold.LocallyLinearEmbedding—Scikit-Learn 0.24.2 documentation. <https://scikit-learn.org/stable/modules/generated/sklearn.manifold.LocallyLinearEmbedding.html> (accessed 14.05.21).

[27] G. Guo, H. Wang, D. Bell, Y. Bi, KNN model-based approach in classification, Aug. 2004.

[28] H. Patel, P. Prajapati, Study and analysis of decision tree based classification algorithms, Int. J. Comput. Sci. Eng. 6 (2018) 74–78. Available from: https://doi.org/10.26438/ijcse/v6i10.7478.

[29] I. Rish, An empirical study of the Naïve Bayes classifier, IJCAI 2001 workshop on empirical methods in artificial intelligence, 3, Jan2001.

[30] A. Natekin, A. Knoll, Gradient boosting machines, a tutorial, Front. Neurorobot. 7 (2013) 21. Available from: https://doi.org/10.3389/fnbot.2013.00021. Dec.

[31] A.A. Borkowski, M.M. Bui, L.B. Thomas, C.P. Wilson, L.A. DeLand, S.M. Mastorides, Lung and colon cancer histopathological image dataset (LC25000), ArXiv Prepr. ArXiv191212142, 2019.

[32] Kaggle, Lung and colon cancer histopathological images. < https://kaggle.com/andrewmvd/lung-and-colon-cancer-histopathological-images> (accessed 14.05.21).

[33] K. He, X. Zhang, S. Ren, J. Sun, Deep residual learning for image recognition, ArXiv151203385 Cs, <http://arxiv.org/abs/1512.03385>, Dec. 2015 (accessed 14.05. 21).

[34] K. Simonyan, A. Zisserman, Very deep convolutional networks for large-scale image recognition, ArXiv Prepr. ArXiv14091556, 2014.

[35] C. Szegedy, V. Vanhoucke, S. Ioffe, J. Shlens, Z. Wojna, Rethinking the inception architecture for computer vision, in: Proceedings of the IEEE conference on computer vision and pattern recognition, 2016, pp. 2818–2826.

[36] A.G. Howard et al., MobileNets: efficient convolutional neural networks for mobile vision applications, ArXiv170404861 Cs, <http://arxiv.org/abs/1704.04861>, Apr. 2017 (accessed 14.05.21).

[37] G. Huang, Z. Liu, L. van der Maaten, K.Q. Weinberger, Densely connected convolutional networks, ArXiv160806993 Cs, <http://arxiv.org/abs/1608.06993> Jan. 2018 (accessed 14.05.21).

[38] Papers with Code, Inception-ResNet-v2 explained. <https://paperswithcode.com/method/inception-resnet-v2> (accessed 14.05.21).

[39] M. Sandler, A. Howard, M. Zhu, A. Zhmoginov, L.-C. Chen, MobileNetV2: inverted residuals and linear bottlenecks, ArXiv180104381 Cs, <http://arxiv.org/abs/1801.04381>, Mar. 2019 (accessed 14.04.21).

[40] S. Ghosh, A. Dasgupta, A. Swetapadma, A study on support vector machine based linear and non-linear pattern classification, in: 2019 International Conference on Intelligent Sustainable Systems (ICISS), Feb. 2019, pp. 24–28. Available from: https://doi.org/10.1109/ISS1.2019.8908018.

[41] M.-C. Popescu, V. Balas, L. Perescu-Popescu, N. Mastorakis, Multilayer perceptron and neural networks, WSEAS Trans. Circuits Syst. 8 (2009). Jul.

[42] S. Sharma, S. Sharma, A. Athaiya, Activation functions in neural networks, Int. J. Eng. Appl. Sci. Technol. 04 (2020) 310–316. Available from: https://doi.org/10.33564/IJEAST.2020.v04i12.054. May.

[43] E. Grossi, M. Buscema, Introduction to artificial neural networks, Eur. J. Gastroenterol. Hepatol. 19 (Jan. 2008). Available from: https://doi.org/10.1097/MEG.0b013e3282f198a0.

[44] S. Nagpal, M. Kumar, M.R. Maruthi, Ayyagari, Kumar, A survey of deep learning and its applications: a new paradigm to machine learning, Arch. Comput. Methods Eng. (2019). Jul.

[45] R. Khandelwal, Convolutional neural network: feature map and filter visualization, Medium (2020). Available from: https://towardsdatascience.com/convolutional-neural-network-feature-map-and-filter-visualization-f75012a5a49c. May 18.

[46] L. Alzubaidi, et al., Review of deep learning: concepts, CNN architectures, challenges, applications, future directions, J. Big Data 8 (1) (2021) 53. Available from: https://doi.org/10.1186/s40537-021-00444-8.

[47] K. Weiss, T.M. Khoshgoftaar, D. Wang, A survey of transfer learning, J. Big Data 3 (1) (2016) 9. Available from: https://doi.org/10.1186/s40537-016-0043-6.

[48] I.T. Jolliffe, J. Cadima, Principal component analysis: a review and recent developments, Philos. Trans. R. Soc. Math. Phys. Eng. Sci. 374 (2065) (2016) 20150202.

CHAPTER

3

Magnetic resonance imagining-based automated brain tumor detection using deep learning techniques

Abhranta Panigrahi[1] *and Abdulhamit Subasi*[2,3]

[1]Department of Biotechnology and Medical Engineering, National Institute of Technology Rourkela, Rourkela, India [2]Institute of Biomedicine, Faculty of Medicine, University of Turku, Turku, Finland [3]Department of Computer Science, College of Engineering, Effat University, Jeddah, Saudi Arabia

OUTLINE

3.1 Introduction

Central nervous system tumor or a CNS tumor refers to excessive growth or accumulation of cells in the CNS that consists of the brain and the spinal cord. An approximated 24,530 adults consisting of 13,840 men and 10,690 women in the United States will be diagnosed with CNS cancer. An estimated 3460 children under the age of 15 will also be

Applications of Artificial Intelligence in Medical Imaging
DOI: https://doi.org/10.1016/B978-0-443-18450-5.00012-8

diagnosed with brain or CNS tumors in 2021. Brain and nervous system cancer is the 10th leading cause of death for men and women. An estimated 18,600 adults will die from primary CNS cancer [1]. Due to the severity of the problem, the early diagnosis of a brain tumor is extremely important for its treatment. Since the invention of X-rays in 1895 by Wilhelm Roentgen, various medical imaging techniques have been used to diagnose brain tumors. German neurosurgeon, Fedor Krause used X-rays extensively to detect brain tumors. Over time, many sophisticated and advanced diagnosis techniques were invented. These included imaging techniques such as computed tomography (CT) scan [2], positron emission tomography scan [3], and various invasive methods such as biopsy. One of the most significant breakthroughs in brain tumor detection came with the development of magnetic resonance imaging (MRI). It is a noninvasive medical imaging technique that uses strong radio magnetic waves to generate an accurate image of the required organ. MRI generates a more detailed picture of the brain and does not involve radiation. Intravenous gadolinium-enhanced MRI is typically used to help create a clearer picture of the brain tumor [4]. With the advent of MRI technology, different variations of MRI imaging started to play a vital role in the image-based diagnosis of brain tumors. Some of the techniques of MRI imaging that are prevalent in modern medicine are fluid attenuated inversion recovery, T1-weighted precontrast, T1-weighted postcontrast (Gd), and T2-weighted.

With the advent of technology, various computational techniques proved to be useful in medical diagnosis. The advent of deep learning [5] has revolutionized the field of computer vision and natural language processing. Various models such as AlexNet [6], ResNet [7], and GoogleNet [8] have given near human-like accuracy in the ImageNet image classification task [6]. The success of convolutional neural networks (CNNs) led the way to their extensive use in diagnosis via medical imaging [9–11]. Since training a CNN from scratch to achieve state of the art performance requires high computational resources, the usage of transfer learning has increased exponentially. This has resulted in the adoption of transfer learning techniques to detect and segment brain tumors [12] and classify [13] them for faster and more efficient diagnosis. In this chapter, we try to provide a detailed assessment of various deep learning methods for the detection of brain tumors and evaluate the various pros and cons of each method, which will enable faster diagnosis of brain tumors. Overview of the MRI-based brain tumor detection is given in Fig. 3.1.

3.2 Literature survey

Extensive work has been done in the field of deep learning and big data analysis and their applications in the diagnosis of brain tumors. Most of the approaches involve the segmentation of the brain tumor via CNNs. Özyurt et al. [14] conducted a study, which proposed a hybrid method using neutrosophy and convolutional neural network. It aimed to classify brain tumors as benign or malignant. CNNs were used to extract features from the segmented brain tumor images and then the tumors were classified as malignant or benign using support vector machine (SVM) and K-nearest neighbor (KNN) classifiers. Jalali and Kaur [15] gave a detailed comparative summary of various methods and works for automatic brain tumor detection using medical imaging. Their work involved the description of various methods such as deep belief networks, recurrent neural networks, KNNs, SVMs, and many others. Deb and Roy [16]

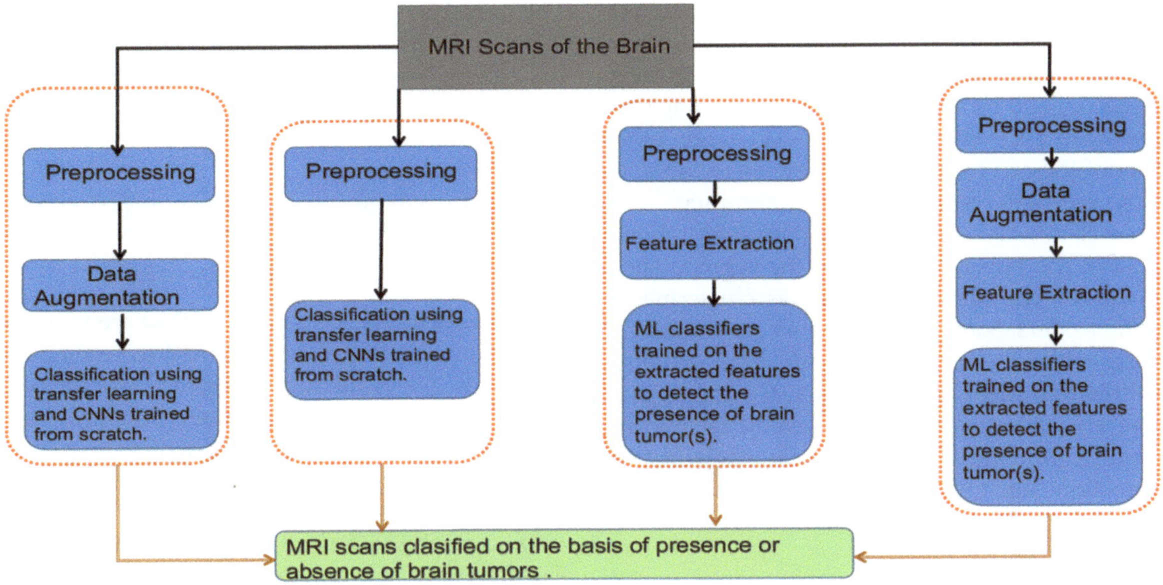

FIGURE 3.1 Overview of the pipelines.

proposed a novel segmentation and classification algorithm for brain tumor detection. They proposed a system that uses an adaptive fuzzy deep neural network with frog leap optimization to detect abnormalities in an image and then the abnormal image was segmented using an adaptive flying squirrel algorithm. Ambily and Suresh [17] proposed an integrated model of CNN and transfer learning automated binary classification of MRI images. They used a 16-layer pretrained network to distinguish normal and abnormal images.

Amin et al. [18] proposed a deep learning model to predict input MR slices as unhealthy/healthy based on the presence of tumors. Multiple image processing techniques were used to make the tumors more prominent. The MR slices were segmented by applying optimal thresholds to cluster the similar pixels together. These segmented slices were then sent to a two-layer stacked sparse autoencoder model. They trained the model on the BRATS dataset and performed a binary classification on the MRI scans and predicted which scans had tumors and which did not.

Vallabhaneni and Rajesh [19] worked on automatic brain tumor detection in noise-corrupted images. Noise in medical imaging can seriously hamper the ability of automated algorithms to correctly detect abnormalities. To tackle this problem, they performed denoising using Edge Adaptive Total Variation Denoising Technique. This technique preserves the edges of an image while denoising. The denoised images were then segmented using mean shift clustering and feature extraction was performed using gray-level cooccurrence matrix. A multiclass SVM classifier was then used to detect the tumor present in the images.

Rai et al. [20] proposed a hybrid deep CNN model for an automated prediction and segmentation of brain tumors from MRI images. They used a dataset with 3929 images, including 1373 images with tumors and 2556 images without tumors. The CNN model was evaluated using Jaccard index, DICE score, F1 score, accuracy, precision, and recall. It was benchmarked against UnetResnet-50 and vanilla U-Net and other state-of-the-art techniques. They were able to show excellent results with 99.7% accuracy.

As it can be seen from the discussion above, most of the papers focused on the segmentation of tumors use algorithms such as clustering or autoencoders. Some of the authors have used CT scans while the most commonly used imaging technique is MRI. In this chapter, we aim to provide an exhaustive study of several techniques ranging from transfer learning to deep feature extraction for the detection of brain tumors. The research goal here was to use a range of techniques to classify MRI scans into two categories—scans with tumors and scans without tumors.

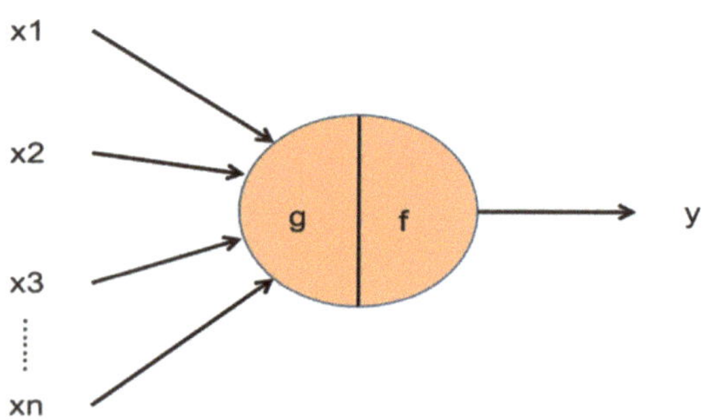

FIGURE 3.2 Artificial neuron.

3.3 Deep learning for disease detection

3.3.1 Artificial neural networks

Artificial neural networks (ANNs) are function approximators that were designed to resemble a primitive idea of the human brain. They are very useful while predicting extremely nonlinear relations. They consist of layers of artificial neurons. Fig. 3.2 shows a simple artificial neuron.

The layers of neurons that accept the input data are called as the input layers (Fig. 3.3A). The layers of neurons that give the final prediction are called as the output layers (Fig. 3.3C), and all the layers in between are called as the hidden layers (Fig. 3.3B). They consist of neurons and synapses. Each neuron has some data (*x*) and some bias (*B*). Each synapse has a weight (*W*) and each layer of neurons has an activation function (*f*). The output (*o*) of each layer is governed by:

$$o = f(x * W + B)$$

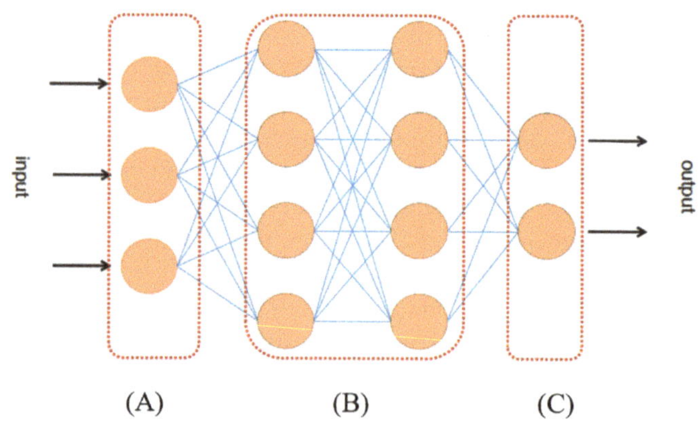

FIGURE 3.3 Overview of a four-layered ANN. *ANN*, Artificial neural network.

Leshno et al. [21] showed that multilayer feedforward networks with a polynomial activation function can approximate any function.

The nonlinear activation function of the neurons is very crucial for an ANN. These differentiable functions are the main reason why neural networks can fit to extremely nonlinear data. There are many such activation function choices such as sigmoid, tanh, ReLU, and many more.

3.3.2 Deep learning

Deep learning refers to a type of machine learning algorithm which uses ANNs to learn. Deep learning consists of steps such as forward propagation, calculation of loss, and backpropagation. This method was formalized by LeCun et al. [5]. ANNs learn by constantly updating the weights and biases of the synapses on an ANN. The weights and biases are updated to decrease the loss. The steps involved in deep learning (Fig. 3.4) are:

1. Forward propagation of data:
   - The data vector is fed through the network, and it is multiplied with the weight matrix and a bias term is added to it.
   - After passing through the activation functions, the networks give a prediction.
2. Calculation of loss:
   - The value of the prediction is compared to the actual label of the data and a loss term is calculated.
   - There are various types of loss such as mean squared error (MSE) loss, cross-entropy loss, etc.
   - Each different type of loss is used in different types of problems. For example, MSE loss is mainly used in regression problems, whereas cross-entropy loss is mainly used in classification problems.
3. Backpropagation of loss:
   - The information of the loss propagates through all the layers starting from the output layer and moving back all the way up to the input layer; hence, the name "backpropagation."
   - The gradient of the loss is calculated with respect to every layer in the network through chain rule of differentiation.

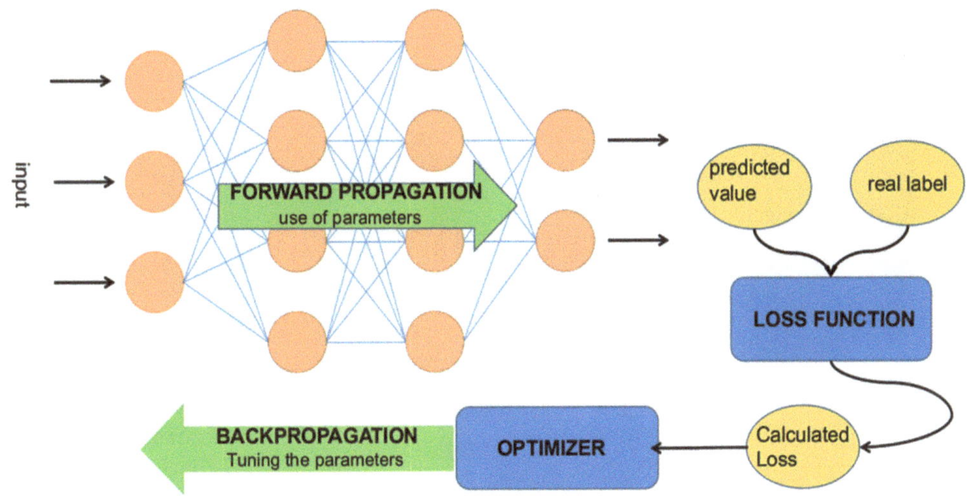

FIGURE 3.4 Deep learning training process.

4. Gradient descent:
   - The motive of the training step is to minimize the loss. For this purpose, the "gradient descent" algorithm is used.
   - The parameters are updated in order to reach the global minima of the loss curve; hence, the name "gradient descent."

```
from sklearn.neural_network import MLPClassifier
# Create the Model
mlp =  MLPClassifier(hidden_layer_sizes=(100, ), learning_rate_init=0.001,
alpha=1, momentum=0.9,max_iter=1000)
# Train the Model with Training dataset
mlp.fit(Xtrain,ytrain)
# Test the Model with Testing dataset
ypred = mlp.predict(Xtest)
```

3.3.3 Convolutional neural networks

Vision is a unique property of life. We rely heavily on out ability of vision to interact with the world and perform daily tasks like walking and driving. In order to give the power of vision to computer, LeCun et al. [22] provided the description of a CNN (Fig. 3.5) for handwritten digit recognition. The popularity of CNNs started after Krizhevsky et al. [6] and created the AlexNet, which was a state-of-the-art image classification network in the ImageNet [23] challenge. Since then, networks such as GoogleNet [8] and ResNet [7] have performed exceedingly well in image classification tasks.

CNNs are a type of ANN which implements the convolution operation on images in each

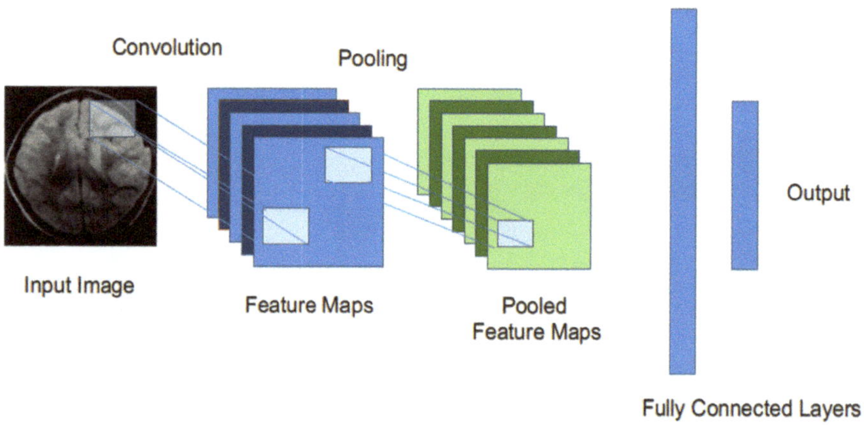

FIGURE 3.5 Overview of a convolutional neural network.

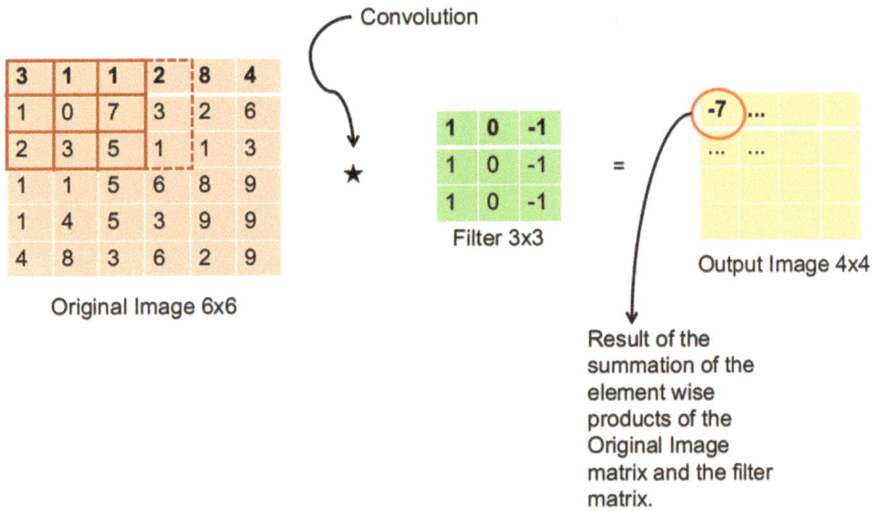

FIGURE 3.6 Convolution operation in matrices.

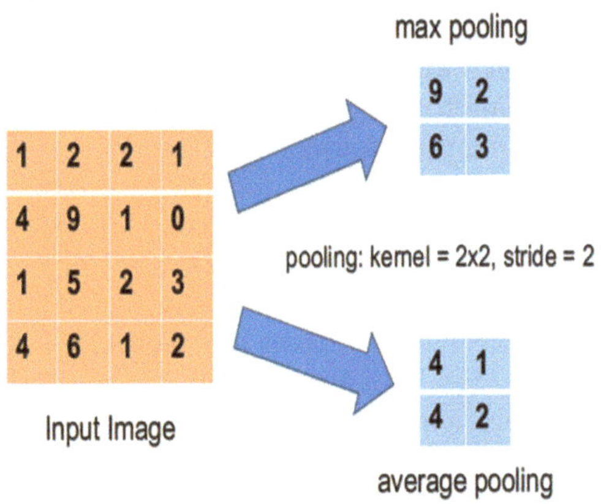

FIGURE 3.7 Pooling operation in matrices.

layer. The networks consist of convolution (Fig. 3.6) layers and pooling (Fig. 3.7) layers and each layer progressively extract the features of an image. The values of the convolution filters are the parameters that get updated with training.

The extracted feature maps are then flattened and fed to a series of dense layers that perform the final prediction. CNNs take the information about a pixel and its neighboring pixels as well, whereas dense networks or fully connected networks flatten the image. This results in loss of the spatial information of an image in case of a dense neural network. Due to this reason, CNNs are better suited for image tasks when compared to densely connected networks. ANNs and deep learning are still in their infancy but the rapid rate of research in these fields has ensured their adoption in various domains. In health care, deep learning techniques have done wonders. In recent years, researchers at Deep-mind have been able to solve the age-old problem of protein folding using their model called as Alphafold [24]. Similarly, many other state-of-the-art algorithms such as U-Net [25] have aided the medical community by providing segmented images for image-based diagnoses.

```
#Creating a basic 2 layered CNN with Keras
model_2 = Sequential()

#Adding a convolution layer, batch-normalization layer, ReLU activation and #maxpooling layer
model_2.add(Conv2D(16,(3,3),padding='same',input_shape=
                              X_train_prep[0].shape))
model_2.add(BatchNormalization())
model_2.add(Activation('relu'))
model_2.add(MaxPooling2D(pool_size=(2,2),strides=2,padding = 'same'))

#Adding a convolution layer, batch-normalization layer, ReLU activation and #maxpooling layer
model_2.add(Conv2D(32,(3,3),padding='same'))
model_2.add(BatchNormalization())
model_2.add(Activation('relu'))
model_2.add(MaxPooling2D(pool_size=(2,2),strides=2,padding = 'same'))

#Flattening and adding a FC layer with sigmoid Activation
model_2.add(Flatten())
model_2.add(Dense(nClasses,activation='sigmoid'))
```

3.4 Disease detection using artificial intelligence

3.4.1 Feature extraction

Feature extraction is an inherent property of neural networks. In Convolutional Neural Networks (CNN), the feature maps of an image are extracted in each layer. After each convolutional layer, features of an image such as edge information, gradient information, etc. are retrieved. These features are then learnt by the network for the required classification task. This is how a computer "sees" (Fig. 3.8).

Machine learning algorithms such as KNN and XGBoost are excellent algorithms for predicting something from a given set of features. They perform well on tabular data where the feature vectors are well defined. In case of images where there are a huge number of not well-defined features, implementing these algorithms can be difficult. To solve that problem, deep feature extraction is used where CNNs are used to extract the features from the input image which are then flattened and used as the training data for machine learning algorithms.

In this chapter, a total of 11 pretrained CNN models have been used for feature extraction (Fig. 3.9), namely, VGG16 [26], VGG19 [26], ResNet50 [7], ResNet101 [7], MobileNet-V2 [27], MobileNet [28], InceptionNet-V3 [29], InceptionResNet-V2 [30], DenseNet169 [31], DenseNet121 [31], and XceptionNet [32]. All the models were pretrained on the ImageNet dataset. The MRI scans were passed through the pretrained networks to get a feature map. A dropout layer was added with a dropout probability of 0.5. This feature map was then flattened and converted into a 64×1 embedding/latent vector. This latent vector was then passed through a batch normalization layer. The resulting feature embedding was used as the training features for six machine learning algorithms: Multi layer perceptron (MLP)

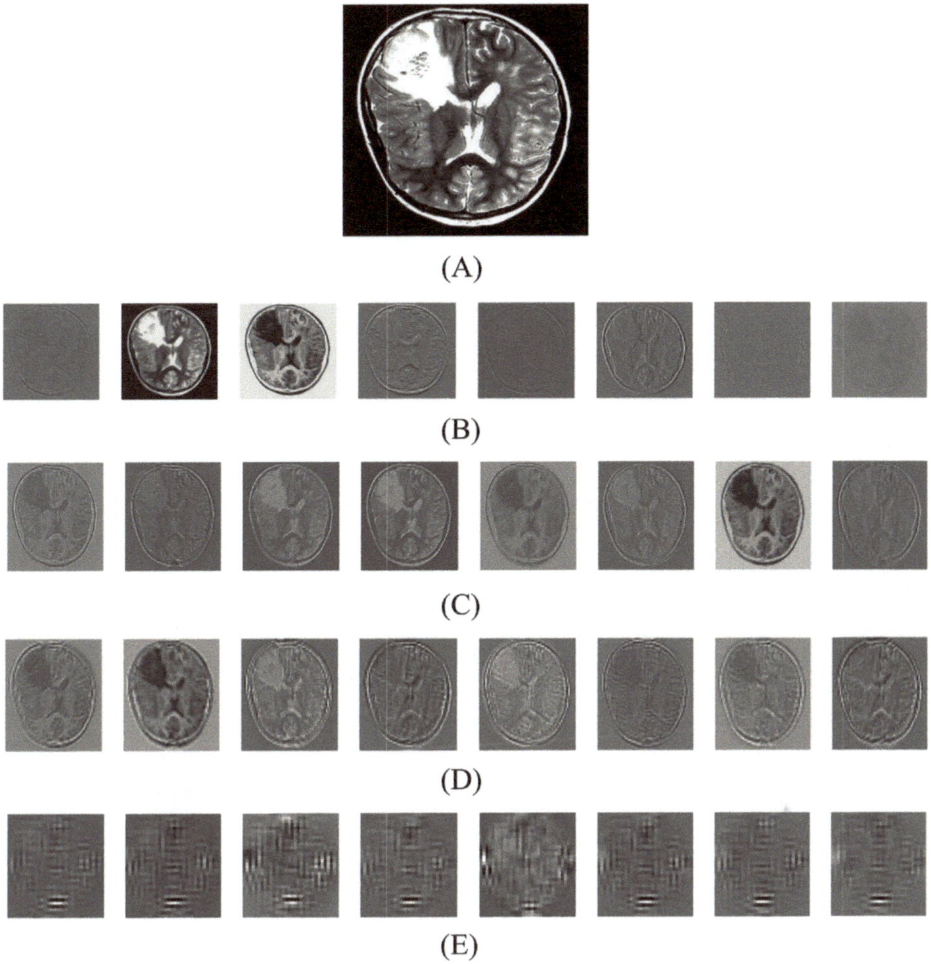

FIGURE 3.8 How a computer "sees"? Visualizing the feature maps of an image retrieved from an ResNet50 model. Fig. 3.8A shows the input image. Fig. 3.8B shows some of the features extracted in the very first layer of the network. Fig. 3.8C shows the features extracted in the 12th layer of the network. Fig. 3.8D shows the features extracted in the 25th layer of the network. Fig. 3.8E shows the features extracted in the last convolutional layer of the network. This visualization clearly shows that the edge information are the most commonly extracted features in CNNs. As the number of layers increase, the features get more abstract.

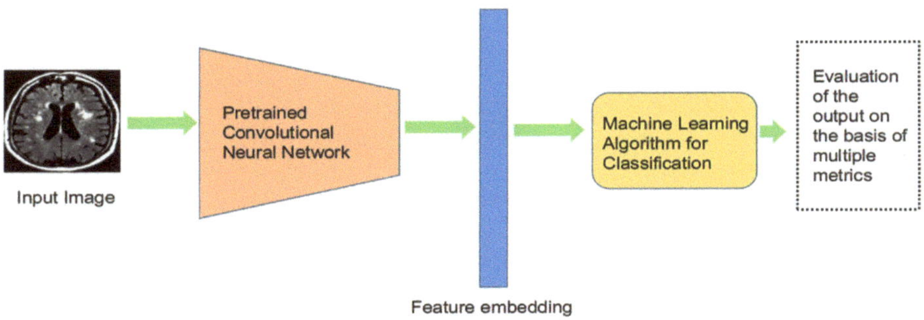

FIGURE 3.9 Overview of deep feature extraction for classification of MRI scans. *MRI*, Magnetic resonance imaging.

classifier/ANN, KNN, support vector machines (SVM), Random Forest classifier, AdaBoost classifier, and XGBoost classifier. The results were then recorded.

```
#Defining a pretrained ResNet50 model as the feature extractor
#The ResNet model was changed to other models for performing further
#experiments
base_model= ResNet50(input_shape=(224,224,3), weights='imagenet',
include_top=False)

#Applying dropout, flatten and batch-normalization to the extracted features
x = base_model.output
x = Dropout(0.5)(x)
x = Flatten()(x)
x = BatchNormalization()(x)
x = Dense(64,kernel_initializer='he_uniform')(x)
x = BatchNormalization()(x)
predictions = Activation('relu')(x)

model_feat = Model(inputs=base_model.input,outputs=predictions)

#Extracting the train, test and validation features from the respective data
train_features = model_feat.predict(X_train_prep)
val_features=model_feat.predict(X_val_prep)
test_features=model_feat.predict(X_test_prep)


#Creating a pipeline to train various classifiers on the features extracted #by
the pretrained networks.
names = ["K Nearest Neighbour Classifier",'SVM',"Random Forest Classifier",
           "AdaBoost Classifier", "XGB Classifier","ANN Classifier"]
classifiers = [
KNeighborsClassifier(),
SVC(probability = True),
RandomForestClassifier(),
AdaBoostClassifier(),
XGBClassifier(),
MLPClassifier()
]
zipped_clf = zip(names,classifiers)
def classifier_summary(pipeline, X_train, y_train, X_val,
y_val,X_test,y_test):
      sentiment_fit = pipeline.fit(X_train, y_train)
      y_pred_train= sentiment_fit.predict(X_train)
      y_pred_val = sentiment_fit.predict(X_val)
      y_pred_test = sentiment_fit.predict(X_test)
      y_pred_train = [1 if x>0.5 else 0 for x in y_pred_train]
      y_pred_val = [1 if x>0.5 else 0 for x in y_pred_val]
      y_pred_test = [1 if x>0.5 else 0 for x in y_pred_test]
```

```
#Calculating the performance metrics on the training data
train_accuracy = np.round(accuracy_score(y_train, y_pred_train),4)*100
train_precision = np.round(precision_score(y_train, y_pred_train,
                                 average='weighted'),4)
train_recall = np.round(recall_score(y_train, y_pred_train,
                                 average='weighted'),4)
train_F1 = np.round(f1_score(y_train, y_pred_train,
                           average='weighted'),4)
train_kappa = np.round(cohen_kappa_score(y_train, y_pred_train),4)

#Calculating the performance metrics on validation data
val_accuracy = np.round(accuracy_score(y_val, y_pred_val),4)*100
val_precision = np.round(precision_score(y_val, y_pred_val,
                                 average='weighted'),4)
val_recall = np.round(recall_score(y_val, y_pred_val,
                                 average='weighted'),4)

val_F1 = np.round(f1_score(y_val, y_pred_val, average='weighted'),4)
val_kappa = np.round(cohen_kappa_score(y_val, y_pred_val),4)

#Calculating the performance metrics on test data
test_accuracy = np.round(accuracy_score(y_test, y_pred_test),4)*100
test_precision = np.round(precision_score(y_test, y_pred_test,
                           average='weighted'),2)
test_recall = np.round(recall_score(y_test, y_pred_test,
                           average='weighted'),2)
test_F1 = np.round(f1_score(y_test, y_pred_test,
                                 average='weighted'),2)
test_kappa = np.round(cohen_kappa_score(y_test, y_pred_test),2)
test_roc_auc = metrics.roc_auc_score(y_test,y_pred_test,
                 multi_class='ovo', average='weighted')
```

3.4.2 Transfer learning

Transfer learning is a machine learning technique in which models trained on a huge dataset for a particular task can be repurposed on a second but related task. It is a technique in which the learning capabilities of deep neural nets are exploited to improve generalization in another setting. Since training a deep neural network from scratch to achieve state-of-the-art performance requires huge computational resources, transfer learning has been gaining popularity. It has enabled faster paced research and, hence, it has resulted in faster adaptation of deep learning. Transfer learning is a type of feature extraction where a pretrained deep network is used to extract features from an image (Fig. 3.10). In a CNN, the filters in the beginning learn low-level features such as edges and contour, which has the filters at the end, learn higher level features. Hence, the top filters of a deep ConvNet trained on a large dataset can generalize very well. Hence, in transfer learning, the weights of the top layers of a pretrained

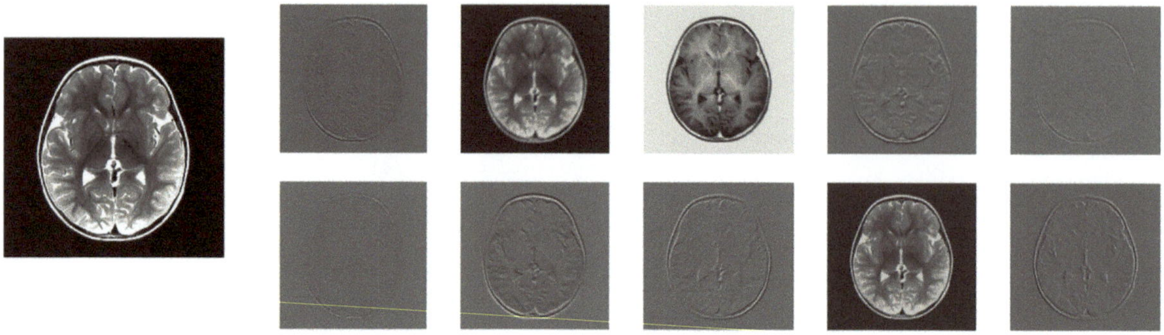

FIGURE 3.10 Visualization of the features extracted for totally different images in the first layer of ResNet50 model. It shows the features extracted for an MRI scan of a brain. As it can be clearly observed, the first layers of the ConvNet mainly extract the basic features such as edge and contour information. Hence, the parameters of the first few layers of a deep neural network can be frozen and the features extracted from these can be used to train smaller and more task-specific neural network to achieve near state-of-the-art performance in various tasks. *MRI*, Magnetic resonance imaging.

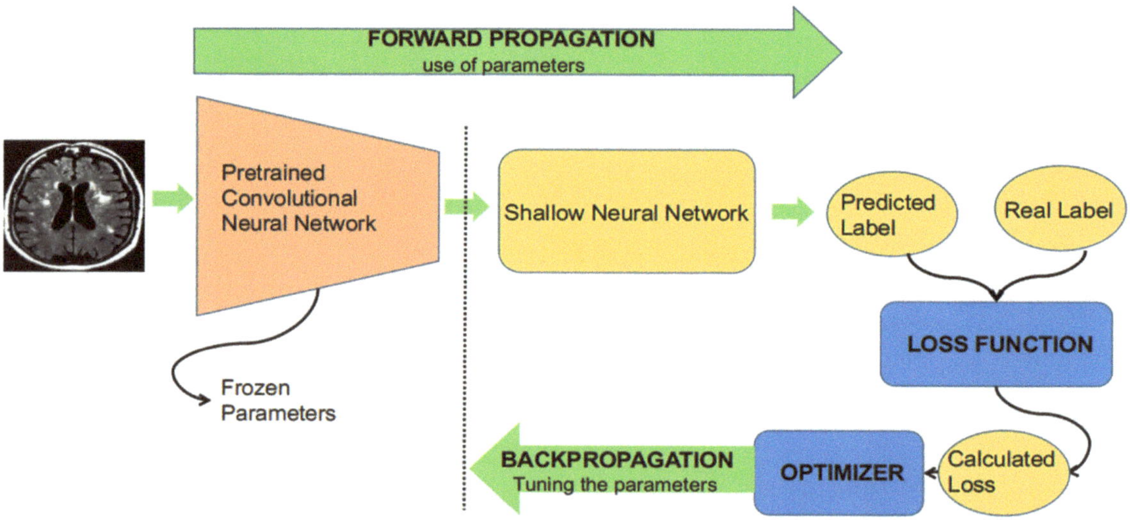

FIGURE 3.11 Overview of transfer learning.

ConvNet are frozen and then a smaller neural network is trained from scratch on the features extracted by the frozen layers to give the final result (Fig. 3.11).

In this chapter, 11 different pretrained CNNs were used for transfer learning. These are the same networks that were mentioned in Section 3.4.1. The parameters of the layers of the networks were frozen and a dense neural net was created that accepted the flattened output of the pretrained CNN. The dense network consisted of a dense layer of dimension 256×1. The output of the dense network was then passed through a batch normalization layer which was then passed through an ReLU activation function. After a dropout layer, the final layer consisted of 1 neuron with a sigmoid activation function to perform binary classification of MRI scans based on the presence or absence of tumors. The pipeline is shown in Fig. 3.12.

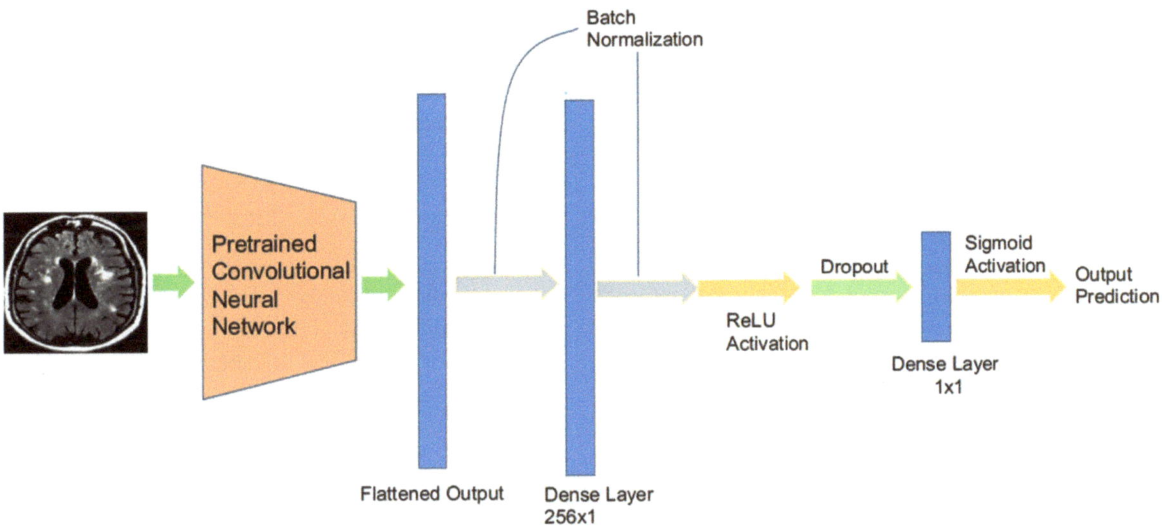

FIGURE 3.12 Pipeline for the experimentation with transfer learning.

```
#Considering a ResNet50 model as the base model
base_Neural_Net= ResNet50(input_shape=(224,224,3),
weights='imagenet', include_top=False)
model=Sequential()

#Adding trainable head to the base network
model.add(base_Neural_Net)
model.add(Flatten())
model.add(BatchNormalization())
model.add(Dense(256,kernel_initializer='he_uniform'))
model.add(BatchNormalization())
model.add(Activation('relu'))
model.add(Dropout(0.5))
model.add(Dense(1,activation='sigmoid'))

#Freezing the parameters of the base network
for layer in base_Neural_Net.layers:
layer.trainable = False

#Compiling the Neural Network
model.compile(loss='binary_crossentropy',optimizer='adam',
                metrics=['accuracy' , 'AUC'])

#Defining the Early stopping parameters
EPOCHS = 30
es = EarlyStopping(monitor='val_acc', mode='max',patience=6)

#Training the model
history = model.fit_generator(train_generator,steps_per_epoch=50,
                    epochs=EPOCHS,
                    validation_data=validation_generator,
                    validation_steps=25,callbacks=[es])
```

3.4.3 Prediction and classification

In this chapter, we have performed binary classification on MRI scans of human brain on the basis of the presence or absence of tumor. Since this was a binary classification problem, a sigmoid activation function was used on the last neuron of an ANN which returned either **0** or **1** as the output.

Three different experiments are carried out. In the first experiment, deep feature extraction was used to extract features from the images. As mentioned earlier in Section 3.4.1, 11 state-of-the-art CNN models were used to extract features from the MRI scans. These features were then flattened and used as training features for various machine learning algorithms such as MLP classifier, KNN classifier, XGBoost classifier, etc. After the classification, the performance of the models was monitored using several performance evaluation metrics, namely, training accuracy, validation accuracy, testing accuracy, F1 score, Cohen's Kappa score, recall, precision, and area under the ROC curve (AUC) score. These performance metrics have been discussed more elaborately in Section 3.4.5.

The second experiment was used for the experiments was transfer learning. Here, the CNN models as described in Section 3.4.1 were used to extract features from the image. The parameters of these networks were frozen, and a shallow ANN was attached to the output of the last layer of the pretrained CNN. The parameters of the shallow network were then updated with the help of backpropagation and gradient descent. The loss function used to train this network was the binary cross-entropy loss. This loss was used as the problem in hand was the binary classification problem. After multiple experimentations, The Adam optimizer was found to give the best result in terms of loss and the monitored performance evaluation metrics which was accuracy and AUC score. More on this method has been discussed in Section 3.4.2.

The third experiment carried out to tackle the problem of training a CNN from scratch. CNNs of various departments ranging from two-layered CNNs to eight-layered CNNs were used for classification. ReLU activation was used for all the hidden layers and a sigmoid neuron was used as the final layer for classification. Appropriate batch normalization layers and pooling layers were added to increase the accuracy of the models.

3.4.4 Experimental data

The dataset that was chosen for the experiment was a publicly available dataset [33]. It consisted of 3000 MRI scans with 1500 of them having a brain tumor and 1500 of them without a brain tumor. The data had compressed JPEG version of the MRI scans. The data was divided into three main folders—Train, Test, and Validation with each of them containing two subfolder—Yes and No. The Yes folder contained the images with the brain tumor and the No folder consisted of the images without the brain tumor. The dataset was then divided into Train, Test, and Validation sets with the Test set containing 300 images of each class, the Train set containing 960 images of each class and the Validation set containing 240 images of each class. Due to the low number of training data, various image augmentation techniques were used which are described in Section 3.4.5. Since the images were of different sizes, resizing them was necessary to train the neural networks. To avoid distortions due to resizing, a few preprocessing steps were taken which are discussed in Section 3.4.5. Sample MRI images are shown in Fig. 3.13.

3.4.5 Experimental setup

The pretrained models that were used in the experiments, accepted 224×224 RGB images. So, these accepted $224 \times 224 \times 3$ tensors. Since the images in the original dataset were of varying dimensions (Fig. 3.14), they needed to be resized. To avoid distortions, some preprocessing steps were required before resizing them.

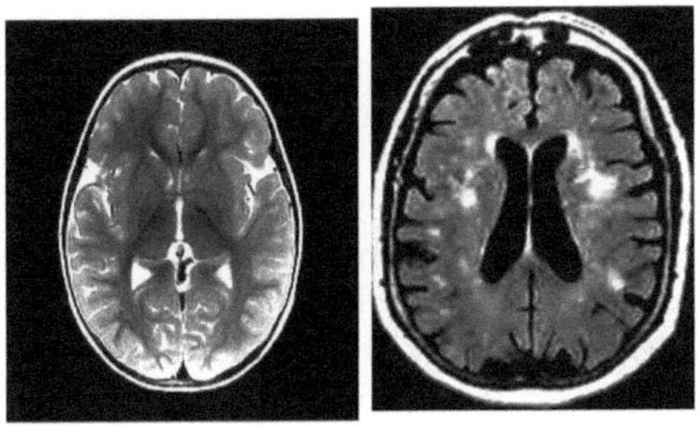

(A)

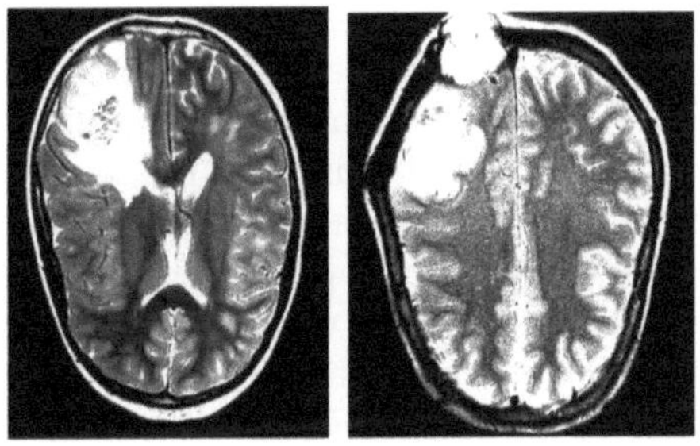

(B)

FIGURE 3.13 (A) shows the images without brain tumor and (B) shows the images with brain tumor.

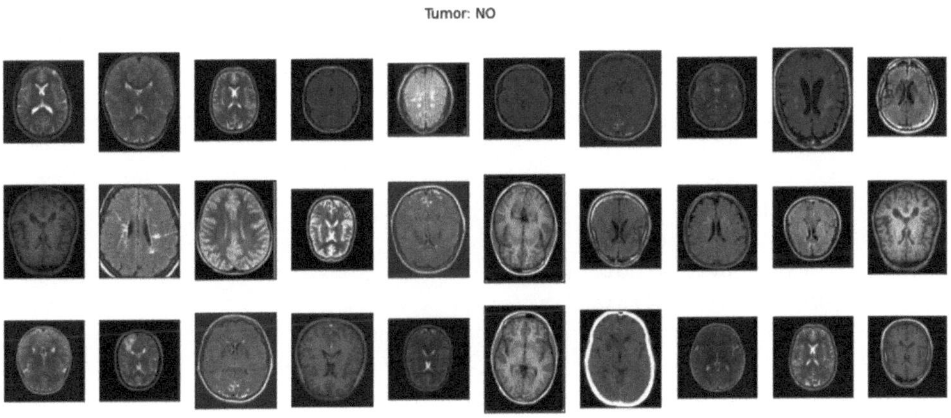

(A)

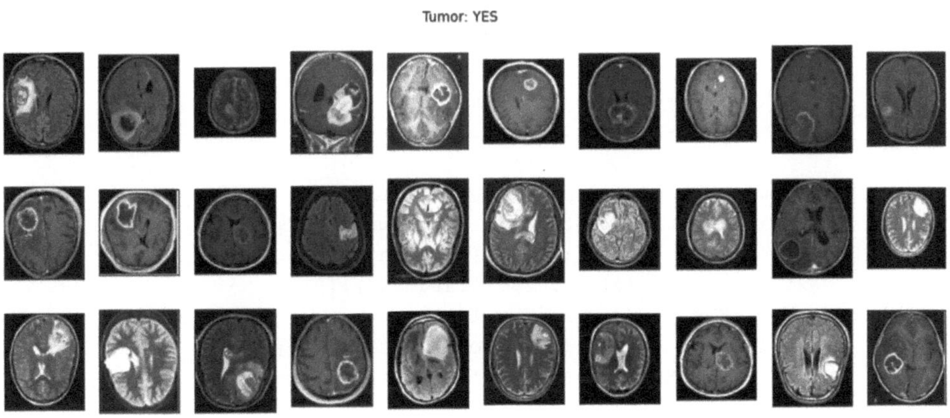

(B)

FIGURE 3.14 (A) shows a batch of images without tumor and (B) shows a batch of images with tumors. This figure shows the varying dimension of the images in the dataset and highlights the requirement of resizing the images.

Before resizing the images, the images were cropped to avoid distortions. As Fig. 3.15 clearly shows, the brain occupied varying areas for different images, that is, the black areas were different across all images. For this reason, the images were cropped to make sure that the models performed well. The first step for cropping the images was to find the contours of the brain in the MRI scan. To do this, first, a Gaussian blur was applied on the image to reduce the noise. Then appropriate thresholding and dilation was used to remove small regions of noisy scans. Then contours were found from these images. After finding the contours, the extreme points in the contours were calculated to get the edge of the brain in the image. Then a bounding box was constructed with the middle of each side being the calculated extreme point of the contour. The image was cropped according to the edges of the bounding box. These steps are clearly shown in Fig. 3.15.

This step was important to make sure that the brain occupies maximum area in the scan. After this step, the images were resized to be of dimension 224×224.

```
def crop_imgs(set_name, add_pixels_value=0):
    """
    Finds the extreme points on the image and crops the rectangular out of
    them
    """
    set_new = []
    for img in set_name:
        gray = cv2.cvtColor(img, cv2.COLOR_RGB2GRAY)
        gray = cv2.GaussianBlur(gray, (5, 5), 0)
        # threshold the image, then perform a series of erosions +
        # dilations to remove any small regions of noise
        thresh = cv2.threshold(gray, 45, 255, cv2.THRESH_BINARY)[1]
        thresh = cv2.erode(thresh, None, iterations=2)
        thresh = cv2.dilate(thresh, None, iterations=2)
        # find contours in thresholded image, then grab the largest one
        cnts = cv2.findContours(thresh.copy(), cv2.RETR_EXTERNAL,
                                cv2.CHAIN_APPROX_SIMPLE)
        cnts = imutils.grab_contours(cnts)
        c = max(cnts, key=cv2.contourArea)
        # find the extreme points
        extLeft = tuple(c[c[:, :, 0].argmin()][0])
        extRight = tuple(c[c[:, :, 0].argmax()][0])
        extTop = tuple(c[c[:, :, 1].argmin()][0])
        extBot = tuple(c[c[:, :, 1].argmax()][0])
        ADD_PIXELS = add_pixels_value
        new_img = img[extTop[1]-ADD_PIXELS:extBot[1]+ADD_PIXELS,
    extLeft[0]-ADD_PIXELS:extRight[0]+ADD_PIXELS].copy()
        set_new.append(new_img)

    return np.array(set_new)
```

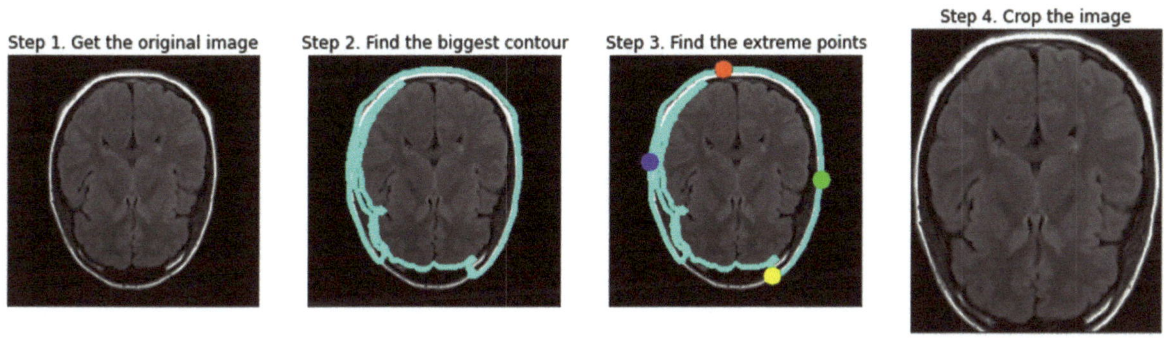

FIGURE 3.15 Overview of the cropping step.

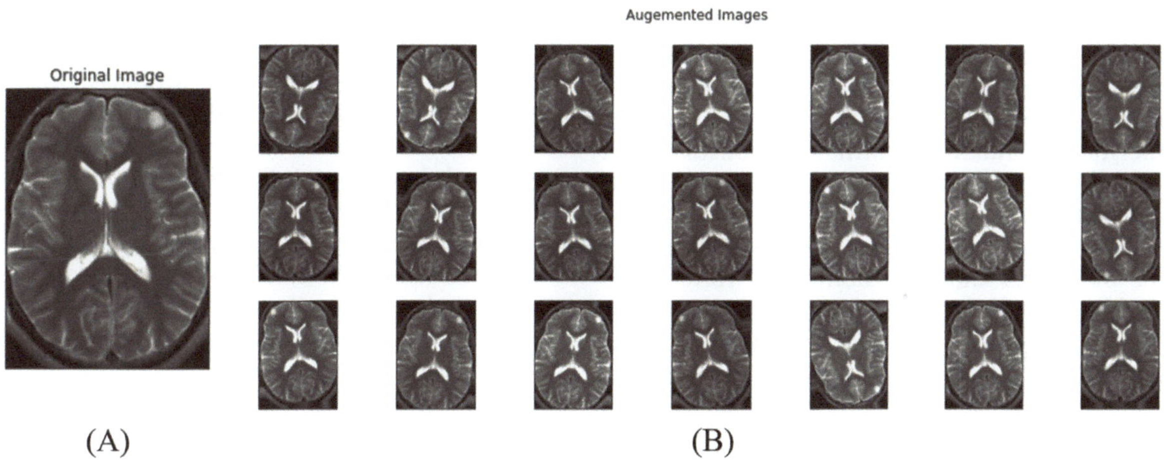

FIGURE 3.16 Sample of an image before augmentation (A) and the resulting images after a few image augmentations are applied (B). This figure shows how image augmentation helps to generate more data which leads to robust models.

Since there are only 960 samples available for training, various image augmentations were used to increase the number of training samples. The augmentations that were performed on the image were: Random rotation in a 15-degree range, shifting the image along the width and the length, rescaling the image, shearing, varying the brightness of the image, and flipping the image vertically and horizontally (Fig. 3.16). All of these augmentations were performed to make the models more robust. A series of experimentations were also performed to compare the performance of the models on the augmented and nonaugmented images to study the effect of number of training samples on the machine learning algorithms. All the observations are showed in Section 3.4.7.

```
train_datagen = ImageDataGenerator(rotation_range=15, width_shift_range=0.1,
                  height_shift_range=0.1, shear_range=0.1,
                  brightness_range=[0.5, 1.5], horizontal_flip=True,
                  vertical_flip=True,preprocessing_function=preprocess_input)

train_generator = train_datagen.flow_from_directory(TRAIN_DIR,
           color_mode='rgb', target_size=IMG_SIZE, batch_size=32,
              class_mode='binary', seed=RANDOM_SEED)
```

3.4.6 Performance evaluation metrics

Performance evaluation metrics are very important while dealing with the problems related to health care. Since different performance metrics mean different things and choosing a wrong performance metric to evaluate the model can result in false claims, which can ultimately lead to a health disaster. Hence, in this chapter, the models were evaluated on the most common performance metrics to get an idea about their robustness and performance.

First, a confusion matrix was plotted for all the models to know the number of correct prediction and wrong prediction. When positive labels are predicted correctly, it is called as a true positive or **TP**. Similarly, when negative labels are predicted correctly, it is called as a true negative or **TN**. When positive labels are misclassified as negative labels or vice-versa, it is called as false negative **(FN)** and false positive **(FP)**, respectively. These values give us a very clear idea about the performance of the model.

After getting the confusion, the accuracy of the prediction was calculated. This is defined as the ratio of correct prediction to the total number of predictions. Since the data was balanced, accuracy is a good indicator of the performance of the model. To get an even greater insight to the performance, the precision, recall, F1 score, Cohen's Kappa score, and the AUC score were calculated and reported. All these metrics, combined together, gave a clear indication about the performance of the models.

```
from sklearn import metrics

 print('Accuracy score is :', np.round(metrics.accuracy_score(y_test,
      predictions),4))
 print('Precision score is :', np.round(metrics.precision_score(y_test,
      predictions, average='weighted'),4))
 print('Recall score is :', np.round(metrics.recall_score(y_test,
      predictions, average='weighted'),4))
 print('F1 Score is :', np.round(metrics.f1_score(y_test, predictions,
      average='weighted'),4))
 print('ROC AUC Score is :', np.round(metrics.roc_auc_score(y_test,
      prob_pred,multi_class='ovo', average='weighted'),4))
 print('Cohen Kappa Score:', np.round(metrics.cohen_kappa_score(y_test,
      predictions),4))
 print('\t\tClassification Report:\n', metrics.classification_report(y_test,
      predictions))
```

3.4.7 Experimental results

While experimenting, many different pretrained architectures were used to perform the classification. But, to check the performance on CNNs trained from scratch, various CNNs with different number of layers were trained as mentioned in Section 3.4.3. As Table 3.1 clearly shows that the overall performance of the models was higher when image augmentations were used to increase the amount of data. It was also seen that larger models fit better to larger datasets as the CNN with seven layers performed the best when image augmentations were applied, whereas the CNN with just three layers performed the best when no image augmentations were applied. The overall accuracy, F1 score, and Kappa score are higher when the number of training samples is higher.

Next series of experiments were with transfer learning. Since the shallow network used to fine-tune the pretrained networks were the same for all the models, the performance difference was solely based on the pretrained model's abilities to extract features from the MRI scans (Table 3.2).

As observed, ResNet101 [7] model performed the best with and without the image augmentations. Since the training accuracy was 100%, it indicated to possible overfitting, but as the test accuracy was also above 99.5%, it was concluded that the models were generalizing well.

For classification using deep feature extraction, as mentioned in Section 3.4.1, 11 different neural networks and 6 different classifiers were used (Table 3.3).

When the features were extracted using VGG16 [26] model, ANN classifier performed the best, but the SVM classifier had the highest test accuracy. However, with the nonaugmented data, XGBoost and Random Forest performed equally well with similar test accuracies and F1 score (Table 3.4).

As observed from Table 3.4, XGBoost classifier had the highest test accuracy, F1 score, Cohen's Kappa score, and AUC score when the features were extracted by VGG19 [26] model. The training accuracy of the XGBoost, Random Forest, and ANN was 100%, while the test accuracy was around 85%. This showed that the classifiers were not generalizing well on the unseen data when the features were being extracted by the VGG19 [26] model (Table 3.5).

As Table 3.5 shows, an ANN classifier performed the best when an ResNet50 [7] model was used for extracting the features. Overall test accuracy, F1 score, Kappa score, and AUC score were higher when the amount of training data was increased with the help of augmentations. In spite of performing the best, these models failed to generalize well and ended up overfitting on the training data (Table 3.6).

As observed from Table 3.6, ANN/MLP classifier performed the best when the features were extracted using an ResNet101 [7] model. When the number of training samples was increased by augmentations, there was a clear increase in the performance of the classifiers with the test accuracy being over 1% higher. When augmentations were not used, SVM classifier performed similar to the ANN/MLP classifier (Table 3.7).

When MobileNet-v2 [27] was used to extract the features for classification, it was observed that the classifiers gave a better result on the nonaugmented data. Although the difference was very small, ANN/MLP classifier performed the best on nonaugmented data with MobilNet-v2 feature extractor (Table 3.8).

As observed from Table 3.8, ANN/MLP classifier again emerged as the winner in terms of test accuracy when the features were extracted from the augmented data using a pretrained MobileNet [28]. When the features were extracted form a smaller nonaugmented dataset; however, Random Forest had the highest test accuracy. It was also observed that all classifiers ended up overfitting on the training data (Table 3.9).

TABLE 3.1 Details of performance of the convolutional neural networks (CNNs) that were trained from scratch on this data.

Classifier	Training accuracy	Validation accuracy	Test accuracy	F1 measure	KAPPA	AUC
(A)						
CNN 2 Layer	0.74	0.72	0.76	0.7479	0.51	0.8671
CNN 3 Layer	0.81	0.79	0.78	0.7803	0.5666	0.8905
CNN 4 Layer	0.5	0.5	0.5	0.33	0	0.5
CNN 5 Layer	0.5	0.5	0.5	0.33	0	0.5
CNN 6 Layer	0.5	0.5	0.5	0.33	0	0.5
CNN 7 Layer	0.97	0.96	0.96	0.9599	0.92	0.99438
CNN 8 Layer	0.94	0.94	0.94	0.9308	0.8766	0.98753
(B)						
CNN 2 Layer	1	0.93	0.9288	0.9282	0.8566	0.9776
CNN 3 Layer	1	0.94	0.94	0.9399	0.88	0.984
CNN 4 Layer	0.5	0.5	0.5	0.33	0	0.5
CNN 5 Layer	0.74	0.71	0.695	0.6673	0.39	0.9236
CNN 6 Layer	0.93	0.86	0.8783	0.8767	0.7566	0.9877
CNN 7 Layer	0.96	0.90	0.93166	0.9314	0.8633	0.99301
CNN 8 Layer	0.92	0.88	0.87166	0.8701	0.7433	0.97942

Table 3.1A shows the performance result of the CNNs on the data after image augmentations were applied and Table 3.1B shows the perforce on the smaller dataset without image augmentations.

When the features of the data are extracted using an Inception-V3 [29] network, Random Forest classifier performed the best on almost all the metrics. The test accuracy was barely higher when augmentations are not performed. It was also observed that all the models tended to overfit on the training data and hence this led to poor generalization (Table 3.10).

Random Forest classifier performed the best in all the parameters when the features were extracted using Inception-ResNet-v2 [30] from the augmented data. With a test accuracy of 85.83%, it was higher by 2%. While ANN/MLP classifiers performed the best in ResNets, the introduction of Inception modules in the network tilted the balance toward tree-based models as seen in Table 3.9 and Table 3.10. The introduction of residual connections increased the overall performance of the models (Table 3.11).

XGBoost classifier emerged as the clear winner in terms of all the metrics when the features are extracted from the augmented data using a DenseNet169 [31] model. When nonaugmented data was used, the overall test accuracy of all the classifiers was decreased by over 1%. With a training accuracy of 100% and a test accuracy of 82.33%, it was very clear that the model was overfitting on the training features (Table 3.12).

When DenseNet-121 [31] was used to extract features from the augmented data, XGBoost model performed the best with 82.17% test accuracy and an F1 score of 0.82. When image

TABLE 3.2 Results of the experiments that were conducted with transfer learning.

Classifier	Training accuracy	Validation accuracy	Test accuracy	F1 measure	KAPPA	AUC
(A)						
ResNet50	1	0.99	0.99	0.9933	0.9867	0.9986
VGG16	1	0.99	0.99	0.9933	0.9867	0.9998
VGG19	1	0.98	0.99	0.9883	0.9766	0.9992
Inception_v3	0.84	0.84	0.8033	0.8032	0.6066	0.883
MobileNet	0.98	0.97	0.97	0.9716	0.9433	0.9972
DenseNet169	0.94	0.93	0.94	0.9328	0.8766	0.9808
DenseNet121	0.96	0.95	0.95	0.9516	0.9033	0.9925
InceptionResNet-V2	0.76	0.76	0.73	0.7253	0.4533	0.8237
MobileNet-V2	0.97	0.95	0.95	0.9516	0.9033	0.9932
ResNet101	1	0.98	0.995	0.9949	0.99	0.9998
(B)						
ResNet50	1	0.97	0.9817	0.9817	0.9633	0.9982
VGG16	1	0.97	0.9567	0.9567	0.9133	0.9907
VGG19	0.99	0.95	0.9733	0.9733	0.9466	0.9869
Inception_v3	0.99	0.88	0.885	0.885	0.77	0.9485
MobileNet	1	0.96	0.9733	0.9733	0.9467	0.9957
DenseNet169	0.94	0.88	0.8967	0.8964	0.7933	0.9657
DenseNet121	0.95	0.9	0.9117	0.9111	0.8233	0.9765
InceptionResNet-V2	0.88	0.8	0.8117	0.8117	0.6233	0.8924
MobileNet-V2	0.97	0.95	0.95	0.9516	0.9033	0.9932
ResNet101	1	0.98	0.995	0.9949	0.99	0.9998

Table 3.2A contains the performance of the networks on the dataset after image augmentations were applied and Table 3.2B contains the performance of the models on the dataset without any image augmentations.

augmentations were not used, the performance of all the models saw a significant drop and ANN/MLP classifier performed the best among all with an 80% test accuracy (Table 3.13).

When a network with depth separable convolution was used, ANN/MLP classifier performed the best with and without any data augmentations. Test accuracy of the models somewhat increased when image augmentations were not used.

3.5 Discussion

The series of experiments revealed the performance of various machine learning algorithms

TABLE 3.3 The performance of various classifiers when the features were extracted using VGG16 [26].

Model	Training accuracy	Validation accuracy	Test accuracy	F1 score	Kappa	Recall	Precision	AUC
(A)								
ANN	100%	84.33%	84.67%	0.85	0.69	0.85	0.85	0.8466
KNN	89.94%	78.83%	80.50%	0.8	0.61	0.8	0.83	0.805
SVM	90.06%	81.00%	8883.00%	0.83	0.66	0.83	0.83	0.8283
Random Forest	100%	81.67%	82.33%	0.82	0.65	0.82	0.82	0.8233
AdaBoost	81.11%	75.15%	73.50%	0.73	0.47	0.74	0.74	0.735
XGBoost	100%	82.17%	83.67%	0.84	0.67	0.84	0.84	0.83666
(B)								
ANN	100%	86.50%	85.67%	0.86	0.71	0.86	0.86	0.8566
KNN	89.94%	82.00%	82.33%	0.82	0.65	0.82	0.85	0.8233
SVM		84.50%	85.33%	0.85	0.71	0.85	0.85	0.8533
Random Forest	100%	84.83%	86.67%	0.87	0.73	0.87	0.87	0.8666
AdaBoost	83.61%	78.67%	75.83%	0.76	0.52	0.76	0.76	0.7583
XGBoost	100%	86.17%	86.67%	0.87	0.87	0.73	0.87	0.8666

Table 3.3A shows the performance when data augmentation was used, and Table 3.3B shows the performance when data augmentation was not used.

TABLE 3.4 Performance metrics of classifiers on the features extracted by a pretrained VGG19 [26] network.

Model	Training accuracy	Validation accuracy	Test accuracy	F1 score	Kappa	Recall	Precision	AUC
(A)								
ANN	100%	87.17%	85.33%	0.85	0.71	0.85	0.85	0.8533
KNN	87.33%	78.33%	79.17%	0.79	0.58	0.79	0.81	0.7916
SVM	91.72%	86.50%	84.33%	0.84	0.69	0.84	0.84	0.8433
Random Forest	100%	85.00%	85.33%	0.85	0.71	0.85	0.85	0.8533
AdaBoost	84.06%	76.00%	77.33%	0.77	0.55	0.77	0.77	0.7733
XGBoost	100%	85.67%	86.00%	0.86	0.72	0.86	0.86	0.86
(B)								
ANN	100%	83.67%	82.33%	0.82	0.65	0.82	0.82	0.8233
KNN	85.28%	78.50%	80.00%	0.8	0.6	0.8	0.83	0.8
SVM	91.78%	81.83%	82.33%	0.82	0.65	0.82	0.82	0.8233
Random Forest	100%	82.17%	81.33%	0.81	0.63	0.81	0.82	0.8133
AdaBoost	81.67%	71.50%	75.00%	0.75	0.5	0.75	0.75	0.7499
XGBoost	100%	84.33%	83.33%	0.83	0.67	0.83	0.83	0.8333

Table 3.4A shows the performance of the classifiers on the dataset with augmentations and Table 3.4B shows the performance on the dataset without augmentations.

TABLE 3.5 Performance of the classifiers based on the features extracted by ResNet50 [7] model.

Model	Training accuracy	Validation accuracy	Test accuracy	F1 score	Kappa	Recall	Precision	AUC
(A)								
ANN	100%	88.33%	89.00%	0.89	0.78	0.89	0.89	0.89
KNN	92.11%	85.17%	85.33%	0.85	0.71	0.85	0.86	0.8533
SVM	92.78%	87.33%	85.00%	0.85	0.7	0.85	0.85	0.85
Random Forest	100%	86.67%	84.50%	0.84	0.69	0.84	0.85	0.845
AdaBoost	85.50%	81.33%	78.33%	0.79	0.58	0.79	0.79	0.7833
XGBoost	100%	87.33%	85.50%	0.85	0.71	0.86	0.86	0.855
(B)								
ANN	100%	90.50%	87.83%	0.88	0.76	0.88	0.88	0.8783
KNN	92.89%	86.33%	87.00%	0.87	0.74	0.87	0.87	0.87
SVM	93.89%	87.83%	85.33%	0.85	0.71	0.85	0.85	0.8533
Random Forest	100%	85.50%	82.83%	0.83	0.66	0.83	0.83	0.8283
AdaBoost	85.67%	82.17%	77.17%	0.77	0.54	0.77	0.77	0.7716
XGBoost	100%	87.33%	85.50%	0.85	0.71	0.86	0.86	0.855

Table 3.5A shows the performance of the classifiers when the number of training samples were increased using image augmentations. Table 3.5B shows the performance of the classifiers when image augmentations were not used.

TABLE 3.6 Performance of various classifiers trained on the features extracted by an ResNet101 [7] model.

Model	Training accuracy	Validation accuracy	Test accuracy	F1 score	Kappa	Recall	Precision	AUC
(A)								
ANN	100%	89.83%	88.83%	0.89	0.78	0.89	0.89	0.8883
KNN	92.89%	87.50%	88.50%	0.88	0.77	0.88	0.89	0.8849
SVM	93.33%	86.17%	87.00%	0.87	0.74	0.87	0.87	0.87001
Random Forest	100%	84.83%	84.33%	0.84	0.69	0.84	0.84	0.8433
AdaBoost	85.06%	77.33%	77.50%	0.77	0.55	0.78	0.78	0.775
XGBoost	100%	88.33%	85.00%	0.85	0.7	0.85	0.85	0.8499
(B)								
ANN	100%	88.50%	87.33%	0.87	0.75	0.87	0.87	0.8733
KNN	92.17%	85.83%	84.67%	0.85	0.69	0.85	0.86	0.8466
SVM	93.11%	87.50%	87.33%	0.87	0.75	0.87	0.88	0.8733
Random Forest	100%	86.83%	86.50%	0.86	0.73	0.86	0.87	0.865
AdaBoost	86.57%	77.50%	75.00%	0.75	0.5	0.75	0.75	0.75
XGBoost	100%	85.50%	85.50%	0.85	0.71	0.86	0.86	0.855

Table 3.6A shows the performance of the classifiers when the features are extracted from the dataset after image augmentations are applied and Table 3.6B shows the performance on the data without any image augmentations.

TABLE 3.7 Performance of various classifiers trained on the features extracted by a MobileNet-v2 [27] model.

Model	Training accuracy	Validation accuracy	Test accuracy	F1 score	Kappa	Recall	Precision	AUC
(A)								
ANN	100%	81.67%	77.50%	0.77	0.55	0.78	0.78	0.775
KNN	84.72%	75.33%	74.67%	0.74	0.49	0.75	0.76	0.7466
SVM	89.61%	79.50%	76.33%	0.76	0.53	0.76	0.76	0.7633
Random Forest	100%	75.83%	76.67%	0.77	0.53	0.77	0.77	0.766
AdaBoost	79.39%	74.00%	69.00%	0.69	0.38	0.69	0.69	0.69
XGBoost	100%	78.00%	79.00%	0.79	0.58	0.79	0.79	0.7899
(B)								
ANN	100%	79.83%	79.83%	0.8	0.6	0.8	0.8	0.7983
KNN	84.67%	76.00%	77.70%	0.77	0.55	0.78	0.79	0.7766
SVM	91.39%	83.50%	78.17%	0.78	0.56	0.78	0.78	0.7816
Random Forest	100%	80.83%	76.50%	0.76	0.53	0.76	0.77	0.765
AdaBoost	81.89%	73.50%	71.17%	0.71	0.42	0.71	0.714	0.71166
XGBoost	100%	79.33%	76.33%	0.76	0.53	0.76	0.76	0.7633

Table 3.7A shows the performance of the classifiers when the features are extracted from the dataset after image augmentations are applied and Table 3.7B shows the performance on the data without any image augmentations.

TABLE 3.8 Performance of various classifiers trained on the features extracted by a MobileNet [28] model.

Model	Training accuracy	Validation accuracy	Test accuracy	F1 score	Kappa	Recall	Precision	AUC
(A)								
ANN	100%	78.83%	79.83%	0.8	0.6	0.8	0.8	0.7983
KNN	83.44%	73.00%	72.67%	0.72	0.45	0.73	0.74	0.7266
SVM	88.89%	77.50%	78.50%	0.78	0.57	0.78	0.79	0.7849
Random Forest	100%	76.17%	77.00%	0.77	0.54	0.77	0.77	0.77
AdaBoost	77.06%	68.83%	67.67%	0.68	0.35	0.38	0.68	0.6766
XGBoost	100%	77.50%	76.67%	0.77	0.53	0.77	0.77	0.7666
(B)								
ANN	100%	82.67%	717.00%	0.77	0.54	0.77	0.7	0.7716
KNN	83.83%	76.50%	75.67%	0.75	0.51	0.76	0.79	0.7566
SVM	87.61%	83.33%	76.50%	0.76	0.53	0.76	0.77	0.765
Random Forest	100%	80.67%	77.50%	0.77	0.55	0.78	0.78	0.775
AdaBoost	79.56%	73.17%	70.17%	0.7	0.4	0.7	0.7	0.7016
XGBoost	100%	77.83%	77.00%	0.77	0.54	0.77	0.77	0.77

Table 3.8A shows the performance of the classifiers when the features are extracted from the dataset after image augmentations are applied and Table 3.8B shows the performance on the data without any image augmentations.

TABLE 3.9 Performance of various classifiers trained on the features extracted by an Inception-V3 [29] model.

Model	Training accuracy	Validation accuracy	Test accuracy	F1 score	Kappa	Recall	Precision	AUC
(A)								
ANN	100%	74.17%	76.83%	0.77	0.54	0.77	0.77	0.7683
KNN	85.33%	71.83%	76.17%	0.76	0.52	0.76	0.76	0.7616
SVM	84.78%	73.83%	74.83%	0.75	0.5	0.75	0.75	0.74833
Random Forest	100%	73.83%	77.83%	0.78	0.56	0.78	0.78	0.7783
AdaBoost	75.61%	60.67%	63.83%	0.64	0.28	0.64	0.64	0.6383
XGBoost	100%	73.67%	75.33%	0.75	0.51	0.75	0.75	0.7533
(B)								
ANN	100%	77.00%	77.33%	0.77	0.55	0.77	0.77	0.7733
KNN	83.83%	74.83%	74.50%	0.74	0.49	0.74	0.75	0.745
SVM	84.78%	74.33%	72.00%	0.72	0.44	0.72	0.72	0.72
Random Forest	100%	76.67%	78.00%	0.78	0.56	0.78	0.78	0.78
AdaBoost	76.72%	64.67%	60.83%	0.61	0.22	0.61	0.61	0.6083
XGBoost	100%	75.83%	73.83%	0.74	0.48	0.74	0.74	0.7383

Table 3.9A shows the performance of the classifiers when the features are extracted from the dataset after image augmentations are applied and Table 3.9B shows the performance on the data without any image augmentations.

TABLE 3.10 Performance of various classifiers trained on the features extracted by an Inception-ResNet-v2 [30] model.

Model	Training accuracy	Validation accuracy	Test accuracy	F1 score	Kappa	Recall	Precision	AUC
(A)								
ANN	80.61%	74.67%	73.00%	0.73	0.46	0.73	0.73	0.73
KNN	83.44%	75.33%	74.33%	0.74	0.49	0.74	0.78	0.7433
SVM	70.11%	68.50%	65.67%	0.65	0.31	0.66	0.67	0.6555
Random Forest	100%	82.33%	85.83%	0.86	0.72	0.86	0.86	0.8583
AdaBoost	79.39%	70.67%	69.50%	0.69	0.39	0.7	0.7	0.695
XGBoost	100%	81.33%	83.83%	0.84	0.68	0.84	0.84	0.8383
(B)								
ANN	79.78%	72.17%	73.00%	0.73	0.46	0.73	0.73	0.73
KNN	83.22%	72.00%	72.17%	0.72	0.44	0.72	0.74	0.7216
SVM	67.89%	65.67%	64.00%	0.64	0.28	0.64	0.64	0.64
Random Forest	100%	80.50%	81.17%	0.81	0.62	0.81	0.81	0.8116
AdaBoost	76.78%	70.17%	68.67%	0.69	0.37	0.69	0.69	0.6866
XGBoost	100%	80.67%	83.00%	0.83	0.66	0.83	0.83	0.8299

Table 3.10A shows the performance of the classifiers when the features are extracted from the dataset after image augmentations are applied and Table 3.10B shows the performance on the data without any image augmentations.

TABLE 3.11 Performance of various classifiers trained on the features extracted by a DenseNet169 [31] model.

(A)

Model	Training accuracy	Validation accuracy	Test accuracy	F1 score	Kappa	Recall	Precision	AUC
ANN	99.39%	81.67%	81.17%	0.81	0.62	0.81	0.82	0.8116
KNN	86.39%	77.17%	81.00%	0.81	0.62	0.81	0.83	0.81
SVM	80.00%	76.17%	76.00%	0.76	0.52	0.76	0.76	0.76
Random Forest	100%	81.00%	82.17%	0.82	0.64	0.82	0.82	0.8216
AdaBoost	79.27%	69.83%	72.50%	0.72	0.45	0.72	0.73	0.7249
XGBoost	100%	82.17%	82.83%	0.83	0.66	0.83	0.83	0.8283
(B)								
ANN	99.61%	80.67%	79.83%	0.81	0.6	0.8	0.81	0.7983
KNN	86.89%	78.50%	78.33%	0.78	0.57	0.78	0.79	0.7833
SVM	79.78%	74.67%	71.83%	0.72	0.44	0.72	0.72	0.7183
Random Forest	100%	79.50%	80.17%	0.8	0.6	0.8	0.8	0.8016
AdaBoost	79.06%	68.33%	69.00%	0.69	0.38	0.69	0.69	0.69
XGBoost	100%	79.17%	81.17%	0.81	0.62	0.81	0.81	0.8116

Table 3.11A shows the performance of the classifiers when the features are extracted from the dataset after image augmentations are applied and Table 3.11B shows the performance on the data without any image augmentations.

TABLE 3.12 Performance of various classifiers trained on the features extracted by a DenseNet121 [31] model.

(A)

Model	Training accuracy	Validation accuracy	Test accuracy	F1 score	Kappa	Recall	Precision	AUC
ANN	99.94%	83.00%	81.17%	0.81	0.62	0.81	0.82	0.8116
KNN	87.50%	81.50%	81.17%	0.81	0.62	0.81	0.82	0.8116
SVM	79.83%	74.83%	74.17%	0.74	0.48	0.74	0.75	0.7416
Random Forest	100%	81.33%	81.50%	0.81	0.63	0.82	0.82	0.815
AdaBoost	78.06%	67.67%	69.50%	0.69	0.39	0.7	0.7	0.695
XGBoost	100%	81.50%	82.17%	0.82	0.64	0.82	0.82	0.8216
(B)								
ANN	100%	82.50%	80.50%	0.8	0.61	0.8	0.81	0.805
KNN	87.33%	74.00%	78.67%	0.78	0.57	0.79	0.8	0.7866
SVM	82.44%	78.50%	72.67%	0.73	0.45	0.73	0.723	0.7266
Random Forest	100%	81.33%	79.67%	0.8	0.59	0.8	0.8	0.7966
AdaBoost	80.50%	69.83%	69.50%	0.69	0.39	0.7	0.7	0.695
XGBoost	100%	83.17%	80.00%	0.8	0.6	0.8	0.8	0.8

Table 3.12A shows the performance of the classifiers when the features are extracted from the dataset after image augmentations are applied and Table 3.12B shows the performance on the data without any image augmentations.

TABLE 3.13 Performance of various classifiers trained on the features extracted by a XceptionNet [32] model.

Model	Training accuracy	Validation accuracy	Test accuracy	F1 score	Kappa	Recall	Precision	AUC
(A)								
ANN	100%	81.33%	81.33%	0.81	0.63	0.81	0.81	0.8133
KNN	86.39%	76.17%	75.50%	0.75	0.51	0.76	0.77	0.7549
SVM	86.11%	79.00%	76.50%	0.76	0.53	0.76	0.77	0.765
Random Forest	100%	79.50%	77.50%	0.77	0.55	0.78	0.78	0.7749
AdaBoost	80.56%	73.50%	71.00%	0.71	0.42	0.71	0.71	0.71
XGBoost	100%	80.83%	81.00%	0.81	0.62	0.81	0.81	0.8099
(B)								
ANN	100%	79.83%	83.50%	0.83	0.67	0.84	0.84	0.835
KNN	84.56%	75.50%	77.50%	0.77	0.55	0.78	0.79	0.775
SVM	86.56%	77.83%	78.50%	0.78	0.57	0.78	0.79	0.785
Random Forest	100%	79.50%	83.17%	0.83	0.66	0.86	0.83	0.8316
AdaBoost	78.50%	69.83%	76.00%	0.76	0.52	0.76	0.76	0.76
XGBoost	100%	78.67%	81.00%	0.81	0.62	0.81	0.81	0.8099

Table 3.13A shows the performance of the classifiers when the features are extracted from the dataset after image augmentations are applied and Table 3.13B shows the performance on the data without any image augmentations.

and deep learning techniques while classifying normal (without tumor) and abnormal (with tumor) brain MRI images. From the observations, it is very clear that on an average, the algorithms perform better for augmented data. This is because, higher number of data points lead to higher knowledge of the model about the data. This is because machine learning algorithms are very data centric and require a huge amount of information to perform tasks.

As evident from Fig. 3.17, all the machine learning classifiers had 1% higher test accuracy on an average when the features were collected from the bigger dataset which was created after applying image augmentations. Since the dataset used was balanced, the "test accuracy" metric holds the most importance. If the dataset were unbalanced, then F1 and AUC scores would be of more importance as the accuracy would give an illusion of performance.

When transfer learning was used to classify the MRI scans, it was observed that the test accuracy was higher when image augmentations were used to increase the number of training samples. This goes on to show how important data is for deep learning models to learn. Without sufficient data, the models fail to extract meaningful patterns and hence the performance decreases. This trend is clearly shown in Fig. 3.18.

The test accuracy of IncpetionNet-v3 and InceptionResnetv-2 is higher when data augmentation is not performed, but their overall performance is lower when compared to other models. These can be outlier cases as in machine learning, it is a general rule that higher amount of data leads to better generalization of models (Fig. 3.19).

When inception modules are introduced in the network, the test accuracy of the model drops to 70%–80%. A similar drop in performance was also observed for all the other performance metrics. This shows that InceptionNets are not well suited for problem. It could also be concluded that further tuning and a larger

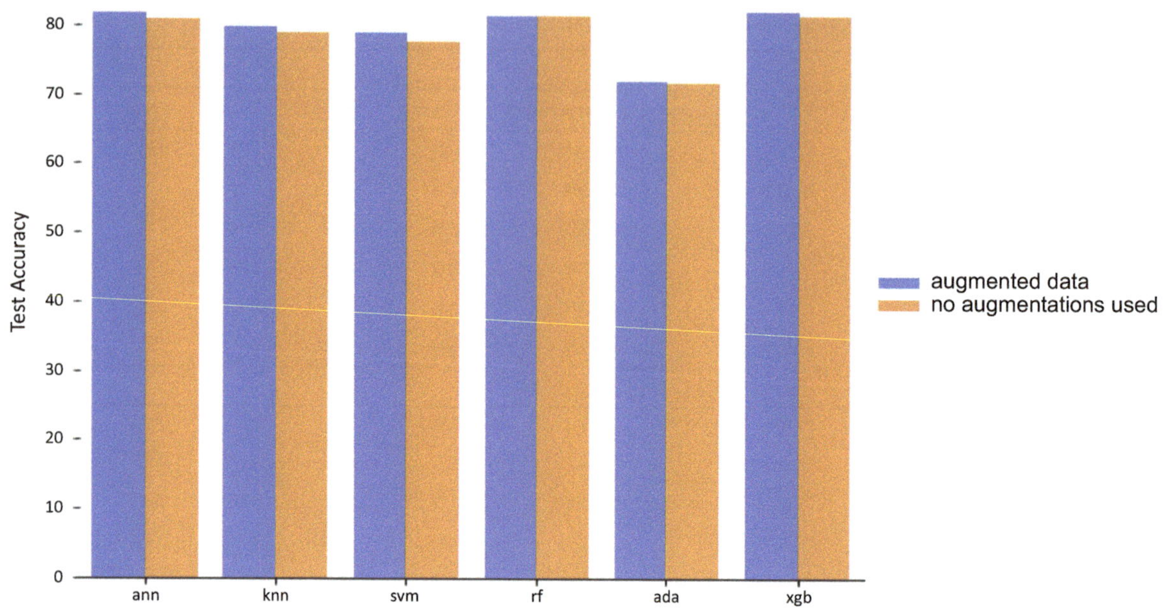

FIGURE 3.17 Comparison of the performance of the models on the features from augmented and nonaugmented data.

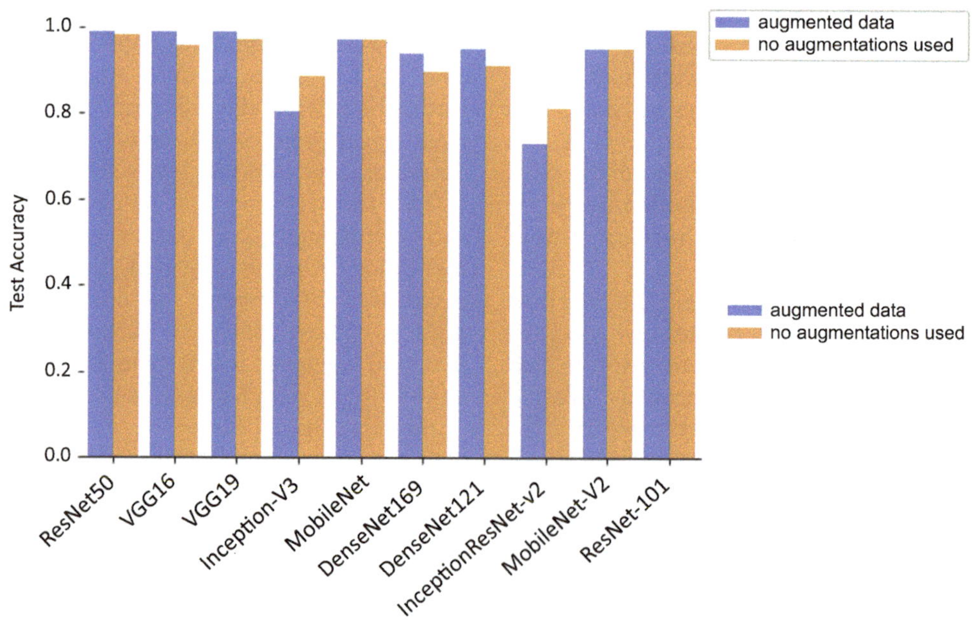

FIGURE 3.18 Comparison of various models for transfer learning on augmented and nonaugmented data.

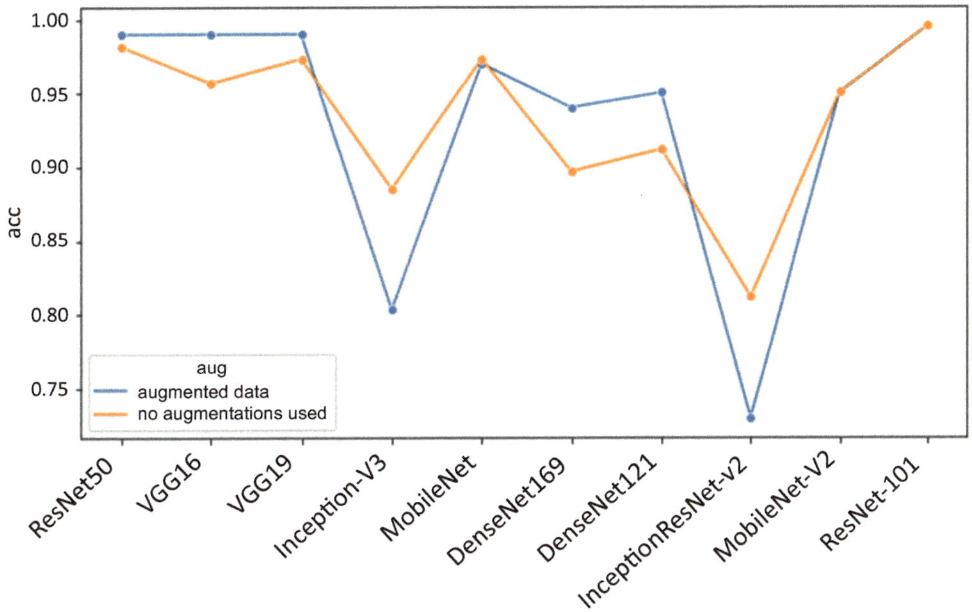

(A)

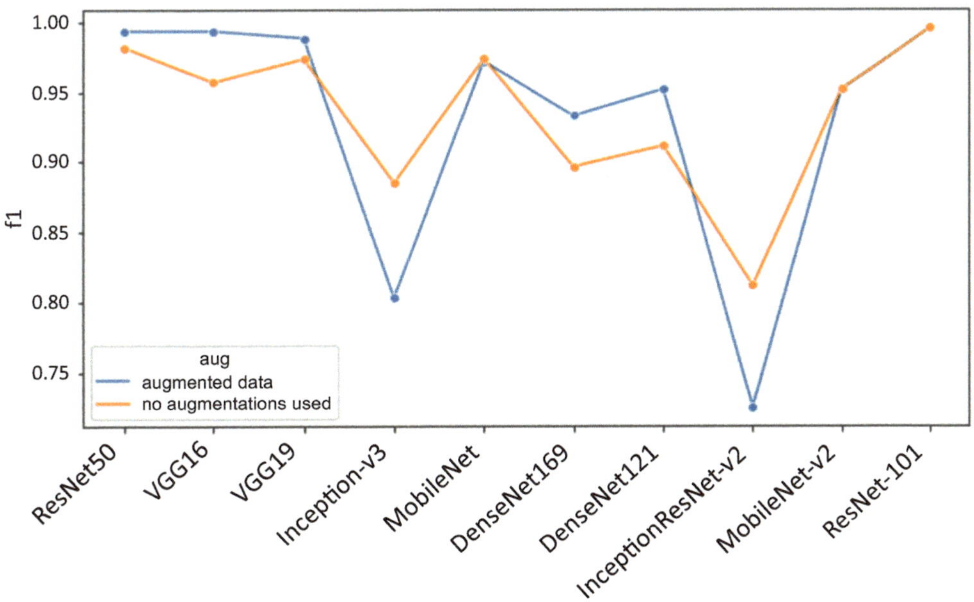

(B)

FIGURE 3.19 Comparison of the test accuracies of various models trained by transfer learning. (A) shows the test accuracies of various models and (B) shows the F1 score of various models.

dataset could have improved the performance of inception nets. ResNets and VGG Nets performed the best with ResNet50 having a test accuracy of 99% and an F1 score of 0.9933 and VGG16 having a test accuracy of 99% and an F1 score of 0.9933. Similarly, the bigger ResNet101 and VGG19 have test accuracies of 99.5% and 99% and F1 score of 0.9949 and 0.9883, respectively. So, networks with residual connections ended up performing the best in transfer learning on this dataset.

Deep feature extraction helped in determining which models preserve the maximum data about the original image in their latent representation. This meant that we could figure out which network maps the original $224 \times 224 \times 3$ dimensional space to a smaller 64×1 dimensional space. Since the same classifiers with the exact same hyperparameters were used to classify the images from the extracted features, the difference in accuracies completely was completely dependent on the features themselves. The features were a mapping of a vector in the high-dimensional space (the image) to a vector in a lower dimensional space (the latent representation / features). It was observed that the classifiers had the highest average test accuracy on the features extracted by ResNets. The average test accuracy of classifiers trained on the features extracted by ResNet101 from the augmented dataset is 85.193%, whereas the test accuracy of the classifiers trained on the features extracted by ResNet50 from the augmented dataset is 84.61%. Fig. 3.20 also shows that the features extracted form a smaller non-augmented dataset. The classifiers still perform the best on the features extracted by ResNets with ResNet50 having an average of 84.27% and ResNet101 having an average of 84.38% (Fig. 3.21).

When CNNs of varying depths were trained on the data from scratch, it was observed that

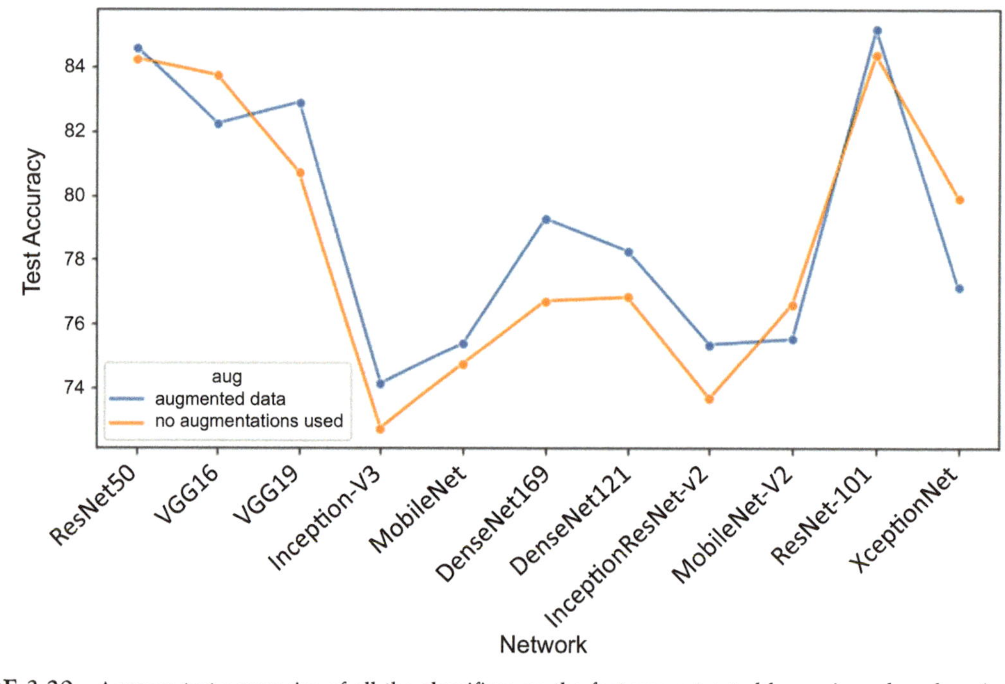

FIGURE 3.20 Average test accuracies of all the classifiers on the features extracted by various deep learning models. This plot shows which network represents the original data the best.

the bigger networks (with more convolutional layers) performed better on the larger dataset (with augmentations). The models of four, five, and six convolutional layers performed the worst and showed no agreement between the test data and labels which was evident from the extremely low Cohen's Kappa score. The maximum test accuracy of 94% was shown by

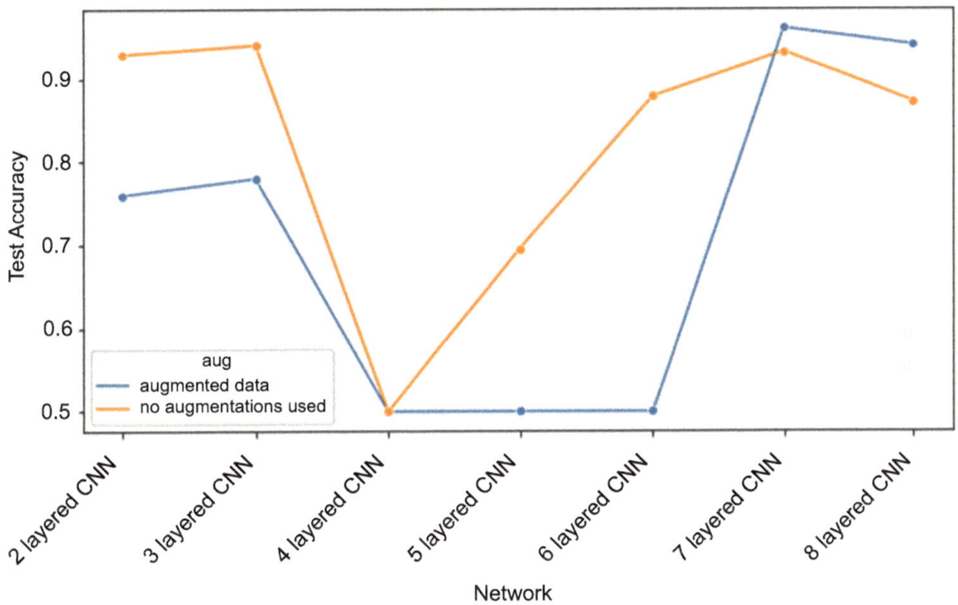

FIGURE 3.21 Test accuracies of various CNNs that were trained from scratch on the dataset. *CNNs*, Convolutional neural networks.

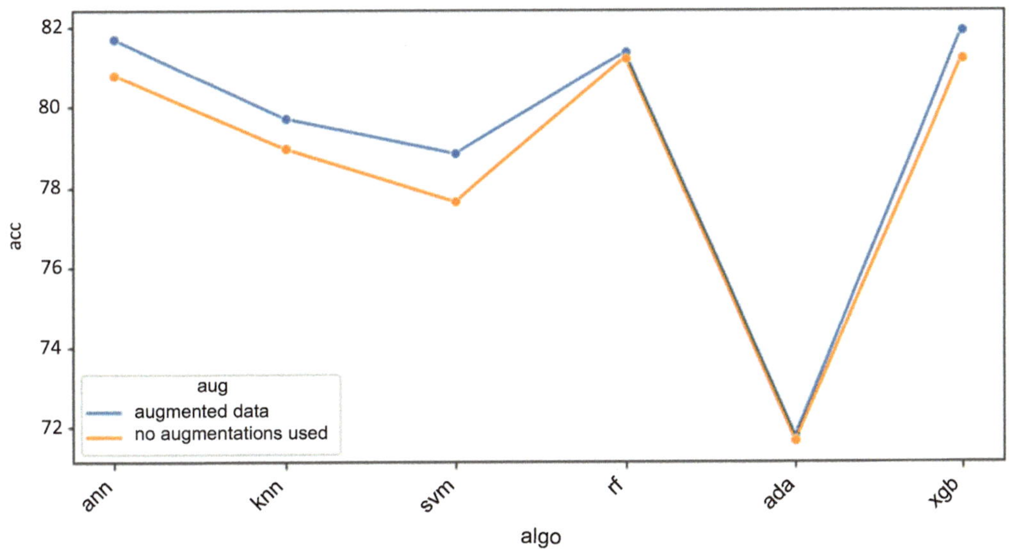

FIGURE 3.22 Average test accuracies of the machine learning classifiers on the features extracted by various convolutional neural networks.

the eight-layered CNN on augmented data. The two-layered CNN shows the maximum accuracy of 92.8% on the smaller nonaugmented data. It was evident that it would take much more data to train the CNNs from scratch and hence, transfer learning and deep feature extraction were superior methods to handle the problem of brain tumor detection.

ANN/MLP classifier and tree-based classifiers perform the best in terms of test accuracies on features extracted by pretrained convolution neural networks. ANN classifier performed the best with 81.69% test accuracy on the augmented data and it was followed by XGBoost classifier with 81.9% test accuracy. As shown in Fig. 3.22, AdaBoost classifier had the least average accuracy on the test set. The training accuracy of ANN, XGBoost, and Random Forest was 100% for most of the features. This indicated that these models tend to overfit on the training data and require a larger dataset to generalize well.

3.6 Conclusion

This chapter deals with the problem of automated brain tumor detection using various deep learning techniques. For this, a plethora of deep learning techniques such as transfer learning, deep feature extraction, and training CNNs from scratch were used. Multiple experiments were conducted to evaluate the performance of these techniques. This chapter tried to show the importance of machine learning in detection of brain tumor and their efficiency of doing that. Detailed comparative studies were carried out to determine the best possible models and techniques. Multiple performance metrics were used in order to avoid falling in accuracy traps and other performance metrics related biases. The findings have revealed that CNNs are able to detect brain tumors with 99% accuracy. We are also focused on creating novel ways to solve the problem of MRI-based automated brain tumor detection.

References

[1] R.L. Siegel, K.D. Miller, H.E. Fuchs, A. Jemal, Cancer statistics, 2021, CA: A Cancer J. Clinicians 71 (1) (2021) 7–33. Available from: https://doi.org/10.3322/caac.21654.

[2] Cancer.Net, Computed tomography (CT) scan, 2012. <https://www.cancer.net/navigating-cancer-care/diagnosing-cancer/tests-and-procedures/computed-tomography-ct-scan> (accessed 14.05.21).

[3] Cancer.Net, Positron emission tomography and computed tomography (PET-CT) scans, 2012. <https://www.cancer.net/navigating-cancer-care/diagnosing-cancer/tests-and-procedures/positron-emission-tomography-and-computed-tomography-pet-ct-scans> (accessed 14.05.21).

[4] Cancer.Net, Brain tumor - diagnosis, 2012. <https://www.cancer.net/cancer-types/brain-tumor/diagnosis> (accessed 14.05.21).

[5] Y. LeCun, Y. Bengio, G. Hinton, Deep learning, Nature 521 (7553) (2015) 436–444. Available from: https://doi.org/10.1038/nature14539.

[6] A. Krizhevsky, I. Sutskever, G.E. Hinton, ImageNet classification with deep convolutional neural networks, Commun. ACM 60 (6) (2017) 84–90. Available from: https://doi.org/10.1145/3065386.

[7] K. He, X. Zhang, S. Ren, J. Sun, Deep residual learning for image recognition, arXiv:1512.03385 [cs], Dec. 2015, <http://arxiv.org/abs/1512.03385> (accessed 15.05. 21).

[8] C. Szegedy et al., Going deeper with convolutions, arXiv:1409.4842 [cs], Sep. 2014, <http://arxiv.org/abs/1409.4842> (accessed 15.05.21).

[9] D.R. Sarvamangala, R.V. Kulkarni, Convolutional neural networks in medical image understanding: a survey, Evol. Intell. 15 (2021). Available from: https://link.springer.com/article/10.1007/s12065-020-00540-3 (accessed 18.05.21).

[10] A.S. Lundervold, A. Lundervold, An overview of deep learning in medical imaging focusing on MRI, Zeitschrift für Medizinische Physik 29 (2019). https://www.sciencedirect.com/science/article/pii/S0939388918301181(accessed May 18, 2021).

[11] M.I. Razzak, S. Naz, A. Zaib, Deep learning for medical image processing: overview, challenges and future, arXiv:1704.06825 [cs], Apr. 2017, <http://arxiv.org/abs/1704.06825> (accessed 18.05.21).

[12] S. Ahuja, B.K. Panigrahi, T. Gandhi, Transfer learning based brain tumor detection and segmentation using superpixel technique, in: 2020 International Conference on Contemporary Computing and Applications (IC3A), Feb. 2020, pp. 244–249. Available from: https://doi.org/10.1109/IC3A48958.2020.233306.

[13] R. Mehrotra, M.A. Ansari, R. Agrawal, R.S. Anand, A Transfer Learning approach for AI-based classification of brain tumors, Mach. Learn. Appl. 2 (2020)

100003. Available from: https://doi.org/10.1016/j.mlwa.2020.100003.

[14] F. Özyurt, E. Sert, E. Avci, E. Dogantekin, Brain tumor detection based on Convolutional Neural Network with neutrosophic expert maximum fuzzy sure entropy, Measurement 147 (2019) 106830. Available from: https://doi.org/10.1016/j.measurement.2019.07.058. Dec.

[15] V. Jalali, D. Kaur, A study of classification and feature extraction techniques for brain tumor detection, Int. J. Multimed. Info Retr. 9 (4) (2020) 271–290. Available from: https://doi.org/10.1007/s13735-020-00199-7.

[16] D. Deb, S. Roy, Brain tumor detection based on hybrid deep neural network in MRI by adaptive squirrel search optimization, Multimed. Tools Appl. 80 (2) (2021) 2621–2645. Available from: https://doi.org/10.1007/s11042-020-09810-9.

[17] N. Ambily, K. Suresh, Classification of brain MRI images using convolution neural network and transfer learning, in: 2020 11th International Conference on Computing, Communication and Networking Technologies (ICCCNT), Jul. 2020, pp. 1–6. Available from: https://doi.org/10.1109/ICCCNT49239.2020.9225504.

[18] J. Amin, et al., Brain tumor detection by using stacked autoencoders in deep learning, J. Med. Syst. 44 (2) (2020) 32. Available from: https://doi.org/10.1007/s10916-019-1483-2.

[19] R.B. Vallabhaneni, V. Rajesh, Brain tumour detection using mean shift clustering and GLCM features with edge adaptive total variation denoising technique, Alex. Eng. J. 57 (4) (2018) 2387–2392. Available from: https://doi.org/10.1016/j.aej.2017.09.011.

[20] H.M. Rai, K. Chatterjee, S. Dashkevich, Automatic and accurate abnormality detection from brain MR images using a novel hybrid UnetResNext-50 deep CNN model, Biomed. Signal. Process. Control. 66 (2021) 102477. Available from: https://doi.org/10.1016/j.bspc.2021.102477.

[21] M. Leshno, V.Ya Lin, A. Pinkus, S. Schocken, Multilayer feedforward networks with a nonpolynomial activation function can approximate any function, Neural Netw. 6 (6) (1993) 861–867. Available from: https://doi.org/10.1016/S0893-6080(05)80131-5.

[22] Y. Lecun, L. Bottou, Y. Bengio, P. Haffner, Gradient-based learning applied to document recognition, Proc. IEEE 86 (11) (1998) 2278–2324. Available from: https://doi.org/10.1109/5.726791.

[23] IEEE, ImageNet: a large-scale hierarchical image database. <https://ieeexplore.ieee.org/document/5206848>, 2009 (accessed 15.08.21).

[24] A.W. Senior, et al., Improved protein structure prediction using potentials from deep learning, Nature 577 (7792) (2020). Available from: https://doi.org/10.1038/s41586-019-1923-7. Art. no. 7792, Jan.

[25] O. Ronneberger, P. Fischer, T. Brox, U-Net: convolutional networks for biomedical image segmentation, arXiv:1505.04597 [cs], <http://arxiv.org/abs/1505.04597>, May 2015 (accessed 26.07.21).

[26] K. Simonyan, A. Zisserman, Very deep convolutional networks for large-scale image recognition, arXiv:1409.1556 [cs], <http://arxiv.org/abs/1409.1556>, Apr. 2015 (accessed 27.06.21).

[27] M. Sandler, A. Howard, M. Zhu, A. Zhmoginov, L.-C. Chen, MobileNetV2: inverted residuals and linear bottlenecks, arXiv:1801.04381 [cs], <http://arxiv.org/abs/1801.04381>, Mar. 2019, (accessed 27.07.21).

[28] A.G. Howard et al., MobileNets: efficient convolutional neural networks for mobile vision applications, arXiv:1704.04861 [cs], <http://arxiv.org/abs/1704.04861>, Apr. 2017 (accessed 27.07.21).

[29] C. Szegedy, V. Vanhoucke, S. Ioffe, J. Shlens, Z. Wojna, Rethinking the inception architecture for computer vision, arXiv:1512.00567 [cs], <http://arxiv.org/abs/1512.00567>, Dec. 2015 (accessed 27.07.21).

[30] C. Szegedy, S. Ioffe, V. Vanhoucke, A. Alemi, Inception-v4, Inception-ResNet and the impact of residual connections on learning, arXiv:1602.07261 [cs], <http://arxiv.org/abs/1602.07261>, Aug. 2016 (accessed 27.07.21).

[31] G. Huang, Z. Liu, L. van der Maaten, K.Q. Weinberger, Densely connected convolutional networks, arXiv:1608.06993 [cs], <http://arxiv.org/abs/1608.06993>, Jan. 2018 (accessed 27.07.21).

[32] F. Chollet, Xception: deep learning with depthwise separable convolutions, arXiv:1610.02357 [cs], <http://arxiv.org/abs/1610.02357>, Apr. 2017 (accessed 27.07.21).

[33] Hamada, Br35H: brain tumor detection 2020. <https://kaggle.com/ahmedhamada0/brain-tumor-detection> (accessed 28.07. 21).

CHAPTER

4

Breast cancer detection from mammograms using artificial intelligence

Abdulhamit Subasi[1,2], *Aayush Dinesh Kandpal*[3], *Kolla Anant Raj*[4] *and Ulas Bagci*[5]

[1]Institute of Biomedicine, Faculty of Medicine, University of Turku, Turku, Finland [2]Department of Computer Science, College of Engineering, Effat University, Jeddah, Saudi Arabia [3]Department of Metallurgical and Materials Engineering, National Institute of Technology Rourkela, Rourkela, Odisha, India [4]Department of Electrical Engineering, Indian Institute of Technology Kharagpur, Kharagpur, West Bengal, India [5]Northwestern University, Chicago, IL, United States

OUTLINE

4.1 Introduction

Breast cancer occurs when cells in the breast region begin to grow out of control. These cells most commonly form a tumor that can often be seen on an X-ray film or felt as a lump. These tumors are categorized as benign or malignant: the tumor is malignant if the cells can grow into (invade) surrounding tissues or spread (metastasize) to distant areas of the body, otherwise called benign. As of 2017, 1 out of every 800 women was reported to have breast cancer in the United

Applications of Artificial Intelligence in Medical Imaging
DOI: https://doi.org/10.1016/B978-0-443-18450-5.00005-0

States. As of 2020, female breast cancer has surpassed lung cancer as the most commonly diagnosed cancer, with an estimated 2.3 million new cases (it constitutes 11.7% of all forms of cancer). It is the fifth leading cause of cancer mortality worldwide, with 685,000 deaths. Among women, breast cancer accounts for 1 in 4 cancer cases and 1 in 6 cancer deaths, ranking first for incidence in the vast majority of countries [1].

Because of the medical importance of breast cancer screening, CAD (computer-aided detection) methods for detecting anomalies such as calcifications, masses, architectural distortion, and bilateral asymmetry have been created [2]. One reason why detection of breast cancer is difficult is because mammography results are highly dependent on the age of the patient, breast density, and the type of lesion present. The density can lead to differences in the contrast of the malignant regions and could lead to wrong conclusion. This field of research is highly sensitive to details in the mammograms of each individual and hence requires accurate and faster methods of processing large amounts of data.

Deep learning techniques are revolutionizing the field of medical image analysis, and hence in this study, we utilized convolutional neural networks (CNNs) for early detection of breast cancer to minimize the overheads of manual analysis [3]. Deep learning has been instrumental in dealing with large amounts of data with relative ease. Preprocessing methods, such as image augmentation, make it possible to train models on selective ROI (region of interest) and extract region-specific information for further model building. In general, complications may arise due to presence of calcium deposits which are also known as calcifications and microcalcification. These granular deposits present themselves in irregular shapes, lines. The presence of such deposits may lead to the magnified and highlighted dark white spots in the mammograms. These white spots could be mistaken for abnormal breast lesions and lead to false conclusions.

Deep learning models can make it possible to work with highly specific regions in mammograms and avoid these. However, even deep learning models are not error free and can often misclassify the benign and malignant classes for one another. The use of deep learning models is governed by the amount of good-quality data that is available for training. Due to the high specificity of the application, the data required for model training is to be of high quality. Very often, it is observed that the datasets compiled by institutions and organizations is highly skewed and imbalanced. This can lead to the propagation of unwanted class-specific bias through the model while in the training process. Through the help of deep learning this can be dealt with by using the help of data augmentation. This allows researchers to increase the amount of data in different classes by using image manipulation techniques such as flipping, rotation of image through certain angles, applying geometric transformations, color manipulation, and images. One such approach of increasing images per class in imbalanced datasets is using GANs (generative adversarial networks). Through GANs similar images are formed by the introduction of random/specific noise vectors to the images. Some new-age techniques such GANs, neural style transfer, and meta-learning make it possible to artificially improve the class distributions and helped remove bias for specific classes to some extent.

Mammography is the mostly used breast cancer screening technique. It is a type of imaging that uses a low-dose X-ray system to examine the breast. It is considered the most reliable method for screening breast abnormalities before they become clinically palpable [2,4,5]. Transfer learning has been used extensively for this purpose, as shown in work proposed by a paper in 2018 [6]. In this chapter, we aim to evaluate and assess the impact of artificial intelligence (AI)-based techniques that can help health professionals screen and diagnose breast cancer early and help save lives [7] (Fig. 4.1).

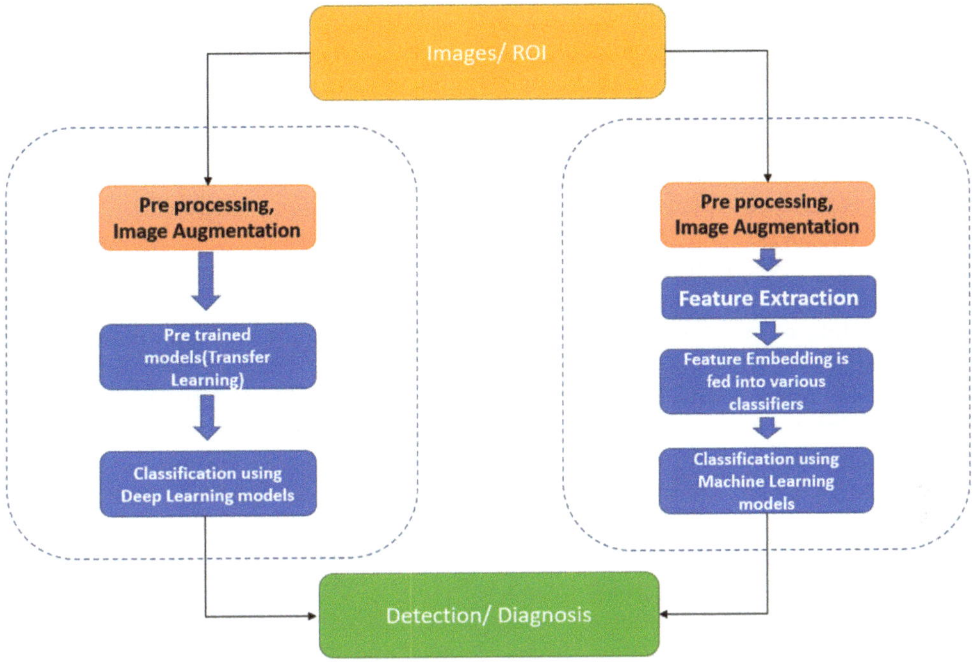

FIGURE 4.1 Overview of the proposed CAD system. *CAD*, Computer-aided detection.

4.2 Background and literature review

Extensive work has been done in this field in the last few decades. However, in recent years deep learning has played a pivotal role in early breast cancer detection through classification and segmentation as object detection models evolve. Segmentation of abnormal masses and classification of these masses into different classes has been the most crucial objective. Deep learning has made it possible to use large pre-trained architectures to retrain mammography data through transfer learning. Recently, the use of an encoder–decoder architecture to classify mammograms into different classes has been proposed. Some researchers have employed deep learning in the literature to detect suspicious breast lesions to improve classification accuracy. Traditional CAD-based systems that have been implemented over the past few decades still required radiologist to specifically manually outline the cancerous regions. This practice in many cases has a high chance of propagating human errors and ultimately lead to false conclusions. Now with the help of deep learning models, outlines and regions can be highlighted automatically. The primary concern that exists with regards to manual outlining of cancerous regions is the fact that radiologists may sometimes not be able to completely analyze the density of different regions in the mammograms and this could lead to false labels and outlines of the cancerous regions. There are many techniques to generate 3D scans of breasts to analyze the images and detect cancerous developments, and mammography data has been widely used compared to other methods. One advantage of mammography over other scanning methods is that mammography exposes patients to relatively low amounts of radiation. Recent studies in this field have been observed to focus on three crucial steps, namely, segmentation/feature extraction, data augmentation, and classification. The segmentation/feature extraction process helps in extracting relevant information (specific features) from the mammograms and includes identifying and labeling the ROIs for further processing. In the next step, various manipulation techniques could be used to clean the data and increase the data artificially using techniques such as GAN [8] and neural style transfer [9]. In the last step, that is, classification, CNN is used to identify the mammograms either as benign or malignant. The segmentation process has seen the rise of CNN architectures especially for medical image segmentation tasks such as U-Net and U^2-Net [10,11]. Some of the examples that have been mentioned below are great examples of how deep learning has helped improve breast cancer detection through segmentation and classification in recent times. In 2019 a method was proposed using the patch-based CNN method to detect breast lesions in full-field digital mammograms [12]. In one study, transfer learning was used to pretrain their CNN model using an extensive public database of digitized mammograms (i.e., CBIS-DDSM dataset), and then it was tested the trained model using the INbreast dataset [13]. The study evaluated breast detection using VGG16, ResNet50, and InceptionV3 as in-depth feature extractors. It was concluded that the model based on InceptionV3 achieved promising results with a true positive rate (TPR) of 98% [12]. In 2018 a CAD system based on the Faster R-CNN detected and classified breast lesions as benign or malignant. The study evaluated and tested the model using the breast dataset, achieving an overall classification accuracy of 95% in area under the ROC curve (AUC) [14].

4.3 Artificial intelligence techniques

4.3.1 Artificial neural networks

Artificial neural networks (ANNs) have been used in many fields for classification tasks, pattern recognition, and predictive modeling and have proved to deliver excellent performance on such data. ANNs have been fast growing for their ability to adapt to different kinds of data. This is done with the combination of external networks, deep networks, and hyperparameter optimization [15]. Our research involves training a classification model to increase the depth and then tuning the hyperparameters to achieve optimal results. ANNs, in particular, have been extensively used in disease classification. ANNs have been used for the classification and segmentation of diseases such as Alzheimer's disease, breast cancer, lung cancer, and brain tumors, among other well-known diseases [16–19]. A comparative study presented in 2020 focused on the strengths and weaknesses of some ANN models such as ML-NN, SNN, SADE, DBN, and PCA-Net. The study reveals that ANN has been very widely used, and the performance achieved by ANN-based models is auspicious and is comparable with the results of state-of-the-art CNN architecture [20]. The results achieved by ANNs in disease classification look promising, and its wide use in this field is rapidly on the rise.

4.3.2 Deep learning

The swift advancement of deep learning continues to aid the medical imaging community in applying advanced techniques to improve the accuracy of cancer screening. Many CAD-based screening techniques have been extensively researched since their inception in the 1990s. Deep learning has vastly helped in improving accuracy in early detection of breast cancer through mammograms. In recent years deep learning has revolutionized fields concerned with object detection and classification, pattern recognition, and many other domains [21]. A breakthrough in the field of image processing was achieved in the year 2012 when a deep learning model (convolution neural network) was able to outperform all other models in the ImageNet Large-Scale Visual Recognition Challenge [22]. Deep learning models have two primary advantages over classical machine learning algorithms that do not use distributed representations. First, learning distributed representations enable generalization to new combinations of the values of learned features beyond those seen during training. Second, deep learning models can learn highly specific information with an increase in depth [23]. Some of the health-care applications of deep learning include biosignal analysis (ECG), the prediction of cardiac arrests, sudden seizures, drug discovery, and analysis of electronic health records [24]. Recently, a 50-year-old protein folding problem was solved using deep learning models. These models were used in accurately predicting protein structures by a group of researchers at DeepMind [25,26]. Deep learning has been very useful to the medical imaging community in not only recognition and detection task but also in segmentation tasks. The U-Net and U^2-Net architectures have been used extensively in segmentation tasks and allow for efficient use of available data in specifically biomedical imaging applications [10,11]. Overall, it can be said that the use of deep learning is still in its infancy stage. The adoption of advanced techniques such as deep learning models is at different stages in different parts of the world. However, deep learning models have shown promising results in the research works and studies done so far. These techniques have helped the health industry in multiple ways and continue to evolve at a rapid rate [27]. A simple deep learning Python code is given below.

```
#Defining our Deep Neural Network Model
# Initiating DNN Model
DNN_model=Sequential()

# Adding Batch Normalization, Dropout and Dense Layers
DNN_model.add(Dense(16, input_dim=128, kernel_initializer = 'uniform',
                                            activation = 'relu'))

DNN_model.add(BatchNormalization())
DNN_model.add(Dropout(0.2))
DNN_model.add(Dense(32, kernel_initializer = 'uniform', activation = 'relu'))

# Adding Batch Normalization, Dropout and Dense Layers
DNN_model.add(BatchNormalization())
DNN_model.add(Dropout(0.2))
DNN_model.add(Dense(64, kernel_initializer = 'uniform', activation = 'relu'))

# Adding Batch Normalization, Dropout and Dense Layers
DNN_model.add(BatchNormalization())
DNN_model.add(Dropout(0.2))
DNN_model.add(Dense(64, kernel_initializer = 'uniform', activation = 'relu'))

# Output Layer
DNN_model.add(Dense(1,activation='sigmoid'))

# Model Summary
DNN_model.summary()
```

4.3.3 Convolutional neural networks

CNNs are specifically designed to perform better on image data. In the past few years, deep learning has achieved excellent performance in various fields, such as visual recognition, speech recognition, and natural language processing. Among different types of deep neural networks, CNNs have been most extensively studied. Leveraging on the rapid growth in the amount of the annotated data and the significant improvements in the strengths of graphics processor units, the research on CNNs has been emerged swiftly and achieved state-of-the-art results on various tasks. CNNs can do this by analyzing the images with a grid-like structure (Fig. 4.2). Multiple convolutional layers create successive feature maps. Various filters are used through which the image is passed to retrieve relevant information such as edge detection, pixel correlations, and pixel-specific information, among other valuable features. These feature maps contain information specific to that image that helps classify images into different categories, such as malignant or benign in our case. Among other techniques that examine the breast, the mammogram is a widely utilized and dependable screening innovation, and in our work, we have used various CNN architectures to detect cancerous tissues in the Mammographic Image Analysis Society (MIAS) [28,29] and CBIS-DDSM [30,31] datasets. During a convolution, the original image size is reduced. To maintain the image size, various padding techniques are used. A CNN consists primarily of three steps: Feature extraction, dimensionality reduction, and classification. The convolutional layers perform the task of feature extraction, pooling layers perform feature size reduction, and finally, the SoftMax layer classifies the image. CNNs tend to outperform dense neural networks (feed-forward networks) as they can extract image-specific information through the use of multiple feature maps. In contrast, densely connected networks simply

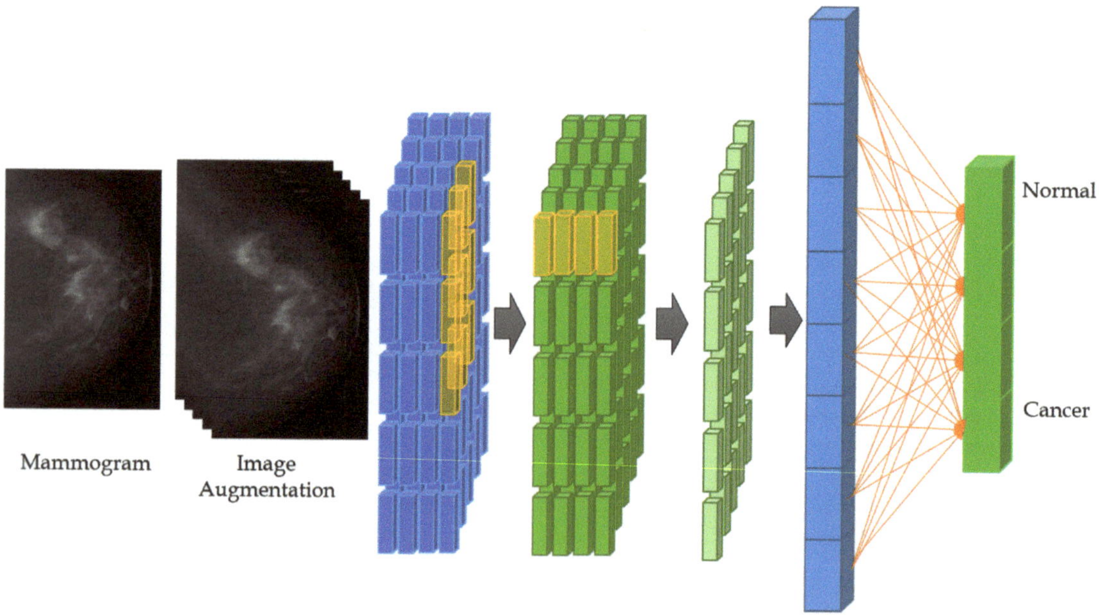

FIGURE 4.2 Simple CNN architecture with image augmentation. *CNN*, Convolutional neural network.

flatten the image and feed the flattened information to successive layers without any image-specific context. A sample Python code is given below.

```
# Simple CNN Model
# Initiating CNN Model
model = Sequential()

# Adding Convolution, MaxPooling and Dropout Layers
model.add(Conv2D(16, (3, 3), padding='same', activation='relu', input_shape=(100,100,3)))
model.add(MaxPooling2D(pool_size= (2, 2)))
model.add(Dropout(0.25))

# Adding Convolution, MaxPooling and Dropout Layers
model.add(Conv2D(32,(3,3),padding='same'))
model.add(Activation('relu'))
model.add(MaxPooling2D(pool_size=(2,2),strides=2,padding = 'same'))

# Adding Convolution, MaxPooling and Dropout Layers
model.add(Conv2D(64,(3,3),padding='same'))
model.add(Activation('relu'))
model.add(MaxPooling2D(pool_size=(2,2),strides=2,padding = 'same'))

# Adding Flattening and Dense Layer
model.add(Flatten())
model.add(Dense(16, activation='relu'))

# Final Output Layer
model.add(Dense(1, activation='sigmoid'))
```

4.4 Breast cancer detection using artificial intelligence

4.4.1 Feature extraction using deep learning

A total of 11 models have been used for feature extraction. The models used were pretrained on the ImageNet dataset, and the corresponding weights were used (Fig. 4.3). Although machine learning is helpful in effectively extracting features for certain tasks, the remaining challenge is deciding which specific features should be extracted to feed into the algorithms for accurate diagnosis. This is made possible with the help of deep learning. In this regard, it provides a basis for developing new and improved algorithms that are better equipped to generalize data by enabling the computer to build complex insights out of simple ideas [23].

The training images were passed through the pretrained models to generate a feature embedding/feature vector of either 16/32/64/128 dimensions. Feature extraction techniques have previously been used to extract specific features from image scans using deep CNNs such as ResNet and VGG architectures [32]. After experimenting with multiple-dimensional outputs, the best feature embedding/feature vector was used for further processing and model building. To obtain a good feature embedding, dropout layers with appropriate dropout rates were used along with batch normalization. Transfer learning was used to retrain the trained models on image data and obtain feature embeddings that contain class and image-specific information. A sample Python code is given below.

```
# Example of Extracting Features using ResNet101 as base model
base_model= ResNet101(input_shape=(100,100,3), weights='imagenet',include_top=False)
x = base_model.output
#x = Dropout(0.5)(x)
x = Flatten()(x)
x = BatchNormalization()(x)
# x = Dense(16,kernel_initializer='he_uniform')(x)
# x = BatchNormalization()(x)
# x = Activation('relu')(x)
# x = Dropout(0.5)(x)
predictions = Dense(128, activation='softmax')(x)

model_feat = Model(inputs=base_model.input,outputs=predictions)

train_features = model_feat.predict(x_train)
val_features=model_feat.predict(x_val)
test_features=model_feat.predict(x_test)

# ML Classification Algorithms are then used on the extracted features
# A custom pipeline is used to train and predict using ML Classifiers
from sklearn.pipeline import make_pipeline
from sklearn.pipeline import Pipeline
names = [
        "K Nearest Neighbour Classifier",
        'SVM',
        "Random Forest Classifier",
        "AdaBoost Classifier",
        "XGB Classifier",
         ]
classifiers = [
    KNeighborsClassifier(),
    SVC(),
    RandomForestClassifier(),
    AdaBoostClassifier(),
    XGBClassifier(),
        ]
zipped_clf = zip(names,classifiers)
```

4.4.2 Prediction and classification

In this work, two datasets are utilized, namely, MIAS and DDSM. The objective in the experiments was to classify images into two classes, namely, benign and malignant. As this was a two-class classification, a sigmoid layer was used for making the prediction and classification of an input image into the two respective classes. The classes in the two datasets were not balanced, and different techniques, such as callbacks, propagating custom weights through the network, and reducing learning rate with epochs, among other techniques, were used to tackle this problem. Various optimizers such as Adam, SGD, RMSprop,

```
def classifier_summary(pipeline, X_train, y_train, X_val, y_val,X_test,y_test):
    sentiment_fit = pipeline.fit(X_train, y_train)

    y_pred_train= sentiment_fit.predict(X_train)
    y_pred_val = sentiment_fit.predict(X_val)
    y_pred_test = sentiment_fit.predict(X_test)

    train_accuracy = np.round(accuracy_score(y_train, y_pred_train),4)*100
    train_precision = np.round(precision_score(y_train, y_pred_train, average='weighted'),4)
    train_recall = np.round(recall_score(y_train, y_pred_train, average='weighted'),4)
    train_F1 = np.round(f1_score(y_train, y_pred_train, average='weighted'),4)
    train_kappa =  np.round(cohen_kappa_score(y_train, y_pred_train),4)


    val_accuracy = np.round(accuracy_score(y_val, y_pred_val),4)*100
    val_precision = np.round(precision_score(y_val, y_pred_val, average='weighted'),4)
    val_recall = np.round(recall_score(y_val, y_pred_val, average='weighted'),4)
    val_F1 = np.round(f1_score(y_val, y_pred_val, average='weighted'),4)
    val_kappa =  np.round(cohen_kappa_score(y_val, y_pred_val),4)


    test_accuracy = np.round(accuracy_score(y_test, y_pred_test),4)*100
    test_precision = np.round(precision_score(y_test, y_pred_test, average='weighted'),2)
    test_recall = np.round(recall_score(y_test, y_pred_test, average='weighted'),2)
    test_F1 = np.round(f1_score(y_test, y_pred_test, average='weighted'),2)
    test_kappa =  np.round(cohen_kappa_score(y_test, y_pred_test),2)
    print('Train Set Metrics')
    print("Accuracy core : {}%".format(train_accuracy))
    cm=confusion_matrix(y_train,y_pred_train)
    cm_plot=plot_confusion_matrix(cm,classes=['Class 1','Class 2',...,'Class n'])

    print('Validation Set Metrics')
    print()
    print("Accuracy score : {}%".format(val_accuracy))
    cm=confusion_matrix(y_val,y_pred_val)
    cm_plot=plot_confusion_matrix(cm,classes=['Class 1','Class 2',...,'Class n'])

    print('Test Set Metrics')
    print()
    print("Accuracy score : {}%".format(test_accuracy))
    print("F1_score : {}".format(test_F1))
    print("Kappa Score : {} ".format(test_kappa))
    print("Recall score: {}".format(test_recall))
    print("Precision score : {}".format(test_precision))
    cm=confusion_matrix(y_test,y_pred_test)
    cm_plot=plot_confusion_matrix(cm,classes=['Class 1','Class 2',...,'Class n'])
```

Adamax, etc., were experimented with. Adam, which showed a great convergence pattern and converged faster than another optimizer, was chosen for final model training. Hence, the Adam optimizer yielding the best results was chosen for final prediction and classification while working with convolution neural networks and transfer learning techniques. Finally, we put a fully connected layer at the end of the network architecture to classify the images. We put raw images and labels into the CNNs with no information on the underlying data, and it almost gives us the state-of-the-art [33]. Similarly, in the case of classifiers trained on features extracted from state-of-the-art models, dense layers of appropriate

```
# Custom plotting Function
def plot_confusion_matrix(cm, classes,
                          normalize=False,
                          title='Confusion matrix',
                          cmap=plt.cm.YlOrRd):
    plt.figure(figsize = (6,6))
    plt.imshow(cm, interpolation='nearest', cmap=cmap)
    plt.title(title)
    plt.colorbar()
    tick_marks = np.arange(len(classes))
    plt.xticks(tick_marks, classes, rotation=90)
    plt.yticks(tick_marks, classes)
    if normalize:
        cm = cm.astype('float') / cm.sum(axis=1)[:, np.newaxis]

    thresh = cm.max() / 2.
    cm = np.round(cm,2)
    for i, j in itertools.product(range(cm.shape[0]), range(cm.shape[1])):
        plt.text(j, i, cm[i, j],
                 horizontalalignment="center",
                 color="white" if cm[i, j] > thresh else "black")
    plt.tight_layout()
    plt.ylabel('True label')
    plt.xlabel('Predicted label')
    plt.show()

def classifier_comparator(X_train,y_train,X_val,y_val,X_test,y_test,classifier=zipped_clf):
        result = []
        for n,c in classifier:
            checker_pipeline = Pipeline([('Classifier', c)])
            print("Fitting {} on input_data ".format(n))

            classifier_summary(checker_pipeline,X_train, y_train, X_val, y_val,X_test,y_test)
```

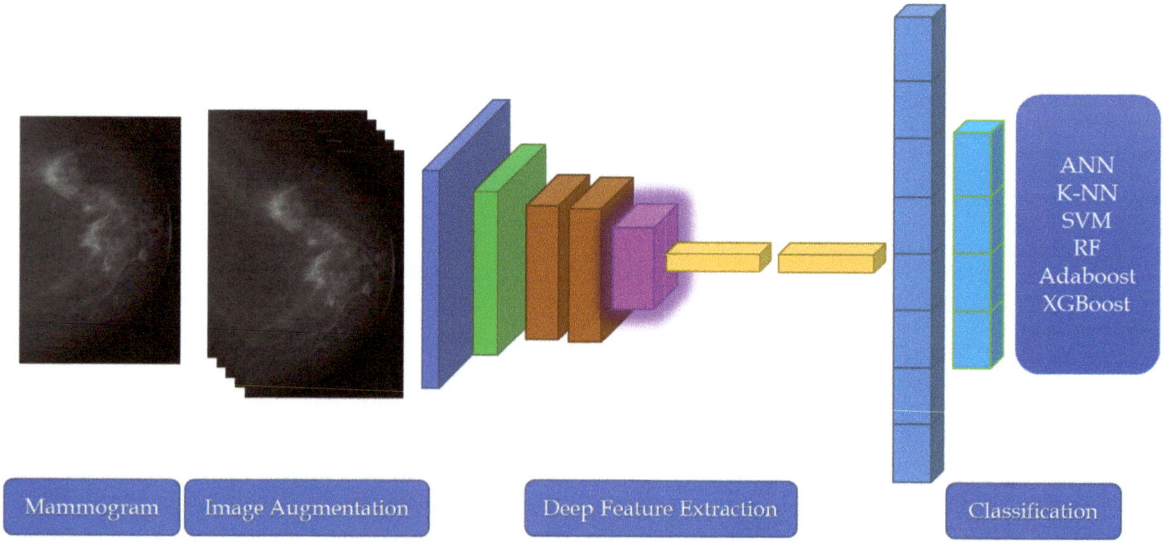

FIGURE 4.3 Deep feature extraction with pretrained model.

dimensions were used with a sigmoid layer. A sigmoid layer is used to map the input only between zero and one. Even if the input values are out of this range, they are mapped to values between zero and one. This way of classification of input data is possible even if input values grow exponentially. After the features have been extracted and selected, they are fed into a classifier to categorize the images into malignant and benign classes. The commonly used classifiers include linear, ANN, Bayesian neural networks, tree-based classifiers, support vector machine (SVM), XGBoost, and AdaBoost [34]. Finally, we place a fully connected layer at the end of the network architecture to classify the images into their respective classes. We put raw images and labels into the CNNs with no information on the underlying data, and it almost gives us state-of-the-art results.

4.4.3 Experimental data

The CBIS-DDSM dataset is the most popular in breast cancer detection using AI techniques. The dataset that we have used is publicly available and consists of images from the DDSM [1] and CBIS-DDSM [3] datasets. The images have been preprocessed and converted to 299 × 299 images by extracting the ROIs. The data is stored as tfrecords files for TensorFlow. The dataset contains 55,890 training examples, of which 14% are positive and the remaining 86% negative, divided into five tfrecords files, CBIS-DDSM dataset [30,31].

Early detection of breast cancer on screening mammography is a challenging classification task because the tumors themselves occupy only a tiny portion of the image of the entire breast scan [6]. We have merged the training and testing dataset in CBIS-DDSM, and the data was further split into training, validation, and testing.

The MIAS is an organization of UK research groups interested in understanding mammograms and has generated a digital mammogram database. The MIAS [28] dataset consists of 322 images of size 1024 × 1024.

Our study primarily focuses on implementing and tuning state-of-the-art models to attain higher accuracy. Transfer learning has played a significant role in this field. Pretrained weights have led to better results, and techniques such as feature extraction make it possible to extract representation vectors/feature embeddings from large amounts of data and maintain high accuracy when combined with machine learning models. Below are some sample images (Figs. 4.4–4.6).

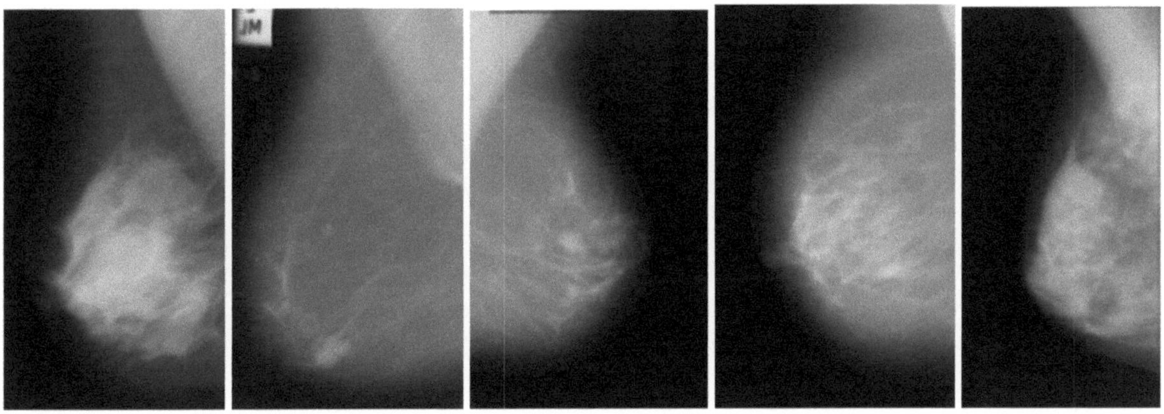

FIGURE 4.4 Sample of benign mammograms from MIAS [28,29] dataset. *MIAS*, Mammographic Image Analysis Society.

```
# Extracting Data from MIAS Dataset
# Image Augmentation
no_angles = 360
url = 'MIAS_Dataset_Kaggle'

def save_dictionary(path,data):
        #open('u.item', encoding="utf-8")
        import json
        with open(path,'w') as outfile:
            json.dump(str(data), fp=outfile)

def read_image():
        info = {}
        for i in range(322):
            if i<9:
                image_name='mdb00'+str(i+1)
            elif i<99:
                image_name='mdb0'+str(i+1)
            else:
                image_name = 'mdb' + str(i+1)
            image_address= url+image_name+'.pgm'
            img = cv2.imread(image_address,1)
            img = cv2.resize(img, (224,224))
            rows, cols,channel = img.shape
            info[image_name]={}
            for angle in range(0,no_angles,8):
                M = cv2.getRotationMatrix2D((cols / 2, rows / 2), angle, 1)
                img_rotated = cv2.warpAffine(img, M, (cols, rows))
                info[image_name][angle]=img_rotated
        return (info)

def read_lable():
    filename = url+'Info.txt'
    text_all = open(filename).read()
    #print(text_all)
    lines=text_all.split('\n')
    info={}
    for line in lines:
        words=line.split(' ')
        if len(words)>3:
            if (words[3] == 'B'):
                info[words[0]] = {}
                for angle in range(0,no_angles,8):
                    info[words[0]][angle] = 0
            if (words[3] == 'M'):
                info[words[0]] = {}
                for  angle in range(0,no_angles,8):
                    info[words[0]][angle] = 1
    return (info)
```

```
import numpy as np
lable_info=read_lable()
image_info=read_image()
ids=lable_info.keys()
del lable_info['Truth-Data:']
X=[]
Y=[]
for id in ids:
    for angle in range(0,no_angles,8):
        X.append(image_info[id][angle])
        Y.append(lable_info[id][angle])
X=np.array(X)
Y=np.array(Y)
Y=to_categorical(Y,2)
x_train, x_test1, y_train, y_test1 = train_test_split(X, Y, test_size=0.3, random_state=42)
x_val, x_test, y_val, y_test = train_test_split(x_test1, y_test1, test_size=0.3, random_state=42)
```

```
# Extracting .tfrecords (CBIS-DDSM Dataset)
images=[]
labels=[]
feature_dictionary = {
    'label': tf.io.FixedLenFeature([], tf.int64),
    'label_normal': tf.io.FixedLenFeature([], tf.int64),
    'image': tf.io.FixedLenFeature([], tf.string)
    }

def _parse_function(example, feature_dictionary=feature_dictionary):
    parsed_example = tf.io.parse_example(example, feature_dictionary)
    return parsed_example

def read_data(filename):
    full_dataset = tf.data.TFRecordDataset(filename,
                    num_parallel_reads=tf.data.experimental.AUTOTUNE)
    full_dataset = full_dataset.cache()
    print("Size of Training Dataset: ", len(list(full_dataset)))

    feature_dictionary = {
    'label': tf.io.FixedLenFeature([], tf.int64),
    'label_normal': tf.io.FixedLenFeature([], tf.int64),
    'image': tf.io.FixedLenFeature([], tf.string)
    }
```

```
    full_dataset = full_dataset.map(_parse_function,
                                              num_parallel_calls=tf.data.experimental.AUTOTUNE)
    print(full_dataset)
    for image_features in full_dataset:
        image = image_features['image'].numpy()
        image = tf.io.decode_raw(image_features['image'], tf.uint8)
        image = tf.reshape(image, [299, 299])
        image=image.numpy()
        image=cv2.resize(image,(100,100))
        image=cv2.merge([image,image,image])
        image
        images.append(image)
        labels.append(image_features['label_normal'].numpy())

filenames=['CBIS-DDSM_Dataset(kaggle).tfrecords']

for file in filenames:
    read_data(file)

X=np.array(images)
y=np.array(labels)

# Splitting Data into Train, Test and Validation sets
x_train, x_test1, y_train, y_test1 = train_test_split(X, y, test_size=0.3, random_state=42,
                                                    shuffle=True,stratify=y)

x_val, x_test, y_val, y_test = train_test_split(x_test1, y_test1, test_size=0.3, random_state=42,
                                              shuffle=True,stratify=y_test1)
```

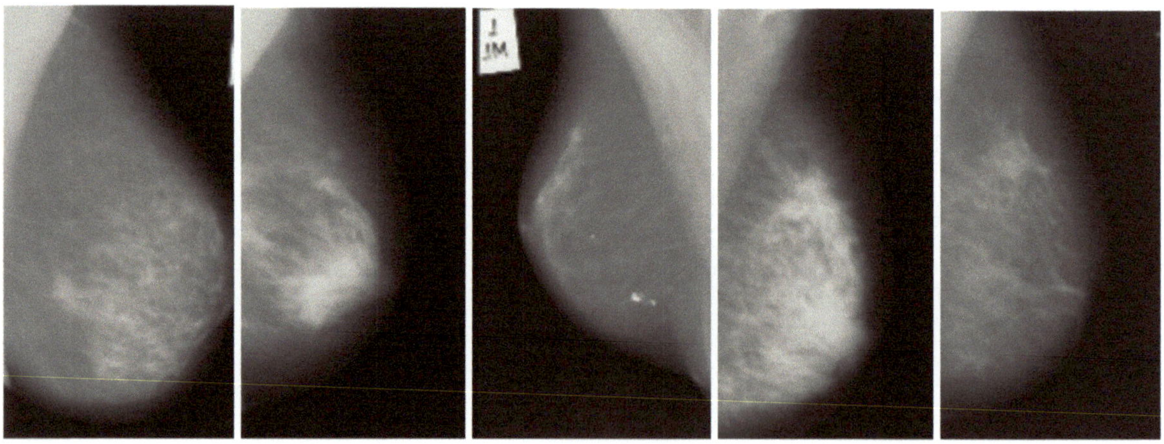

FIGURE 4.5 Sample of malignant mammograms from MIAS [28,29] dataset. *MIAS*, Mammographic Image Analysis Society.

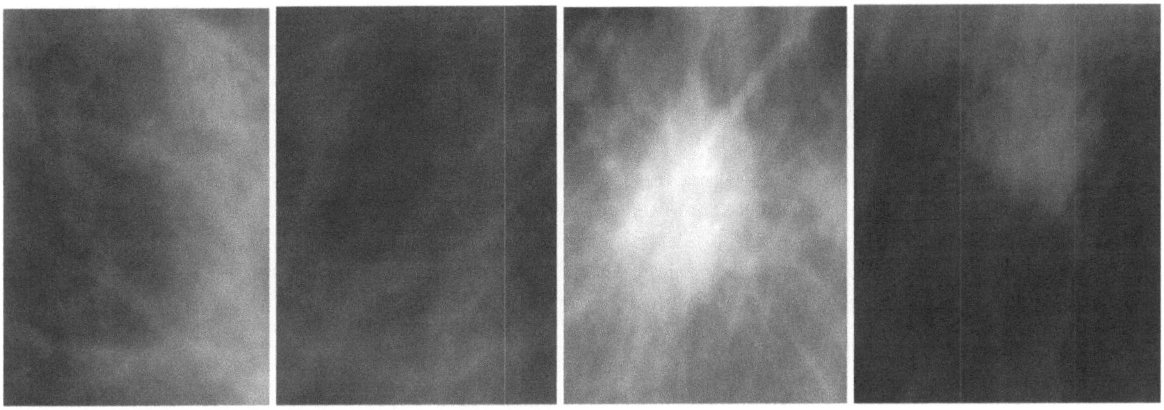

FIGURE 4.6 Samples from DDSM [30,31] dataset.

4.4.4 Performance evaluation measures

The MIAS dataset was randomly split as follows for training, validation, and testing purposes (train size = 70%, validation size = 21%, and test size = 9%). Similarly, the CBIS-DDSM dataset was split as follows (train size = 70%, validation size = 20% of train size, and test size = 30%). The splits were stratified on the labels to maintain the class ratios throughout the experimentation process. The evaluation metrics are to choose carefully when dealing with medical data. Even the slightest of errors could lead to major changes in applying and adapting deep learning models in real time since medical data is susceptible to individual classes. F1 score metric, Kappa score, area under the ROC curve (AUC), recall, and precision scores are some popularly known and most widely used metrics to interpret the performance of multiclass models built with the help of deep learning and machine learning techniques.

To evaluate the results of our classification model, we require some values such as TP (true positive), FN (false negative), FP (false positive), and TN (true negative).

True positives (TP)—These are the correctly predicted positive values that mean that the actual class's value is yes, and the value of the predicted class is also yes.

True negatives (TN)—These are the correctly predicted negative values that mean that the actual class's value is no, and value of the predicted class is also no.

False positive and false negative values occur when your actual class contradicts the predicted class.

False positives (FP)—When the actual class is no, and the predicted class is yes.

False negatives (FN)—When actual class is yes, but predicted class is no.

Accuracy—It is the ratio of correctly predicted observations to the total number of observations.

$$\text{Accuracy} = \frac{\text{TP} + \text{TN}}{(\text{TP} + \text{FP} + \text{FN} + \text{TN})}$$

Precision—Precision is the ratio of correctly predicted positive observations to the total predicted positive observations.

$$\text{Precision} = \frac{\text{TP}}{\text{TP} + \text{FP}}$$

Recall (sensitivity)—Recall is the ratio of correctly predicted positive observations to all observations in the actual class.

$$\text{Recall} = \frac{\text{TP}}{\text{TP} + \text{FP}}$$

F1—F1 score is the weighted average of precision and recall. Therefore this Score takes both false positives and false negatives into account.

$$\text{F1 Score} = 2 * (\text{Recall} * \text{Precision}) / (\text{Recall} + \text{Precision})$$

Cohen's Kappa coefficient—This statistic is used to measure the reliability between evaluators (and also the reliability within evaluators) for qualitative (categorical) items. Cohen's Kappa is based on the idea of measuring the correspondence between the predicted and the actual labels, which are considered to be two random categorical variables [35]. The kappa statistic is the most widely used metric for measuring categorical data where there is no objective way of determining the likelihood of a coincidental encounter between two or more observers. Values closer to 1 are considered good, and values closer to 0 are uncertain. Cohen [36] defined the Kappa statistic as an agreement index and defined as the following:

$$k = \frac{Po - Pe}{1 - Pe}$$

Po has observed agreement, and Pe measures the agreement expected by chance [37].

AUC score—ROC (receiver operating characteristic) curve is an evaluation metric for binary classification problems. It is a probability curve that plots the TPR versus the false positive rate (FPR) at different thresholds. The area under the curve (AUC) is a measure of a classifier's ability to distinguish between classes and is used to summarize the ROC curve [38]. The higher the value of this metric, the better the performance of the classification model. It is evaluated by calculating the area under the ROC curve [39].

A sample Python code for performance metrics is given below.

```
# Performance Evaluation Metrics
from sklearn import metrics
print('Accuracy score is :', np.round(metrics.accuracy_score(y_test, y_pred),4))
print('Precision score is :', np.round(metrics.precision_score(y_test, y_pred,
                                                            average='weighted'),4))
print('Recall score is :', np.round(metrics.recall_score(y_test,y_pred, average='weighted'),4))
print('F1 Score is :', np.round(metrics.f1_score(y_test, y_pred, average='weighted'),4))
print('ROC AUC Score is :', np.round(metrics.roc_auc_score(y_test, y_pred,multi_class='ovo',
                                                            average='weighted'),4))
print('\t\tClassification Report:\n', metrics.classification_report(y_test, y_pred,
                                                            target_names=target))
print('Cohen Kappa Score:', np.round(metrics.cohen_kappa_score(y_test, y_pred),4))
```

4.4.5 Experimental results

This section contains the experimental results. We have worked on two datasets, namely, MIAS and DDSM. The experimental results have been presented below:

4.4.5.1 Mammographic Image Analysis Society dataset

It can be seen from Table 4.1 that pretrained models were able to show good results. VGG16 was able to outperform other models on all our performance evaluation metrics. Despite the imbalance class data, pretrained models were able to identify the test data into correct classes accurately.

It can be seen from Table 4.2 that multilayered CNN architectures outperformed the pretrained models. The models appear to overfit on the train data, but the model performs equally well on the validation and test data.

TABLE 4.1 Pretrained networks (transfer learning).

Model	Training accuracy	Validation accuracy	Test accuracy	F1 measure	KAPPA	ROC area
ResNet50	97.8	94.11	93.99	0.9399	0.8777	0.9394
VGG16	98.3	96.04	95.71	0.957	0.9122	0.9545
VGG19	98.3	95.22	94.64	0.9464	0.8907	0.9456
Inception_v3	88	78.01	77.04	0.7702	0.5311	0.7651
MobileNet	97	90.71	90.99	0.9099	0.8165	0.9087
DenseNet169	89.5	77.92	78.11	0.7818	0.557	0.7805
DenseNet121	89.4	82.24	82.19	0.8226	0.6406	0.8236
InceptionResNetV2	69.6	66.42	67.81	0.6674	0.321	0.6551
MobileNetV2	96	86.38	86.48	0.8638	0.7212	0.8565
ResNet101	97.6	93.74	94.21	0.9421	0.882	0.9412

TABLE 4.2 Convolutional neural networks (CNNs) with different number of layers.

Classifier	Training accuracy	Validation accuracy	Test accuracy	F1 measure	KAPPA	ROC area
CNN 2 Layer	0.9699	0.9731	0.9678	0.9677	0.9344	1.0
CNN 3 Layer	0.9731	0.9719	0.971	0.9709	0.9411	0.9967
CNN 4 Layer	0.9712	0.9849	0.9836	0.9836	0.9667	0.9978
CNN 5 Layer	0.9851	0.9844	0.9932	0.9932	0.9863	0.999
CNN 6 Layer	0.9719	0.9918	0.9945	0.9945	0.9889	0.9998
CNN 7 Layer	0.99	1.00	0.999	0.999	0.998	1
CNN 8 Layer	0.993	0.9989	0.9974	0.9974	0.9948	1

As the number of layers increases, the accuracy of the model increases.

It can be seen from Table 4.3 that except for the SVM classifier, all the other classifiers were overfitting on training data. The accuracy scores were consistently in the range of 51–59; similar scores were seen in recall and precision.

It can be seen from Table 4.4 that although XGBoost and Random Forest were overfitting on training data, they showed the best overall performance. Accuracy scores were consistently between 51–59, and a similar trend was seen in recall and precision scores.

It can be seen from Table 4.5 that the Random Forest classifier achieved the highest test accuracy. Besides SVM, other classifiers were overfitting on training data. The test accuracy, recall, and precision scores were found to be in the range of 51–56.

It can be seen from Table 4.6 that XGBoost classifier has the best overall performance. While it overfits training data, it outperformed other models in all performance measures. Random Forest also, while it overfits, was able to perform well on test data. SVM did not overfit on train data, unlike other classifiers.

TABLE 4.3 Performance of VGG16 deep feature extraction with different machine learning models.

Model	Train accuracy	Val accuracy	Test accuracy	F1 score	Kappa	Recall	Precision
ANN	59.69	53.26	52.57	0.5154	0.0234	0.5258	0.5166
KNN	70.65	52.34	54.51	0.54	0.07	0.55	0.54
SVM	56.74	55.84	56.22	0.45	0.05	0.56	0.59
Random Forest	78.99	55.37	56.87	0.52	0.08	0.57	0.56
AdaBoost	60.3	54.37	53	0.5	0.01	0.53	0.51
XGBoost	99.09	55.75	51.5	0.51	0	0.52	0.51

TABLE 4.4 Performance of VGG19 deep feature extraction with different machine learning models.

Model	Training accuracy	Validation accuracy	Test accuracy	F1 measure	Kappa	Recall	Precision
ANN	58.3	51.51	51.5	0.4922	−0.01	0.515	0.4983
KNN	70.54	51.51	53	0.52	0.04	0.53	0.53
SVM	55.55	55.01	54.51	0.39	0	0.55	0.48
Random Forest	77.72	54.74	58.58	0.54	0.12	0.59	0.59
AdaBoost	59.53	53.54	57.5	0.55	0.11	0.58	0.57
XGBoost	98.87	56.67	57.94	0.58	0.14	0.58	0.58

TABLE 4.5 Performance of ResNet50 deep feature extraction with different machine learning models.

Model	Training accuracy	Validation accuracy	Test accuracy	F1 measure	Kappa	Recall	Precision
ANN	56.59	51.33	53.21	0.5229	0.0379	0.5322	0.524
KNN	70.24	52.07	54.72	0.55	0.08	0.55	0.55
SVM	55.83	55.11	54.08	0.31	-0.01	0.54	0.49
Random Forest	72.89	54.92	55.79	0.48	0.05	0.56	0.55
AdaBoost	60.35	52.16	51.07	0.48	-0.02	0.51	0.49
XGBoost	99.03	52.81	51.72	0.51	0.02	0.52	0.51

It can be seen from Table 4.7 that the Random Forest classifier overfit on train data showed the best performance on all our performance evaluation measures. The accuracy, recall, and precision scores were consistently found to be in the range of 51–61.

It can be seen from Table 4.8 that the Random Forest classifier overfit on train data showed the best performance on all our performance evaluation measures. The accuracy, recall, precision scores were consistently found to be in the range of 49–58.

TABLE 4.6 Performance of ResNet101 deep feature extraction with different machine learning models.

Model	Training accuracy	Validation accuracy	Test accuracy	F1 measure	Kappa	Recall	Precision
ANN	59.82	52.8	53.64	0.5254	0.0443	0.5365	0.5276
KNN	71.07	54.28	52.86	0.54	0.07	0.54	0.54
SVM	55.83	54.83	54.08	0.42	0	0.54	0.49
Random Forest	78.6	56.76	59.87	0.55	0.15	0.6	0.61
AdaBoost	59.61	54.19	53.43	0.52	0.03	0.53	0.52
XGBoost	99.33	55.01	61.37	0.61	0.22	0.61	0.61

TABLE 4.7 Performance of MobileNetV2 deep feature extraction with different machine learning models.

Model	Training accuracy	Validation accuracy	Test accuracy	F1 measure	Kappa	Recall	Precision
ANN	59.22	53.9	56.65	0.5324	0.0864	0.5665	0.5594
KNN	70.84	51.7	52.79	0.52	0.04	0.53	0.52
SVM	58.09	55.11	56.44	0.51	0.07	0.56	0.56
Random Forest	76.89	55.47	59.87	0.56	0.15	0.6	0.61
AdaBoost	60.71	53.82	56.44	0.53	0.08	0.56	0.56
XGBoost	99.12	52.53	54.08	0.54	0.07	0.54	0.54

TABLE 4.8 Performance of MobileNet deep feature extraction with different machine learning models.

Model	Training accuracy	Validation accuracy	Test accuracy	F1 easure	Kappa	Recall	Precision
ANN	57.28	53.26	51.5	0.4845	− 0.01	0.515	0.4943
KNN	69.52	53.16	54.29	0.54	0.07	0.54	0.54
SVM	55.63	55.11	53.86	0.4	− 0.01	0.54	0.47
Random Forest	76.92	53.36	58.15	0.54	0.11	0.58	0.58
AdaBoost	60.41	54.83	56.22	0.53	0.08	0.56	0.55
XGBoost	98.95	52.71	52.58	0.52	0.03	0.53	0.52

It can be seen from Table 4.9 that all the models performed equally well, and the scores were consistently in a specific range. XGBoost, Random Forest, and KNN classifier were overfitting on train data.

It can be seen from Table 4.10 that SVM classifier showed the best overall performance. XGBoost and KNN classifiers were overfitting on training data.

It can be seen from the Table 4.11 that KNN, XGBoost classifier showed the best performance. Except for the SVM classifier, all classifiers were overfitting on training data.

TABLE 4.9 Performance of InceptionV3 deep feature extraction with different machine learning models.

Model	Training accuracy	Validation accuracy	Test accuracy	F1 measure	Kappa	Recall	Precision
ANN	55.13	53.17	56.22	0.4618	0.0481	0.5622	0.5777
KNN	68.61	51.98	51.29	0.51	0.02	0.51	0.51
SVM	54.91	54.45	55.15	0.4	0.01	0.55	0.64
Random Forest	62.51	53.73	54.94	0.45	0.02	0.55	0.53
AdaBoost	57.21	53.16	54.94	0.47	0.03	0.55	0.53
XGBoost	98.26	52.99	56.44	0.56	0.12	0.56	0.56

TABLE 4.10 Performance of InceptionResNetV2 deep feature extraction with different machine learning models.

Model	Training accuracy	Validation accuracy	Test accuracy	F1 measure	Kappa	Recall	Precision
ANN	55.57	54.53	54.93	0.3993	0.007	0.5494	0.572
KNN	64.85	52.53	50.21	0.47	0	0.5	0.48
SVM	55.33	54.65	55.15	0.4	0.01	0.55	0.64
Random Forest	57.59	54.92	54.94	0.41	0.01	0.55	0.55
AdaBoost	55.76	54.65	54.94	0.4	0.01	0.55	0.57
XGBoost	75.65	55.57	50.43	0.48	− 0.04	0.5	0.48

TABLE 4.11 Performance of DenseNet169 deep feature extraction with different machine learning models.

Model	Training accuracy	Validation accuracy	Test accuracy	F1 measure	Kappa	Recall	Precision
ANN	57.26	54.51	54.5	0.5153	0.0452	0.5451	0.5318
KNN	69.49	52.44	54.29	0.54	0.07	0.54	0.54
SVM	55.66	55.2	53.43	0.4	−0.02	0.53	0.43
Random Forest	80.78	55.38	55.36	0.52	0.06	0.55	0.54
AdaBoost	60.33	53.08	53	0.52	0.03	0.53	0.52
XGBoost	98.32	52.07	54.08	0.54	0.07	0.54	0.54

It can be seen from Table 4.12 that ANN, KNN, Random Forest, XGBoost, and AdaBoost classifiers were overfitting on training data. KNN classifier showed the best performance out of all.

It can be seen from the Table 4.13 that XGBoost classifier showed the best overall performance. XGBoost, KNN, and Random Forest classifiers were overfitting on training data.

4.4.5.2 *CBIS-DDSM dataset*

It can be observed from Table 4.14 that simple CNN models were able to achieve high

TABLE 4.12 Performance of DenseNet121 deep feature extraction with different machine learning models.

Model	Training accuracy	Validation accuracy	Test accuracy	F1 measure	Kappa	Recall	Precision
ANN	58.25	51.12	49.57	0.4946	−0.02	0.4957	0.4938
KNN	69.71	50.87	52.15	0.52	0.03	0.52	0.52
SVM	56.07	54.92	53.22	0.41	−0.02	0.53	0.46
Random Forest	78.13	54.55	52.15	0.46	−0.02	0.52	0.49
AdaBoost	60.35	54.459	50.21	0.47	−0.04	0.5	0.48
XGBoost	99	51.52	48.71	0.48	−0.04	0.49	0.48

TABLE 4.13 Performance of Xception deep feature extraction with different machine learning models

Model	Training accuracy	Validation accuracy	Test accuracy	F1 measure	Kappa	Recall	Precision
ANN	56.57	56.02	55.36	0.5544	0.1029	0.5536	0.5556
KNN	69.46	52.25	51.29	0.51	0.01	0.51	0.51
SVM	55.6	56.03	56.22	0.44	0.04	0.56	0.62
Random Forest	67.34	56.85	57.94	0.52	0.1	0.58	0.59
AdaBoost	59.94	57.41	56.87	0.52	0.08	0.57	0.56
XGBoost	98.48	58.6	60.73	0.6	0.2	0.61	0.6

TABLE 4.14 Convolutional neural network (CNN) with different number of layers.

Model	Training accuracy	Validation accuracy	Test accuracy	F1 measure	Kappa	ROC area
CNN 2 Layer	89.29	88.52	88.53	0.8521	0.2474	0.8159
CNN 3 Layer	91.55	87.76	87.99	0.8329	0.129	0.911
CNN 4 Layer	87.3	86.5	87.01	0.8096	0	0.5
CNN 5 Layer	88.08	91.53	91.55	0.9079	0.5635	0.9221
CNN 6 Layer	93.32	91.81	91.46	0.899	0.4989	0.9286
CNN 7 Layer	92.51	92.42	92.15	0.9118	0.5739	0.9378
CNN 8 Layer	90.68	91.59	91.84	0.9086	0.5593	0.9228

performance overall. However, CNN 4 layer was not able to give good performance and was overfitting on specific class data, which resulted in a significant drop in ROC and Kappa values. CNN 7 Layer has the best performance overall in other networks.

It can be observed from Table 4.15 that pretrained networks were able to perform very well. MobileNet of 95.73% accuracy achieved the highest test accuracy, and it also has the highest Kappa score, which indicates the reliability of the model. Overall, it was seen that

TABLE 4.15 Pretrained networks (transfer learning).

Model	Training accuracy	Validation accuracy	Test accuracy	F1 measure	KAPPA	ROC area
ResNet50	96.35	91.63	91.98	0.91071	0.57214	0.73626
VGG16	94.03	92.83	93.06	0.92384	0.76656	0.63551
VGG19	93.36	93.32	93.57	0.932	0.68526	0.80854
Inception_v3	97.77	91.85	92.22	0.92735	0.70546	0.91058
MobileNet	99.07	95.73	95.86	0.95665	0.8009	0.86699
DenseNet169	95.59	79.7	80.56	0.83352	0.45	0.85053
DenseNet121	98.39	90.2	90.53	0.88391	0.41763	0.64961
InceptionResNetV2	94.92	93.44	93.86	0.94036	0.74515	0.89553
MobileNetV2	99.47	85.11	85.61	0.87472	0.56102	0.90523
ResNet101	97.57	91.77	92.127	91.023	0.56419	0.72465

TABLE 4.16 Performance of VGG16 deep feature extraction with different machine learning models.

Model	Training accuracy	Validation accuracy	Test accuracy	F1_score	Kappa	Recall	Precision
ANN	86.11	86.24	86.18	0.8484	0.28	0.8618	0.8408
KNN	88.97	86.17	86.28	0.83	0.14	0.86	0.82
SVM	87.41	87.08	87.16	0.82	0.06	0.87	0.84
Random Forest	100	87.25	87.67	0.84	0.17	0.88	0.85
AdaBoost	87.41	87.14	87.51	0.84	0.17	0.88	0.85
XGBoost	94.13	86.96	87.32	0.84	0.21	0.87	0.84

pretrained networks outperformed all the simple CNN models.

It can be observed from Table 4.16 that Random Forest classifier was observed to give the best performance. XGBoost and Random Forest classifiers were overfitting on training data. The ANN model achieved the highest kappa score.

It can be observed from Table 4.17 that Random Forest classifier showed the best overall performance. ANN model was seen to have the highest reliability index. Random Forest and XGBoost classifiers were seen to overfit on training data.

It can be observed from Table 4.18 that XGBoost classifier showed the best overall performance. ANN model was observed to have the highest reliability index. XGBoost classifier was overfitting on training data.

It can be observed from Table 4.19 that all the classifiers were observed to have identical performance. ANN and XGBoost models were observed to have high-reliability scores of about 0.45. The test accuracies, F1 Score, recall, and precision scores were consistently in the same range.

It can be observed from Table 4.20 that all the classifiers were observed to have identical

TABLE 4.17 Performance of VGG19 deep feature extraction with different machine learning models.

Model	Training accuracy	Validation accuracy	Test accuracy	F1_score	Kappa	Recall	Precision
ANN	84.43	83.69	84.37	0.8365	0.25	0.8437	0.8307
KNN	88.01	85.69	86.28	0.82	0.1	0.86	0.81
SVM	86.98	86.95	86.92	0.81	0	0.87	0.76
Random Forest	99.99	87.58	87.71	0.83	0.13	0.88	0.86
AdaBoost	87.18	87.04	87.51	0.83	0.12	0.88	0.85
XGBoost	98.19	88.61	88.89	0.87	0.37	0.89	0.87

TABLE 4.18 Performance of ResNet50 deep feature extraction with different machine learning models.

Model	Training accuracy	Validation accuracy	Test accuracy	F1_score	Kappa	Recall	Precision
ANN	89.26	88.42	88.62	0.8824	0.4654	0.8863	0.8796
KNN	89.69	85.69	86.6	0.84	0.23	0.87	0.84
SVM	87.68	87.4	87.57	0.83	0.09	0.88	0.87
Random Forest	87.98	87.17	87.3	0.82	0.05	0.87	0.87
AdaBoost	88.25	87.91	87.92	0.86	0.35	0.88	0.86
XGBoost	99.09	89.59	89.76	0.89	0.46	0.9	0.89

TABLE 4.19 Performance of ResNet101 deep feature extraction with different machine learning models

Model	Training accuracy	Validation accuracy	Test accuracy	F1_score	Kappa	Recall	Precision
ANN	89.28	88.56	88.81	0.8817	0.453	0.881	0.8782
KNN	89.66	86.78	86.88	0.84	0.2	0.87	0.84
SVM	88.66	88.09	88.23	0.84	0.19	0.88	0.88
Random Forest	100	87.99	88.17	0.84	0.17	0.88	0.88
AdaBoost	88.68	88.28	88.11	0.87	0.36	0.88	0.86
XGBoost	99.1	89.77	89.6	0.89	0.45	0.9	0.88

performance. ANN and XGBoost models were observed to have high-reliability scores. The test accuracies, F1 score, recall, and precision scores were consistently in the same range.

It can be observed from Table 4.21 that ANN model was able to outperform all other classifiers. XGBoost model was overfitting on training data.

It can be observed from Table 4.22 that XGBoost classifier was observed to have the best performance overall. The remaining models showed identical results.

TABLE 4.20 Performance of MobileNetV2 deep feature extraction with different machine learning models.

Model	Training accuracy	Validation accuracy	Test accuracy	F1_score	Kappa	Recall	Precision
ANN	89.17	88.77	88.6	0.8688	0.3596	0.8861	0.8679
KNN	89.32	86.87	86.22	0.83	0.15	0.86	0.82
SVM	88.03	87.82	87.57	0.83	0.12	0.88	0.86
Random Forest	100	88.39	88.41	0.85	0.23	0.88	0.87
AdaBoost	87.89	88	87.65	0.86	0.29	0.88	0.85
XGBoost	97.63	89.35	89.28	0.88	0.39	0.89	0.88

TABLE 4.21 Performance of MobileNetV2 deep feature extraction with different machine learning models.

Model	Training accuracy	Validation accuracy	Test accuracy	F1_score	Kappa	Recall	Precision
ANN	88.55	88.31	88.68	97.68	0.4179	0.8869	0.8731
KNN	89.87	87.11	86.88	0.84	0.21	0.87	0.84
SVM	87.93	87.63	87.46	0.82	0.08	0.87	0.86
Random Forest	89.09	88.1	87.85	0.83	0.13	0.88	0.88
AdaBoost	88.17	88.01	87.67	0.86	0.3	0.88	0.85
XGBoost	98.08	89.47	89.72	0.88	0.43	0.9	0.88

TABLE 4.22 Performance of MobileNet deep feature extraction with different machine learning models.

Model	Training accuracy	Validation accuracy	Test accuracy	F1_score	Kappa	Recall	Precision
ANN	87.2	87.17	86.79	0.8185	0.0598	0.868	0.8174
KNN	87.69	86.4	86.26	0.82	0.08	0.86	0.81
SVM	87.23	87.28	86.96	0.82	0.05	0.87	0.82
Random Forest	88.53	87.52	87.2	0.82	0.07	0.87	0.84
AdaBoost	87.34	87.32	86.92	0.82	0.06	0.87	0.82
XGBoost	94.86	89.21	89.36	0.88	0.42	0.89	0.88

It can be observed from Table 4.23 that XGBoost classifier was observed to have the best performance overall. The remaining models showed identical results.

It can be observed from Table 4.24 that all the models performed equally well. XGBoost and Random Forest classifiers were found to overfit on training data.

It can be observed from Table 4.25 that XGBoost classifier outperformed other classifiers. Random Forest and XGBoost classifiers were overfitting on training data.

TABLE 4.23 Performance of InceptionResNetV2 deep feature extraction with different machine learning models.

Model	Training accuracy	Validation accuracy	Test accuracy	F1_score	Kappa	Recall	Precision
ANN	86.75	86.66	86.79	0.8066	0	0.8679	0.7533
KNN	88.07	84.58	86.2	0.83	0.16	0.86	0.82
SVM	86.77	86.75	86.79	0.81	0	0.87	0.75
Random Forest	99.99	85.04	86.4	0.83	0.15	0.86	0.82
AdaBoost	86.83	86.66	86.59	0.81	0.01	0.87	0.79
XGBoost	97.17	85.51	85.7	0.82	0.08	0.86	0.8

TABLE 4.24 Performance of DenseNet169 deep feature extraction with different machine learning models.

Model	Training accuracy	Validation accuracy	Test accuracy	F1_score	Kappa	Recall	Precision
ANN	83.33	82.84	82.8	0.8426	0.3823	0.828	0.8657
KNN	89.16	86	86.56	0.84	0.21	0.87	0.83
SVM	87.07	87.03	87.06	0.81	0.02	0.87	0.87
Random Forest	99.99	88.96	88.99	0.86	0.32	0.89	0.88
AdaBoost	87.84	87.58	87.99	0.85	0.21	0.88	0.86
XGBoost	98.06	89.77	89.98	0.89	0.48	0.9	0.89

TABLE 4.25 Performance of DenseNet121 deep feature extraction with different machine learning models.

Model	Training accuracy	Validation accuracy	Test accuracy	F1_score	Kappa	Recall	Precision
ANN	85.93	85.65	85.88	84.55	0.2676	0.8588	0.8376
KNN	88.93	85.53	85.86	0.83	0.14	0.86	0.82
SVM	87.06	87.1	87.12	0.82	0.04	0.87	0.85
Random Forest	99.91	88.08	87.95	0.84	0.2	0.88	0.86
AdaBoost	87.12	86.79	86.96	0.82	0.09	0.87	0.83
XGBoost	98.09	89.2	89.54	0.88	0.43	0.9	0.88

It can be observed from Table 4.26 that XGBoost classifier outperformed other classifiers.

4.5 Discussion

After experimenting with various AI techniques to train our classification model to classify mammograms into B (Benign), M (Malignant), transfer learning was found to have the best performance. In our experimentation with the MIAS dataset, VGG16 had the best performance (Train accuracy: 98.3, Test accuracy: 95.71, Validation accuracy: 96.04, F1 score: 0.957, Kappa score: 0.9122, ROC score: 0.9545).

TABLE 4.26 Performance of Xception deep feature extraction with different machine learning models.

Model	Training accuracy	Validation accuracy	Test accuracy	F1_score	Kappa	Recall	Precision
ANN	84.64	84	85.06	0.8413	0.263	0.8507	0.8344
KNN	88.5	86.2	86.68	0.84	0.2	0.87	0.83
SVM	86.96	86.95	86.96	0.81	0	0.87	0.76
Random Forest	91.81	87.81	87.93	0.84	0.2	0.88	0.86
AdaBoost	87.01	86.93	87.02	0.81	0.03	0.87	0.83
XGBoost	97.25	89.2	89.07	0.88	0.41	0.89	0.88

The custom CNN *n*-layered (CNN 6-layered achieved a test accuracy of only 61.59) models and classifiers (ANN, KNN, XGBoost, SVM, and Random Forest, AdaBoost) trained on extracted features were outperformed by the transfer learning models (XGBoost classifier trained on features extracted from ResNet-101 model achieved a test accuracy of only 61.37). And, in our experimentation with the CBIS-DDSM dataset, MobileNet had the best performance (Train accuracy: 99.07, Test accuracy: 95.73, Validation accuracy: 95.86, F1 score: 0.956, Kappa score: 0.80, ROC score: 0.89). The custom CNN *n*-Layered (CNN 5-Layered, CNN 8-Layered models achieved a test accuracy of about 92% each, respectively) models and classifiers (ANN, KNN, XGBoost, SVM, Random Forest, AdaBoost) trained on extracted features were slightly outperformed by the transfer learning models (XGBoost classifier trained on features extracted from DenseNet169 model achieved a test accuracy of only 89.98). When trained on various classifiers, the feature extracted from the CBIS-DDSM dataset performed relatively well compared to the MIAS dataset. Transfer learning was able to generalize to a terrific extent the features it had learned from the training data and effectively use it to classify test mammograms. The exploration of network architectures has been a part of neural network research since their initial discovery. The recent resurgence in the popularity of neural networks has also revived this research domain. The increasing number of layers in modern networks amplifies the differences between architectures and motivates exploring different connectivity patterns and revisiting old research ideas [40].

4.6 Conclusion

In this chapter, we have highlighted the various techniques of image classification in the field of breast cancer detection. Techniques such as transfer learning, deep feature extraction, model tuning, and hyperparameter optimization were used to obtain the best results. An in-depth analysis of different models has been presented in the work and the various performance evaluation metrics used. Popularly known datasets in breast cancer, such as DDSM and MIAS datasets, were used to showcase the impact of how different AI techniques can be used in this field. This chapter consists of a comparative study of various models such as simple CNNs, transfer learning through pretrained models, deep feature extraction, and using traditional machine learning models over these pretrained models. Our findings have shown that the techniques mentioned above can detect breast cancer from mammograms with accuracies as high as 96%. These findings indicate that many improvements could be made by tuning existing architectures or building new architectures that could lead to further improvements in this field of breast cancer detection from mammograms using deep learning and machine learning techniques.

References

[1] H. Sung, et al., Global cancer statistics 2020: GLOBOCAN estimates of incidence and mortality worldwide for 36 cancers in 185 countries, CA. Cancer J. Clin. 71 (3) (2021). Available from: https://doi.org/10.3322/caac.21660.

[2] J. Tang, R.M. Rangayyan, J. Xu, I.E. Naqa, Y. Yang, Computer-aided detection and diagnosis of breast cancer with mammography: recent advances, IEEE Trans. Inf. Technol. Biomed. 13 (2) (2009) 236–251. Available from: https://doi.org/10.1109/TITB.2008.2009441.

[3] S. Hadush, Y. Girmay, A. Sinamo, G. Hagos, Breast cancer detection using convolutional neural networks, ArXiv200307911 Cs, <http://arxiv.org/abs/2003.07911> Aug. 2020 (accessed 30.04.21).

[4] P. Xi, C. Shu, R. Goubran, Abnormality detection in mammography using deep convolutional neural networks, ArXiv180301906 Cs, <http://arxiv.org/abs/1803.01906>, Mar. 2018 (accessed 30.04.21)

[5] A. Jalalian, S.B.T. Mashohor, H.R. Mahmud, M.I.B. Saripan, A.R.B. Ramli, B. Karasfi, Computer-aided detection/diagnosis of breast cancer in mammography and ultrasound: a review, Clin. Imaging 37 (3) (2013) 420–426. Available from: https://doi.org/10.1016/j.clinimag.2012.09.024.

[6] L. Shen, L.R. Margolies, J.H. Rothstein, E. Fluder, R.B. McBride, W. Sieh, Deep Learning to improve breast cancer early detection on screening mammography, Sci. Rep. 9 (1) (2019) 12495. Available from: https://doi.org/10.1038/s41598-019-48995-4.

[7] J.R. Burt, et al., Deep learning beyond cats and dogs: recent advances in diagnosing breast cancer with deep neural networks, Br. J. Radiol. (2018) 20170545. Available from: https://doi.org/10.1259/bjr.20170545. Apr.

[8] I.J. Goodfellow et al., Generative adversarial networks, ArXiv14062661 Cs Stat, <http://arxiv.org/abs/1406.2661>, Jun. 2014 (accessed 27.05.21).

[9] L.A. Gatys, A.S. Ecker, M. Bethge, "A neural algorithm of artistic style," ArXiv150806576 Cs Q-Bio, Sep. 2015, <http://arxiv.org/abs/1508.06576>, (accessed 27.05.21).

[10] O. Ronneberger, P. Fischer, T. Brox, U-Net: convolutional networks for biomedical image segmentation, ArXiv150504597 Cs, <http://arxiv.org/abs/1505.04597>, May 2015 (accessed 29.04.21)

[11] X. Qin, Z. Zhang, C. Huang, M. Dehghan, O.R. Zaiane, M. Jagersand, U^2-Net: going deeper with nested u-structure for salient object detection, Pattern Recognit. 106 (2020) 107404. Available from: https://doi.org/10.1016/j.patcog.2020.107404.

[12] R. Agarwal, O. Diaz, X. Lladó, M.H. Yap, R. Martí, Automatic mass detection in mammograms using deep convolutional neural networks, J. Med. Imaging 6 (03) (2019) 1. Available from: https://doi.org/10.1117/1.JMI.6.3.031409.

[13] I.C. Moreira, I. Amaral, I. Domingues, A. Cardoso, M. J. Cardoso, J.S. Cardoso, INbreast: Toward a Full-field Digital Mammographic Database, Acad. Radiol. 19 (2) (2012) 236–248. Available from: https://doi.org/10.1016/j.acra.2011.09.014.

[14] M.A. Al-Antari, S.-M. Han, T.-S. Kim, Evaluation of deep learning detection and classification towards computer-aided diagnosis of breast lesions in digital X-ray mammograms, Comput. Methods Prog. Biomed. 196 (2020) 105584. Available from: https://doi.org/10.1016/j.cmpb.2020.105584.

[15] J. Schmidhuber, Deep Learning in Neural Networks: An Overview, Neural Netw. 61 (2015) 85–117. Available from: https://doi.org/10.1016/j.neunet.2014.09.003. Jan.

[16] A. Nawaz, S. M. Anwar, R. Liaqat, J. Iqbal, U. Bagci, M. Majid, Deep convolutional neural network based classification of Alzheimer's disease using MRI data, ArXiv210102876 Cs Eess, <http://arxiv.org/abs/2101.02876>, Jan. 2021 (accessed 03.05.21).

[17] O. Ozdemir, R.L. Russell, A.A. Berlin, A 3D probabilistic deep learning system for detection and diagnosis of lung cancer using low-dose CT scans, ArXiv190203233 Cs, <http://arxiv.org/abs/1902.03233>, Jan. 2020 (accessed 03.05.21).

[18] J. Wen, et al., Convolutional neural networks for classification of Alzheimer's Disease: overview and reproducible evaluation, Med. Image Anal. 63 (2020) 101694. Available from: https://doi.org/10.1016/j.media.2020.101694.

[19] D.-P. Fan et al., "Inf-Net: automatic COVID-19 lunginfection segmentation from CT Images," ArXiv200414133 Cs Eess, May 2020 (accessed 29.04.21).

[20] S. Bharati, P. Podder, M.R.H. Mondal, Artificial neural network based breast cancer screening: a comprehensive review, ArXiv200601767 Cs Eess Math, <http://arxiv.org/abs/2006.01767> May 2020 (accessed 12.05.21)

[21] E.-K. Kim, et al., Applying data-driven imaging biomarker in mammography for breast cancer screening: preliminary study, Sci. Rep. 8 (1) (2018) 2762. Available from: https://doi.org/10.1038/s41598-018-21215-1.

[22] A. Krizhevsky, I. Sutskever, G.E. Hinton, ImageNet classification with deep convolutional neural networks, Commun. ACM 60 (6) (2017) 84–90. Available from: https://doi.org/10.1145/3065386.
[23] Y. LeCun, Y. Bengio, G. Hinton, Deep learning, Nature 521 (7553) (2015) 436–444. Available from: https://doi.org/10.1038/nature14539.
[24] A.S. Lundervold, A. Lundervold, An overview of deep learning in medical imaging focusing on MRI, Z. Für Med. Phys. 29 (2) (2019) 102–127. Available from: https://doi.org/10.1016/j.zemedi.2018.11.002.
[25] A.W. Seni, et al., Improved protein structure prediction using potentials from deep learning, Nature, 2020, p. 577. Available from: https://www.nature.com/articles/s41586-019-1923-7.
[26] A. Kryshtafovych, T. Schwede, M. Topf, K. Fidelis, J. Moult, Critical assessment of methods of protein structure prediction (CASP)—Round XIII, Proteins Struct. Funct. Bioinforma. 87 (12) (2019) 1011–1020. Available from: https://doi.org/10.1002/prot.25823.
[27] M. Ghassemi, T. Naumann, P. Schulam, A. L. Beam, I. Y. Chen, R. Ranganath, A review of challenges and opportunities in machine learning for health, ArXiv180600388 Cs Stat, <http://arxiv.org/abs/1806.00388> Dec. 2019 (accessed 13.05.21).
[28] The mini-MIAS database of mammograms. <http://peipa.essex.ac.uk/info/mias.html> (accessed 12.05.21).
[29] MIAS Mammography. <https://kaggle.com/kmader/mias-mammography> (accessed 14.05.21).
[30] R. Sawyer-Lee, F. Gimenez, A. Hoogi, D. Rubin, Curated breast imaging subset of DDSM, Cancer Imaging Archive (2016). Available from: https://doi.org/10.7937/K9/TCIA.2016.7O02S9CY.
[31] DDSM Mammography. <https://kaggle.com/skooch/ddsm-mammography> (accessed 12.05.21).
[32] A. Boyd, A. Czajka, K. Bowyer, Deep learning-based feature extraction in iris recognition: use existing models, fine-tune or train from scratch?, ArXiv200208916 Cs, <http://arxiv.org/abs/2002.08916> Feb. 2020. (accessed 06.05.21).
[33] A. Subasi, A. Mitra, F. Özyurt, T. Tuncer, Automated COVID-19 detection from CT images using deep learning, Computer-aided Design and Diagnosis Methods for Biomedical Applications, Taylor & Francis, 2021, pp. 153–176. Available from: https://doi.org/10.1201/9781003121152-7.
[34] Y. Jiménez-Gaona, M.J. Rodríguez-Álvarez, V. Lakshminarayanan, Deep learning based computer-aided systems for breast cancer imaging: a critical review, ArXiv201000961 Cs Eess, <http://arxiv.org/abs/2010.00961>, Sep. 2020 (accessed 11.05.21).
[35] P. Ranganathan, C. Pramesh, R. Aggarwal, Common pitfalls in statistical analysis: measures of agreement, Perspect. Clin. Res. 8 (4) (2017) 187. Available from: https://doi.org/10.4103/picr.PICR_123_17.
[36] J. Cohen, A coefficient of agreement for nominal scales, Educ. Psychol. Meas. 20 (1) (1960) 37–46.
[37] Z. Yang, M. Zhou, Kappa statistic for clustered physician–patients polytomous data, Comput. Stat. Data Anal. 87 (2015) 1–17.
[38] K. Feng, H. Hong, K. Tang, J. Wang, Decision making with machine learning and ROC curves, ArXiv190502810 Cs Econ Q-Fin Stat, <http://arxiv.org/abs/1905.02810>, May 2019 (accessed 12.05.21).
[39] J.B. Brown, Classifiers and their metrics quantified, Mol. Inform. 37 (1–2) (2018) 1700127. Available from: https://doi.org/10.1002/minf.201700127.
[40] G. Huang, Z. Liu, L. van der Maaten, K.Q. Weinberger, Densely connected convolutional networks, ArXiv160806993 Cs, <http://arxiv.org/abs/1608.06993>, Jan. 2018 (accessed 29.04.21).

CHAPTER

5

Breast tumor detection in ultrasound images using artificial intelligence

Omkar Modi[1] *and Abdulhamit Subasi*[2,3]

[1]Indian Institute of Technology, Kharagpur, West Bengal, India [2]Institute of Biomedicine, Faculty of Medicine, University of Turku, Turku, Finland [3]Department of Computer Science, College of Engineering, Effat University, Jeddah, Saudi Arabia

OUTLINE

5.1 Introduction

Breast cancer begins when healthy cells in the breast grow out of control, forming a mass or sheet of cells called a tumor. Female breast cancer has surpassed lung cancer as the most commonly diagnosed cancer, with an estimated 2.3 million new cases [1]. Mortality can be reduced by early detection and therapy. There are various predominant tests to find or diagnose breast cancer mostly biopsy is the only the certain way for the doctor to know if an area of the body has cancer. In a biopsy, the doctor takes a small sample of tissue for testing in a laboratory. Other diagnoses include image testing such as diagnostic mammography, MRI, and ultrasound [2].

Ultrasound imaging of the breast uses sound waves to produce images of the internal structures of the breast. It is used to diagnose

Applications of Artificial Intelligence in Medical Imaging
DOI: https://doi.org/10.1016/B978-0-443-18450-5.00003-7

breast lumps and is complementary to mammograms or breast MRIs. Ultrasound is a safe, noninvasive, and radiation-free procedure. Breast ultrasonography can assist to detect whether an abnormality is solid (such as a benign cyst), fluid-filled (such as a noncancerous lump of tissue), or both cystic and solid (such as a malignant tumor) [3]. In recent years, with the development in artificial intelligence (AI), especially in the field of deep learning (DL) networks and its outperforming results in the field of image recognition tasks, we can leverage the technology for ultrasound tests of breast cancer for early detection whether the tumor is benign or malignant.

In this chapter, we will review the task of categorizing the tumor with various models and compare them. We applied DL models involving convolutional neural networks (CNNs) and transfer learning for feature extraction and classified it on various DL architectures and traditional machine learning (ML) algorithms. We compared all the models on various metrics.

5.2 Background/literature review

Breast self-examination is a screening method that is performed by the individual themselves. It is feasible to notice any differences or changes in the breasts by palming them at different angles and at varied pressures. Breast inspection, on the other hand, is the least reliable method of cancer detection. Mammography has evolved as a viable alternative and is now commonly utilized in medicine. However, only relying on mammograms carries a considerable risk of false positives, which frequently result in unneeded biopsies and procedures [4].

Due to the development of AI and computer vision, there has been massive research in using AI for automation in the field of medicine. Recently, AI technology has made great progress in the automated analysis of medical images for anomaly detection. The same is true for breast images for possible breast cancer detection [5,6]. Conventional methods or traditional machine algorithms such as K-nearest neighbors (KNN), support vector machine (SVM), and Random Forest showed moderate performances. Emerging of DL algorithms, which process images and extract features, has shown remarkable results. The CNN model is used for training the data very often in medical image diagnosis, analysis, and its applications. In fact, medical imaging in CAD systems becomes successful because of the use of CNN [7]. CNN exploits the spatial data among the image pixels. CNN has helped researchers map important features localizing it from the scan images of a breast and classifying them into various kinds of abnormalities. It is evident that DL requires large data, the pretraining method provided a solution to classify for incomplete data or large data was not available [8].

In Ref. [9], the authors compared the classification results of KNN, SVM, Random Forest, and Decision Tree techniques. The Wisconsin Breast Cancer dataset was utilized, which was downloaded from the UCI repository. KNN was the top classifier in simulations, followed by SVM, Random Forest, and Decision Tree.

In Ref. [10], the authors presented a unique approach for detecting breast cancer using ML techniques such as the Naive Bayes classifier, SVM classifier, bi-clustering AdaBoost techniques, RCNN classifier, and bidirectional recurrent neural networks (HA-BiRNN). A comparison of ML techniques and the proposed methodology [deep neural network (DNN) with support value] was conducted, and the simulated results revealed that the DNN algorithm was superior in terms of performance, efficiency, and image quality, all of which are critical in today's medical systems, whereas the other techniques failed to perform as expected.

MuñozMeza and Gómez [11] used ultrasound images to classify breast tumors using three M-dimensional sets of characteristics and principal component analysis using shared information. The authors initially addressed image segmentation in ultrasound images of breasts using a watershed transformation mechanism and an important features extraction method for breast cancer classification that is based on some statistical tests and shared information.

Kwok [12] used four distinct state-of-the-art pre-trained models for breast cancer classification on histological images (VGG19, InceptionV3, InceptionV4, and ResNetV2). To increase prediction accuracy, several data augmentation procedures are used. The maximum accuracy was reached by the Inception and ResNetV2 models, according to the assessment findings. For binary and multiclass classification issues, the models achieved 91% and 79% accuracy, respectively.

Nawaz et al. [13] used a fine-tuned AlexNet model for breast cancer categorization. On the dataset, the researchers used two ways to train the model: patch-wise and image-wise of size 512×512 pixels. On the dataset, the model obtained 75.73% and 81.25% accuracy.

Dhungel et al. [14] advocated the use of CNNs in microcalcification detection, Mordang et al. [15] proposed the use of CNNs in breast density estimation, and Ahn et al. [16] proposed the use of CNNs in breast density estimation. Huynh et al. [17] proposed using a transfer learning technique for ultrasound breast image categorization in breast ultrasound imaging.

Despite the continual advancement of ML techniques, the performance of these applications has not improved significantly. Meanwhile, DL has proved successful in visual object recognition and classification in a variety of domains, since it learns representations from data and supports the learning of successive layers of increasingly relevant representations.

5.3 Artificial intelligence techniques

5.3.1 Artificial neural networks

An artificial neural network (ANN) is a brain neural system inspired algorithm that consists of layers with connected nodes and is included in ML. It has input and output layers as well as hidden layers. The first layer contains input values, whereas the last layer has labeled values. The value of each node is learnt during training by parameterizing weights using learning algorithms such as backpropagation. Each node's weights are tuned in the direction to reduce losses and hence improve accuracy [18].

ANNs have been applied to a variety of data classification and pattern recognition tasks in medical image processing and have proven to be a promising classification tool in breast cancer [19]. The application of neural networks in image and signal processing increased dramatically in the early 1980s. The key advantage was the decrease in manipulation time owing to neural networks' parallel-distributed processing nature [20]. The network was then widely employed in popular image processing techniques as vector quantization, eigenvector extraction, 2D pulse code modulation, and 2D filtering. The function of an ANN is similar to that of a biological neuron, and it is made up of neurons with different layers that are connected by numeric weights; these weights may be modified as the network learns to get closer to the best output. The number of neurons in image processing applications is usually proportional to the number of pixels in the input picture, and the number of layers is determined by the processing stages [21].

Image segmentation has been broadly applied for cancer diagnosis and categorization. ANNs have been used to train a variety of image segmentation algorithms, including histogram features, edge detection, region growth, and pixel classification. ANNs have been used for the classification and segmentation of diseases such as Alzheimer's, breast cancer, lung cancer, and brain tumors, among other diseases and have vast potential in the medical and health sector.

5.3.2 Deep learning

DL is a subfield of ML. DL learns from the data. The data may be unstructured or unlabeled. Researchers have recently advanced DL to expand ANN into DNN by stacking many hidden layers with linked nodes between the input and output layers. By combining basic decisions between layers, the multilayer can handle increasingly complicated challenges. In prediction tasks such as classification and regression, DNN often outperforms the shallow layered network. To avoid learning converging at the local minimum or overcome overfitting difficulties, each layer of the DNN improved its weights using the unsupervised restricted Boltzmann machine. Recently, residual neural networks used skip connections to avoid vanishing gradient problems. Furthermore, the introduction of big data and graphics processing units has the potential to solve complicated issues and reduce computing time. There are several hidden layers between the input layer and the output layer. The nodes known as neurons are found in each layer. The difference between ML and DL is that DL is closer to its goal than ML and can extract features automatically [18].

Accordingly, the DL algorithm gets a lot of attention these days to solve various problems in the medical imaging field. In this chapter, we have developed and compared various DL models on ultrasound images to detect breast cancer. DL has helped to reduce false-positive rates and decrease assessment time and unnecessary biopsies.

5.3.3 Convolutional neural networks

CNN is widely used in computer vision for both supervised and unsupervised learning. CNN is used to classify the images. It takes the images of the breast cancer dataset as input that and neurons are associated with their corresponding weights. The weights are adjusted to minimize the error and enhance the performance. The extracted information includes facial features, edge detection, object recognition, and other relevant features present in the image. These features help distinguish one class of images from another. This is what sets the CNNs different from other DL techniques. The compositions of CNN are convolutional, pooling, and fully connected layers (Fig. 5.1). In the convolution layer, a feature map is used to extract the features of the given image and makes the original image more compact. The pooling layer is used to reduce the dimensions of the image. Rectified linear unit (ReLU) layer is used as an activation function in which it checks the value

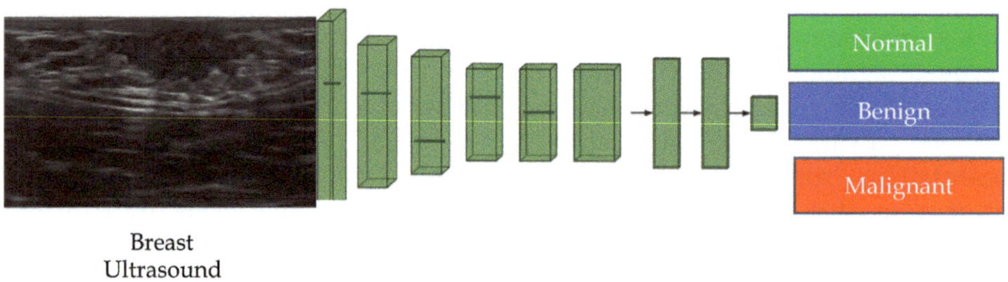

FIGURE 5.1 CNN for breast cancer detection.

of the activation function lies in a given range or not. The fully connected layers are joined at the end of the CNN architecture and offer a final decision for classification or regression tasks. A loss is calculated during training by comparing labeled and predicted values [18]. Python codes for different CNN architectures are given below.

```
from keras.models import Sequential
from keras.layers import Conv2D
from keras.layers import MaxPooling2D
from keras.layers import Dense
from keras.layers import Flatten
from keras.optimizers import SGD
from keras.optimizers import Adam

def cnn_2():
  # Create the model
  model = Sequential()

  # Add convolutional layers
  model.add(Conv2D(32, (3, 3), activation='relu', kernel_initializer='he_uniform',
                                                    padding='valid',input_shape=img_shape))
  model.add(MaxPooling2D((2,2)))

  model.add(Conv2D(32, (3, 3), activation='relu', kernel_initializer='he_uniform',
                                                                        padding='valid'))
  model.add(MaxPooling2D((2,2)))
  # Add dropout layer
  model.add(Dropout(0.5))
  # Add flatten layer
  model.add(Flatten())

  # Add dense layers
  model.add(Dense(128,activation='relu'))
  model.add(Dense(3,activation='softmax'))

  model.compile(optimizer='adam',loss='categorical_crossentropy',metrics=['acc'])

  return model

def cnn_3():
  # Create the model
  model = Sequential()
```

```
    # Add convolutional layers
    model.add(Conv2D(32, (3, 3), activation='relu', kernel_initializer='he_uniform',
                                          padding='valid', input_shape=img_shape))
    model.add(MaxPooling2D((2,2)))

    model.add(Conv2D(32, (3, 3), activation='relu', kernel_initializer='he_uniform',
                                                                padding='valid'))
    model.add(MaxPooling2D((2,2)))

    model.add(Conv2D(64, (3, 3), activation='relu', kernel_initializer='he_uniform',
                                                                padding='valid'))
    model.add(MaxPooling2D((2,2)))

    # Add dropout layer
    model.add(Dropout(0.25))
    # Add flatten layer
    model.add(Flatten())

    # Add dense layers
    model.add(Dense(1024,activation='relu'))
    model.add(Dense(128,activation='relu'))
    model.add(Dense(3,activation='softmax'))

    model.compile(optimizer='adam',loss='categorical_crossentropy',metrics=['acc'])

    return model

def cnn_4():
    # Create the model
    model = Sequential()

    # Add convolutional layers
    model.add(Conv2D(32, (3, 3), padding='valid', strides=(1, 1),input_shape=img_shape))
    model.add(Activation('relu'))
    model.add(MaxPooling2D(pool_size=(2, 2)))
```

```
    model.add(Conv2D(32, (3, 3), padding='same', strides=(1, 1)))
    model.add(Activation('relu'))
    model.add(MaxPooling2D(pool_size=(2, 2),strides=2))
    # Add dropout layer
    model.add(Dropout(0.2))

    model.add(Conv2D(64, (3, 3), padding='same', strides=(1, 1)))
    model.add(Activation('relu'))
    model.add(MaxPooling2D(pool_size=(2, 2),strides=2))

    model.add(Conv2D(64, (3, 3), padding='same', strides=(1, 1)))
    model.add(Activation('relu'))
    model.add(MaxPooling2D(pool_size=(2, 2),strides=2))

    # Add dropout layer
    model.add(Dropout(0.2))

    # Add flatten layer
    model.add(Flatten())
    # Add dense layers
    model.add(Dense(3))
    model.add(Activation('softmax'))

    model.compile(optimizer='adam',loss='categorical_crossentropy',metrics=['acc'])

    return model

def cnn_5():
    # Create the model
    model = Sequential()

    # Add convolutional layers
    model.add(Conv2D(32, (3, 3), padding='same', strides=(1, 1),input_shape=img_shape))
    model.add(Activation('relu'))
    model.add(MaxPooling2D(pool_size=(2, 2)))
    # Add dropout layer
    model.add(Dropout(0.2))
```

```
    model.add(Conv2D(32, (3, 3), padding='same', strides=(1, 1)))
    model.add(Activation('relu'))
    model.add(MaxPooling2D(pool_size=(2, 2),strides=2))

    model.add(Conv2D(64, (3, 3), padding='same', strides=(1, 1)))
    model.add(Activation('relu'))
    model.add(MaxPooling2D(pool_size=(2, 2),strides=2))

    model.add(Conv2D(64, (3, 3), padding='same', strides=(1, 1)))
    model.add(Activation('relu'))
    model.add(MaxPooling2D(pool_size=(2, 2),strides=2))

    model.add(Conv2D(128, (3, 3), padding='same', strides=(1, 1)))
    model.add(Activation('relu'))
    model.add(MaxPooling2D(pool_size=(2, 2),strides=2))

    # Add dropout layer
    model.add(Dropout(0.2))
    # Add flatten layer
    model.add(Flatten())

    # Add dense layers
    model.add(Dense(256))
    model.add(Dense(3))
    model.add(Activation('softmax'))
    model.compile(optimizer='adam',loss='categorical_crossentropy',metrics=['acc'])

    return model

def cnn_6():
    # Create the model
    model = Sequential()
    # Add convolutional layers
    model.add(Conv2D(16, (3, 3), padding='same', strides=(1, 1),input_shape=img_shape))
    model.add(Activation('relu'))
    model.add(MaxPooling2D(pool_size=(2, 2)))
    # Add dropout layer
    model.add(Dropout(0.2))
```

```
    model.add(Conv2D(32, (3, 3), padding='same', strides=(1, 1)))
    model.add(Activation('relu'))
    model.add(MaxPooling2D(pool_size=(2, 2),strides=2))

    model.add(Conv2D(32, (3, 3), padding='same', strides=(1, 1)))
    model.add(Activation('relu'))
    model.add(MaxPooling2D(pool_size=(2, 2),strides=2))

    model.add(Conv2D(64, (3, 3), padding='same', strides=(1, 1)))
    model.add(Activation('relu'))
    model.add(MaxPooling2D(pool_size=(2, 2),strides=2))

    model.add(Conv2D(64, (3, 3), padding='same', strides=(1, 1)))
    model.add(Activation('relu'))
    model.add(MaxPooling2D(pool_size=(2, 2),strides=2))

    model.add(Conv2D(128, (3, 3), padding='same', strides=(1, 1)))
    model.add(Activation('relu'))
    model.add(MaxPooling2D(pool_size=(2, 2),strides=2))

    # Add dropout layer
    model.add(Dropout(0.2))
    # Add flatten layer
    model.add(Flatten())

    # Add dense layers
    model.add(Dense(256,activation = 'relu'))
    model.add(Dense(3))
    model.add(Activation('softmax'))
    model.compile(optimizer='adam',loss='categorical_crossentropy',metrics=['acc'])

    return model

def cnn_7():
    # Create the model
    model = Sequential()
```

```
# Add convolutional layers
model.add(Conv2D(16, (3, 3), padding='same',kernel_initializer='he_uniform',
                                        strides=(1,1),input_shape=img_shape))
model.add(Activation('relu'))

model.add(Conv2D(32, (3, 3), padding='same', strides=(1, 1)))
model.add(Activation('relu'))
model.add(MaxPooling2D(pool_size=(2, 2),strides=2))
# Add dropout layer
model.add(Dropout(0.2))

model.add(Conv2D(32, (3, 3), padding='same', strides=(1, 1)))
model.add(Activation('relu'))
model.add(MaxPooling2D(pool_size=(2, 2),strides=2))
# Add dropout layer
model.add(Dropout(0.2))

model.add(Conv2D(64, (3, 3), padding='same', strides=(1, 1)))
model.add(Activation('relu'))
model.add(MaxPooling2D(pool_size=(2, 2),strides=2))
# Add dropout layer
model.add(Dropout(0.2))

model.add(Conv2D(64, (3, 3), padding='same', strides=(1, 1)))
model.add(Activation('relu'))
model.add(MaxPooling2D(pool_size=(2, 2),strides=2))
# Add dropout layer
model.add(Dropout(0.2))

model.add(Conv2D(128, (3, 3), padding='same', strides=(1, 1)))
model.add(Activation('relu'))
model.add(MaxPooling2D(pool_size=(2, 2),strides=2))
# Add dropout layer
model.add(Dropout(0.2))

model.add(Conv2D(128, (3, 3), padding='same', strides=(1, 1)))
model.add(Activation('relu'))
model.add(MaxPooling2D(pool_size=(2, 2),strides=2))
```

```
    # Add dropout layer
    model.add(Dropout(0.2))
    # Add flatten layer
    model.add(Flatten())

    # Add dense layers
    model.add(Dense(3))
    model.add(Activation('softmax'))
    model.compile(optimizer='adam',loss='categorical_crossentropy',metrics=['acc'])
    return model

def cnn_8():
    # Create the model
    model = Sequential()

    # Add convolutional layers
    model.add(Conv2D(16, (3, 3), padding='same',kernel_initializer='he_uniform',
                                                 input_shape=img_shape))
    model.add(Activation('relu'))
    model.add(MaxPooling2D(pool_size=(2,2),strides=1,padding = 'same'))
    model.add(Conv2D(32,(3,3),padding='same'))
    model.add(Activation('relu'))
    model.add(MaxPooling2D(pool_size=(2,2),strides=2,padding = 'same'))
    # Add dropout layer
    model.add(Dropout(0.2))

    model.add(Conv2D(32,(3,3),padding='same'))
    model.add(Activation('relu'))
    model.add(MaxPooling2D(pool_size=(2,2),strides=2,padding = 'same'))
    # Add dropout layer
    model.add(Dropout(0.2))

    model.add(Conv2D(64,(3,3),padding='same'))
    model.add(Activation('relu'))
    model.add(MaxPooling2D(pool_size=(2,2),strides=2,padding = 'same'))
    # Add dropout layer
    model.add(Dropout(0.2))
```

```
model.add(Conv2D(128,(3,3),padding='same'))
model.add(Activation('relu'))
model.add(MaxPooling2D(pool_size=(2,2),strides=2,padding = 'same'))
# Add dropout layer
model.add(Dropout(0.2))

model.add(Conv2D(256,(3,3),padding='same'))
model.add(Activation('relu'))
model.add(MaxPooling2D(pool_size=(2,2),strides=1,padding = 'same'))
# Add dropout layer
model.add(Dropout(0.2))

model.add(Conv2D(128,(3,3),padding='same'))
model.add(Activation('relu'))
model.add(MaxPooling2D(pool_size=(2,2),strides=1,padding = 'same'))
# Add dropout layer
model.add(Dropout(0.2))

model.add(Conv2D(64,(3,3),padding='same'))
model.add(Activation('relu'))
model.add(MaxPooling2D(pool_size=(2,2),strides=2,padding = 'same'))
# Add dropout layer
model.add(Dropout(0.2))
 # Add flatten layer
model.add(Flatten())

# Add dense layer
model.add(Dense(3))
model.add(Activation('softmax'))

model.compile(optimizer='adam',loss='categorical_crossentropy',metrics=['acc'])

 return model
```

5.4 Breast tumor detection using artificial intelligence

5.4.1 Feature extraction using deep learning

Feature extraction is a step in the image processing, which divides and reduces a large collection of raw data into smaller groupings. As a result, processing will be easier. When you have a huge data collection and need to decrease the number of resources without sacrificing any vital or relevant information, extracting the features might help. Feature extraction aids in the reduction of unnecessary data in data collection. The reduction of data makes it easier for the computer to develop the model with less effort, and it also speeds up the learning and generalization processes in the ML process [22].

In our research, we have extracted featured through multilayered CNN layers. CNNs provide automatic feature extraction. The specified input data is initially forwarded to a feature extraction network, and then the resultant extracted features are forwarded to a classifier network after applying a fully connected layer. Max and average pooling layers were introduced for dimension reduction that helps significantly in reducing computing costs. To obtain a better feature, vector dropout and batch normalization were used which also helped to prevent high variance in data. We have trained various n-layered CNN models (n varying from 2 to 10) and compared the quality of features extracted after passing to classifier neural network through performance matrix. Apart from this, a total of 11 pretrained CNN architectures have been used for feature extraction and trained on various ML models. Model weight was initialized with ImageNet weights which contains more than 14 million images that belong to more than 20,000 classes. This usage of models with pretrained weights is known as transfer learning that helped to save time and resources from having to train multiple ML models from scratch to complete similar tasks. Python codes for deep feature extraction are given below.

```
#VGG16 PreTrained Model for Feature Extraction
base_model = tf.keras.applications.VGG16(
  include_top=False,
  weights="imagenet",
  input_tensor=None,
  input_shape=img_shape,
  pooling=None
)
feature_extraction(base_model)
#VGG19 PreTrained Model for Feature Extraction
base_model = tf.keras.applications.VGG19(
  include_top=False,
  weights="imagenet",
  input_tensor=None,
  input_shape=img_shape,
  pooling=None
)
feature_extraction(base_model)
#ResNet50 PreTrained Model for Feature Extraction
base_model = tf.keras.applications.ResNet50(
    include_top=False,
    weights="imagenet",
    input_tensor=None,
    input_shape=img_shape,
    pooling=None
  )
  feature_extraction(base_model)
#ResNet101 PreTrained Model for Feature Extraction
base_model = tf.keras.applications.ResNet101(
  include_top=False,
  weights="imagenet",
  input_tensor=None,
  input_shape=img_shape,
  pooling=None
)
feature_extraction(base_model)
```

```
#MobileNetV2 PreTrained Model for Feature Extraction
base_model = tf.keras.applications.MobileNetV2(
  include_top=False,
  weights="imagenet",
  input_tensor=None,
  input_shape=img_shape,
  pooling=None
)
feature_extraction(base_model)

#MobileNet PreTrained Model for Feature Extraction
base_model = tf.keras.applications.MobileNet(
  include_top=False,
  weights="imagenet",
  input_tensor=None,
  input_shape=img_shape,
  pooling=None
)
feature_extraction(base_model)

#InceptionV3 PreTrained Model for Feature Extraction
base_model = tf.keras.applications.InceptionV3(
  include_top=False,
  weights="imagenet",
  input_tensor=None,
  input_shape=img_shape,
  pooling=None
)
feature_extraction(base_model)

#InceptionResNetV2 PreTrained Model for Feature Extraction
base_model = tf.keras.applications.InceptionResNetV2(
  include_top=False,
  weights="imagenet",
  input_tensor=None,
  input_shape=img_shape,
  pooling=None
)
feature_extraction(base_model)
```

```
#DenseNet169 PreTrained Model for Feature Extraction
base_model = tf.keras.applications.DenseNet169(
  include_top=False,
  weights="imagenet",
  input_tensor=None,
  input_shape=img_shape,
  pooling=None
)
feature_extraction(base_model)

#DenseNet121 PreTrained Model for Feature Extraction
base_model = tf.keras.applications.DenseNet121(
  include_top=False,
  weights="imagenet",
  input_tensor=None,
  input_shape=img_shape,
  pooling=None
)
feature_extraction(base_model)

#Xception PreTrained Model for Feature Extraction
base_model = tf.keras.applications.Xception(
  include_top=False,
  weights="imagenet",
  input_tensor=None,
  input_shape=img_shape,
  pooling=None
)
feature_extraction(base_model)
```

5.4.2 Prediction and classification

The dataset used in our research consisted of three classes of labels associated with each scanned ultrasound image. Extracted feature vectors processed by CNN layers and various pretrained architecture were passed to a dense neural network consisting of a varied number of hidden layers with ReLU activations. The last layer that is output layer was a softmax layer that classified the feature vectors into three classes. To overcome high variance, dropout layers and batch normalization layers were introduced. The model was compiled using Adam optimizer and the categorial loss function was used to evaluate the loss. This optimization technique showed good convergence in the dataset. Data was stratified to deal with class imbalance problems and early stopping was introduced to avoid overfitting and to improve the learner's performance on data outside of the training set. The Python codes for Transfer learning models are given below.

```
#VGG16
base_model = tf.keras.applications.VGG16(
  include_top=False,
  weights="imagenet",
  input_tensor=None,
  input_shape=img_shape,
  pooling=None
)

for l in base_model.layers:
  l.trainable = False

def VGG16():
  model = Sequential()
  model.add(base_model)
  model.add(Flatten())
  model.add(BatchNormalization())
  model.add(Dense(512,activation='relu'))
  model.add(Dropout(0.5))
  model.add(Dense(256,activation='relu'))
  model.add(Dropout(0.5))
  model.add(Dense(128,activation='relu'))
  model.add(Dropout(0.5))
  model.add(Dense(64,activation='relu'))
  model.add(Dropout(0.5))
  model.add(Dense(3,activation='softmax'))

  model.compile(optimizer='adam',loss='categorical_crossentropy',metrics=['acc'])

  return model
#VGG 19
base_model = tf.keras.applications.VGG19(
  include_top=False,
  weights="imagenet",
  input_tensor=None,
  input_shape=img_shape,
  pooling=None
)
```

```
for l in base_model.layers:
  l.trainable = False

def VGG19():
  model = Sequential()
  model.add(base_model)
  model.add(Flatten())
  model.add(BatchNormalization())
  model.add(Dense(128,activation='relu'))
  model.add(Dropout(0.5))
  model.add(Dense(32,activation='relu'))
  model.add(Dropout(0.5))
  model.add(Dense(3,activation='softmax'))

  model.compile(optimizer='adam',loss='categorical_crossentropy',metrics=['acc'])

  return model

#ResNet50
base_model = tf.keras.applications.ResNet50(
            include_top=False,
            weights="imagenet",
            input_tensor=None,
            input_shape=img_shape,
            pooling=None,
        )
for layer in base_model.layers:
  layer.trainable = False
```

```
def resnet():
  model = Sequential()
  model.add(base_model)
  model.add(Flatten())
  model.add(BatchNormalization())
  model.add(Dense(256,activation='relu'))
  model.add(Dropout(0.5))
  model.add(Dense(128,activation='relu'))
  model.add(Dropout(0.5))
  model.add(Dense(64,activation='relu'))
  model.add(Dropout(0.5))
  model.add(Dense(3,activation='softmax'))
  model.compile(optimizer='adam',loss='categorical_crossentropy',metrics=['acc'])

  return model

#InceptionV3
base_model = tf.keras.applications.InceptionV3(
  include_top=False,
  weights="imagenet",
  input_tensor=None,
  input_shape=img_shape,
  pooling=None
)
for l in base_model.layers:
  l.trainable = False
```

```
def Inception():
  model = Sequential()
  model.add(base_model)
  model.add(Flatten())
  model.add(BatchNormalization())
  model.add(Dense(128,activation='relu'))
  model.add(Dropout(0.5))
  model.add(Dense(128,activation='relu'))
  model.add(Dropout(0.5))
  model.add(Dense(64,activation='relu'))
  model.add(Dropout(0.5))
  model.add(Dense(3,activation='softmax'))

  model.compile(optimizer='adam',loss='categorical_crossentropy',metrics=['acc'])

  return model

#MobileNet
base_model = tf.keras.applications.MobileNet(
              alpha = 0.75,
              include_top=False,
              weights="imagenet",
              input_tensor=None,
              input_shape=img_shape,
              pooling=None,
          )
for layer in base_model.layers:
  layer.trainable = False
```

```
def MobileNet():
  model = Sequential()
  model.add(base_model)
  model.add(Flatten())
  model.add(BatchNormalization())
  model.add(Dense(512,activation='relu'))
  model.add(Dropout(0.5))
  model.add(Dense(256,activation='relu'))
  model.add(Dropout(0.5))
  model.add(Dense(128,activation='relu'))
  model.add(Dropout(0.5))
  model.add(Dense(64,activation='relu'))
  model.add(Dropout(0.5))
  model.add(Dense(3,activation='softmax'))

  model.compile(optimizer='adam',loss='categorical_crossentropy',metrics=['acc'])

  return model

#DenseNet121
base_model = tf.keras.applications.DenseNet121(
              include_top=False,
              weights="imagenet",
              input_tensor=None,
              input_shape=img_shape,
              pooling=None,
          )
for layer in base_model.layers:
  layer.trainable = False

def DenseNet121():
  model = Sequential()
  model.add(base_model)
  model.add(Flatten())
  model.add(BatchNormalization())
  model.add(Dense(512,activation='relu'))
  model.add(Dropout(0.5))
  model.add(Dense(256,activation='relu'))
```

```
  model.add(Dropout(0.5))
  model.add(Dense(128,activation='relu'))
  model.add(Dropout(0.5))
  model.add(Dense(64,activation='relu'))
  model.add(Dropout(0.5))
  model.add(Dense(3,activation='softmax'))

  model.compile(optimizer='adam',loss='categorical_crossentropy',metrics=['acc'])

  return model

#DenseNet169
base_model = tf.keras.applications.DenseNet169(
             include_top=False,
             weights="imagenet",
             input_tensor=None,
             input_shape=img_shape,
             pooling=None,
         )
for layer in base_model.layers:
  layer.trainable = False

def DenseNet169():
  model = Sequential()
  model.add(base_model)
  model.add(Flatten())
  model.add(BatchNormalization())
  model.add(Dense(512,activation='relu'))
  model.add(Dropout(0.5))
  model.add(Dense(256,activation='relu'))
  model.add(Dropout(0.5))
  model.add(Dense(128,activation='relu'))
  model.add(Dropout(0.5))
  model.add(Dense(64,activation='relu'))
  model.add(Dropout(0.5))
  model.add(Dense(3,activation='softmax'))

  model.compile(optimizer='adam',loss='categorical_crossentropy',metrics=['acc'])

  return model
```

```
#InceptionResNetV2
base_model = tf.keras.applications.InceptionResNetV2(
              include_top=False,
              weights="imagenet",
              input_tensor=None,
              input_shape=img_shape,
              pooling=None,
          )
for layer in base_model.layers:
  layer.trainable = False

def InceptionResNetV2():
  model = Sequential()
  model.add(base_model)
  model.add(Flatten())
  model.add(BatchNormalization())
  model.add(Dense(128,activation='relu'))
  model.add(Dropout(0.5))
  model.add(Dense(128,activation='relu'))
  model.add(Dropout(0.5))
  model.add(Dense(64,activation='relu'))
  model.add(Dropout(0.5))
  model.add(Dense(3,activation='softmax'))

  model.compile(optimizer='adam',loss='categorical_crossentropy',metrics=['acc'])

  return model

#MobileNetV2
base_model2 = tf.keras.applications.MobileNetV2(
              alpha = 0.75,
              include_top=False,
              weights="imagenet",
              input_tensor=None,
              input_shape=img_shape,
              pooling=None,
          )
for layer in base_model2.layers:
  layer.trainable = False
```

```
def MobileNet2():
  model = Sequential()
  model.add(base_model2)
  model.add(Flatten())
  model.add(BatchNormalization())
  model.add(Dense(512,activation='relu'))
  model.add(Dropout(0.5))
  model.add(Dense(128,activation='relu'))
  model.add(Dropout(0.5))
  model.add(Dense(128,activation='relu'))
  model.add(Dropout(0.5))
  model.add(Dense(128,activation='relu'))
  model.add(Dropout(0.5))
  model.add(Dense(3,activation='softmax'))

  model.compile(optimizer='adam',loss='categorical_crossentropy',metrics=['acc'])

  return model

#ResNet101
base_model = tf.keras.applications.ResNet101(
              include_top=False,
              weights="imagenet",
              input_tensor=None,
              input_shape=img_shape,
              pooling=None,
         )
for layer in base_model.layers:
  layer.trainable = False

def resnet101():
  model = Sequential()
  model.add(base_model)
  model.add(MaxPooling2D((2,2),strides = 2))
  model.add(Flatten())
  model.add(BatchNormalization())
```

```
    model.add(Dense(512,activation='relu'))
    model.add(Dropout(0.5))
    model.add(Dense(256,activation='relu'))
    model.add(Dropout(0.5))
    model.add(Dense(128,activation='relu'))
    model.add(Dropout(0.5))
    model.add(Dense(64,activation='relu'))
    model.add(Dense(3,activation='softmax'))

    model.compile(optimizer='adam',loss='categorical_crossentropy',metrics=['acc'])

    return model

#AlexNet
def AlexNet():
    AlexNet = Sequential()

    #1st Convolutional Layer
    AlexNet.add(Conv2D(filters=96, input_shape=img_shape, kernel_size=(11,11),
    strides=(4,4), padding='same'))
    AlexNet.add(BatchNormalization())
    AlexNet.add(Activation('relu'))
    AlexNet.add(MaxPooling2D(pool_size=(2,2), strides=(2,2), padding='same'))

    #2nd Convolutional Layer
    AlexNet.add(Conv2D(filters=256, kernel_size=(5, 5), strides=(1,1), padding='same'))
    AlexNet.add(BatchNormalization())
    AlexNet.add(Activation('relu'))
    AlexNet.add(MaxPooling2D(pool_size=(2,2), strides=(2,2), padding='same'))

    #3rd Convolutional Layer
    AlexNet.add(Conv2D(filters=384, kernel_size=(3,3), strides=(1,1), padding='same'))
    AlexNet.add(BatchNormalization())
    AlexNet.add(Activation('relu'))

    #4th Convolutional Layer
    AlexNet.add(Conv2D(filters=384, kernel_size=(3,3), strides=(1,1), padding='same'))
    AlexNet.add(BatchNormalization())
    AlexNet.add(Activation('relu'))
```

```
#5th Convolutional Layer
AlexNet.add(Conv2D(filters=256, kernel_size=(3,3), strides=(1,1), padding='same'))
AlexNet.add(BatchNormalization())
AlexNet.add(Activation('relu'))
AlexNet.add(MaxPooling2D(pool_size=(2,2), strides=(2,2), padding='same'))

#Passing it to a Fully Connected layer
AlexNet.add(Flatten())


# 1st Fully Connected Layer
AlexNet.add(Dense(4096, input_shape=(32,32,3,)))
AlexNet.add(BatchNormalization())
AlexNet.add(Activation('relu'))
# Add Dropout to prevent overfitting
AlexNet.add(Dropout(0.4))


#2nd Fully Connected Layer
AlexNet.add(Dense(4096))
AlexNet.add(BatchNormalization())
AlexNet.add(Activation('relu'))
#Add Dropout
AlexNet.add(Dropout(0.4))


#3rd Fully Connected Layer
AlexNet.add(Dense(1000))
AlexNet.add(BatchNormalization())
AlexNet.add(Activation('relu'))
#Add Dropout
AlexNet.add(Dropout(0.4))

#Output Layer
AlexNet.add(Dense(3))
AlexNet.add(BatchNormalization())
AlexNet.add(Activation('softmax'))

AlexNet.compile(optimizer='adam',loss='categorical_crossentropy',metrics=['acc'])

return AlexNet
```

In the second part of our experiment, various classifiers were trained on features extracted from the pretrained models. The region of interest captured in the extracted features is fed into various classifiers, to classify the images into benign and malignant. The classifiers used were ANNs, KNN classifier, SVM classifier, Random Forest, AdaBoost, Bagging, XGBoost, LSTM, and Bi-LSTM classifier (Fig. 5.2).

```
from sklearn.svm import SVC
from sklearn.ensemble import RandomForestClassifier
from sklearn.ensemble import AdaBoostClassifier
from sklearn.neighbors import KNeighborsClassifier
from xgboost import XGBClassifier
from sklearn.ensemble import BaggingClassifier
from sklearn.neural_network import MLPClassifier
from tensorflow.keras.layers import LSTM,Bidirectional,Reshape
from sklearn.metrics import accuracy_score,recall_score,precision_score,f1_score,
                                                        roc_auc_score,cohen_kappa_score

def eval(classifier_name,y_train,y_train_pred,y_val,y_val_pred,y_true,y_pred):
  y_train = np.argmax(y_train,axis=1)
  y_val = np.argmax(y_val,axis=1)
  y_true = np.argmax(y_true,axis=1)

  train_accuracy = round(accuracy_score(y_train,y_train_pred),4)
  val_accuracy = round(accuracy_score(y_val,y_val_pred),4)
  test_accuracy = round(accuracy_score(y_true,y_pred),4)
  f1_measure = round(f1_score(y_true,y_pred,average='weighted'),4)
  kappa_score = round(cohen_kappa_score(y_true,y_pred),4)
  recall = round(recall_score(y_true,y_pred,average='weighted'),4)
  precision = round(precision_score(y_true,y_pred,average='weighted'),4)

  score={"classifier":classifier_name,"train_accuracy":train_accuracy ,
                              "val_accuracy":val_accuracy,"test_accuracy":test_accuracy,"f1_
      measure":f1_measure,"kappa_score":kappa_score,"recall":recall,"precision":precision}


  for e,a in score.items():
      print(e,a)
  print("--"*20)
```

```
def classifier_eval(classifier,classifier_name,X_train,y_train,X_val,y_val,X_test,
                                                                                  y_test):

  classifier.fit(X_train,np.argmax(y_train,axis=1))

  y_train_pred = classifier.predict(X_train)
  y_val_pred = classifier.predict(X_val)
  y_test_pred = classifier.predict(X_test)

  eval(classifier_name,y_train,y_train_pred,y_val,y_val_pred,y_test,y_test_pred)

names = ['SVM',
  'Random Forest',
  'AdaBoost',
  'KNN',
  'XGBoost',
  'Bagging',
  'ANN'
      ]

classifier = [
  SVC(),
  RandomForestClassifier(),
  AdaBoostClassifier(),
  KNeighborsClassifier(),
  XGBClassifier(),
  BaggingClassifier(),
  MLPClassifier(max_iter = 400),
]

cls_list = zip(names,classifier)
clsm_list = zip(names,classifier)
```

```
def feature_extraction(base_model):
 X_feat_out = base_model.output
 X_feat_flatten = Flatten()(X_feat_out)

 X_feat_model = Model(inputs = base_model.input,outputs = X_feat_flatten)
 X_feat_train = X_feat_model.predict(X_train)
 X_feat_val = X_feat_model.predict(X_val)
 X_feat_test = X_feat_model.predict(X_test)

 Xm_feat_train = X_feat_model.predict(Xm_train)
 Xm_feat_val = X_feat_model.predict(Xm_val)
 Xm_feat_test = X_feat_model.predict(Xm_test)

 for n,c in cls_list:
     classifier_eval(c,n,X_feat_train,y_train.toarray(),X_feat_val,y_val.toarray(),
                                                  X_feat_test,y_test.toarray())

 for l in base_model.layers:
     l.trainable = False

#LSTM
lstm_model = Sequential()
lstm_model.add(base_model)
lstm_model.add(Reshape((base_model.output.shape[1]*base_model.output.shape[2],
                                               base_model.output.shape[3])))

lstm_model.add(LSTM(128, dropout=0.5,recurrent_dropout=0.5))
lstm_model.add(Dense(3,activation='softmax'))

lstm_model.compile(optimizer='adam',loss='categorical_crossentropy',metrics=['acc'])

history = lstm_model.fit(X_train,y_train.toarray(),epochs = 20,validation_data = (X_val,
                                                                  y_val.toarray()))

lstm_train_predict = np.argmax(lstm_model.predict(X_train),axis=1)
lstm_val_predict = np.argmax(lstm_model.predict(X_val),axis=1)
lstm_test_predict = np.argmax(lstm_model.predict(X_test),axis=1)
eval("LSTM",y_train,lstm_train_predict,y_val,lstm_val_predict,y_test,lstm_test_predict)
```

```
#Bi-LSTM
bidir_model = Sequential()
bidir_model.add(base_model)
bidir_model.add(Reshape((base_model.output.shape[1]*base_model.output.shape[2],
base_model.output.shape[3])))
bidir_model.add(Bidirectional(LSTM(128, dropout=0.5,recurrent_dropout=0.5)))
bidir_model.add(Dense(3,activation='softmax'))

bidir_model.compile(optimizer='adam',loss='categorical_crossentropy',metrics=['acc'])

history = bidir_model.fit(X_train,y_train.toarray(),epochs = 20,validation_data =
(X_val,y_val.toarray()))
bidir_train_predict = np.argmax(bidir_model.predict(X_train),axis=1)
bidir_val_predict = np.argmax(bidir_model.predict(X_val),axis=1)
bidir_test_predict = np.argmax(bidir_model.predict(X_test),axis=1)
eval("Bi-dir",y_train,bidir_train_predict,y_val,bidir_val_predict,y_test,
bidir_test_predict)
```

5.4.3 Experimental data

In our study, we have used the dataset of breast ultrasound images [23]. Breast ultrasound dataset is categorized into three classes: normal, benign, and malignant images. The data collected at baseline include breast ultrasound images among women between ages 25 and 75 years old. The number of patients is 600 female patients. The dataset consists of 780 images with an average image size of 500×500 pixels. The data samples are illustrated in Fig. 5.3.

Furthermore, each image has its own ground truth (masked image) as shown in Fig. 5.4.

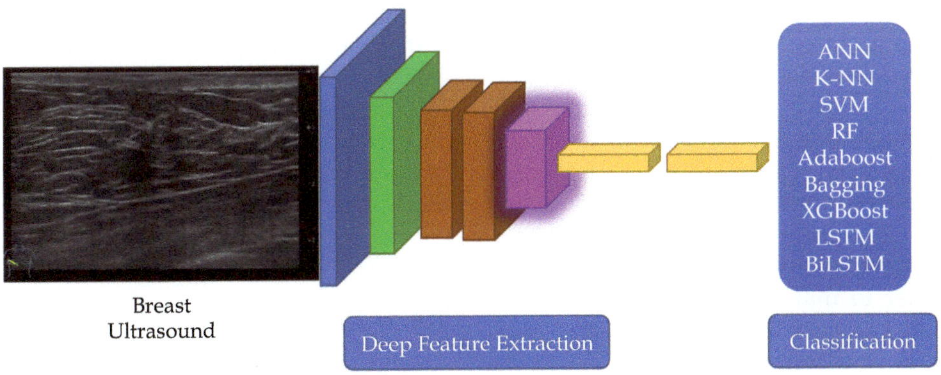

FIGURE 5.2 Classification of breast ultrasound images using deep feature extraction.

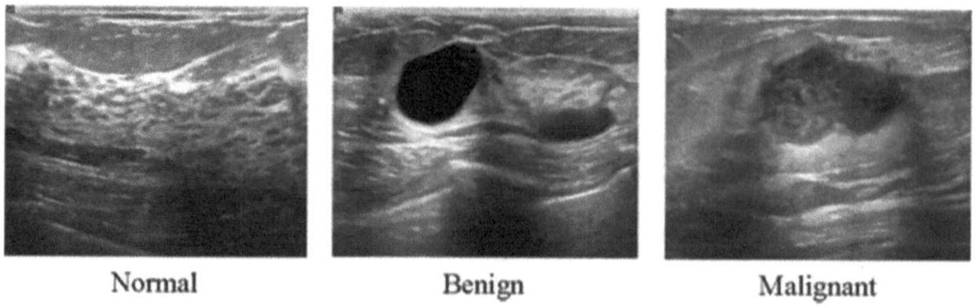

FIGURE 5.3 Data sample for each class.

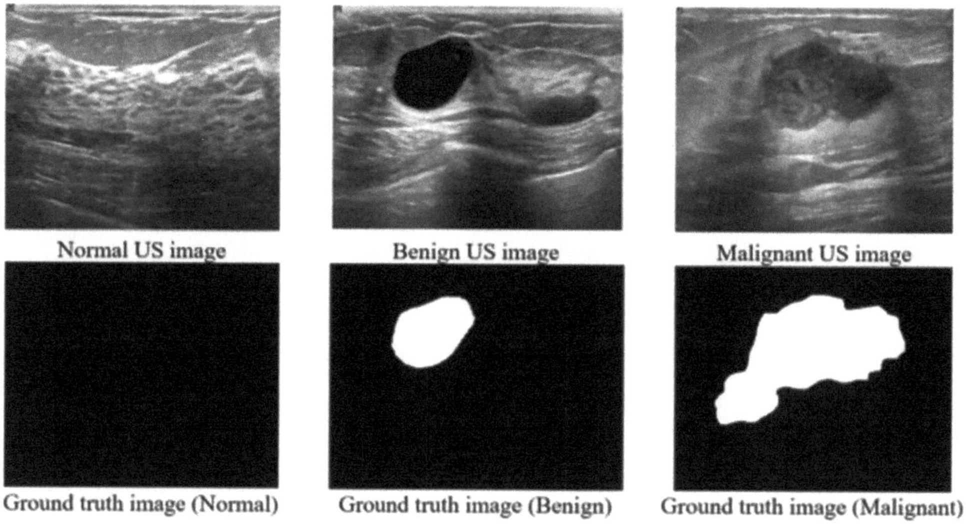

FIGURE 5.4 Ground truth (masked) image example for their respective original image.

At Baheya Hospital, grayscale ultrasound pictures were gathered and saved in a DICOM format. They were preprocessed and improved after being annotated. The number of ultrasound images was decreased to 780 once the dataset was refined. Normal, benign, and malignant images are separated into three categories (cases). To remove unnecessary and irrelevant borders from the images, they were all cropped to various sizes. MATLAB was used to perform ground truth (image boundaries) in order to make the ultrasound dataset more useful. For each image, a freehand segmentation is created independently. As a result, there is a masked image for each image, and we trained the model on both images individually [23].

Each image was changed to a grayscale image with a target shape of size $(128 \times 128 \times 1)$. The dataset consisted of {"benign": 891, "normal": 266, "malignant": 421} labels. The data was split into stratified to maintain class ratios uniformly through the process. The training set consisted of 76.5%, the validation set consisted of 8.5% and the test set consisted of 15% of the total dataset.

5.4.4 Performance evaluation measures

Evaluation metrics are used to measure the performance of the statistical model. In our experimentation work, we used different the performance metrics, such as accuracy, F1 score, Cohen's Kappa score, area under the curve (AUC), precision, and recall. These are widely used metrics to interpret the performance of the model on multiclass data.

Precision: Precision is the ratio of correctly predicted positive observations to the total predicted positive observations.
Recall (Sensitivity): Recall is the ratio of correctly predicted positive observations to all observations in the actual class.

	PREDICTED BENIGN	PREDICTED NORMAL	PREDICTED MALIGNANT
BENIGN	True Benign	False normal	False malignant
NORMAL	False Benign	True normal	False malignant
MALIGNANT	False Benign	False normal	True malignant

The above table is a sample of classification table for multiclass classification. The values in blue are correctly predicted and values in red are wrongly predicted. Depending upon this value other metrics are measured.

True Benign = Total number of observations that are benign and machine has predicted it as benign.
False Benign = Total number of observations that are normal or malignant and machine has predicted it as benign.
True normal = Total number of observations that are normal and machine has predicted it as normal
False normal = Total number of observations that are not normal and machine has predicted it as normal.
True malignant = Total number of observations that are malignant and machine has predicted it as malignant.
False malignant = Total number of observations that are not malignant and machine has predicted it as malignant.
Accuracy: It is the ratio of correctly predicted observations to the total number of observations.

F1: F1 score is the weighted average of precision and recall. Therefore this score takes both false positives and false negatives into account.

1 $Accuracy = \frac{Tp + Tn}{Tp + Tn + Fp + Fn}$
2 $Precision = \frac{Tp}{Tp + Fp}$
3 $Recall = \frac{Tp}{Tp + Tn}$
4 $F1 = 2 * \frac{Precision * Recall}{Precision + Recall}$

5.4.4.1 Cohen's Kappa coefficient

The kappa statistic is frequently used to test model reliability. The importance of model reliability lies in the fact that it represents the extent to which the data collected in the study are correct representations of the variables measured. The kappa can range from -1 to $+1$, where 0 represents the amount of agreement that can be expected from random chance, and 1 represents perfect agreement between the raters. values ≤ 0 as indicating no agreement and 0.01–0.20 as none to slight, 0.21–0.40 as fair, 0.41–0.60 as moderate, 0.61–0.80 as substantial, and 0.81–1.00 as almost perfect agreement [24].

$$\kappa = \frac{Pr(a) - Pr(e)}{1 - Pr(e)}$$

Where *Pr*(*a*) represents the actual observed agreement, and *Pr*(*e*) represents chance agreement.

5.4.4.2 Area under the curve score

The receiver operator characteristic (ROC) curve is a classification problem evaluation metric. It is a probability curve that compares the true positive rate to the false-positive rate at various threshold levels, thereby separating the "signal" from the "noise." The AUC is a summary of the ROC curve that measures a classifier's ability to discriminate between classes. The higher the AUC, the better the performance of the model at distinguishing between the positive and negative classes. When AUC = 1, the classifier is capable of successfully distinguishing between all positive and negative class points. If the AUC was 0, however, the classifier would expect all negatives to be positives and all positives to be negatives. When the AUC reaches 0.5, there is a good likelihood that the classifier will be able to tell the difference between positive and negative class values. Because the classifier can recognize more true positives and true negatives than false negatives and false positives, this is the case. The classifier is unable to discriminate between positive and negative class points when AUC = 0.5. In other words, the classifier predicts a random or constant class for all data points [25]. During classification training, many generative classifiers use accuracy as a criterion for selecting the best answer. However, the accuracy has various flaws, including reduced uniqueness, discriminability, informativeness, and bias to data from the majority class [26].

Simple Python implementation function is shown below.

```
from sklearn.metrics import f1_score, roc_auc_score, cohen_kappa_score

def evaluation(model, X_train, y_train, X_val, y_val, X_test, y_test, history):
                     train_loss, train_acc = model.evaluate(X_train, y_train.toarray())
                     val_loss, val_acc = model.evaluate(X_val, y_val.toarray())
                     test_loss_value, test_accuracy = model.evaluate(X_test,
                     y_test.toarray())

                     y_pred = model.predict(X_test)
                     y_pred_label = np.argmax(y_pred, axis=1)
                     y_true_label = np.argmax(y_test, axis=1)

                     f1_measure = f1_score(y_true_label, y_pred_label, average="weighted")
                     roc_score = roc_auc_score(y_test.toarray(), y_pred)
                     kappa_score = cohen_kappa_score(y_true_label, y_pred_label)

                     print("Train accuracy = " + str(train_acc))
                     print("Validation accuracy = " + str(val_acc))
                     print("Test accuracy = " + str(test_accuracy))
                     print("f1_measure = " + str(f1_measure))
                     print("KAPPA = " + str(kappa_score))
                     print("roc_area = " + str(roc_score))
```

5.4.5 Experimental results

The models were trained on images and masked images separately from the dataset. The following table consists of various different architectures and their respective performance metrics.

i. Multilayered CNN networks on raw image

We trained simple n-layered CNN network and model seemed to be overfit on training data but failed to generalize on test data, 7-layered CNN showed better performance compared to other networks (Table 5.1).

ii. Multilayered CNN networks on masked image

The model performed well on masked images although the model appeared to be overfitting on the training set. Also, dense layered CNN architecture performed comparatively better than shallow layered CNN (Table 5.2).

iii. Pretrained networks (transfer learning) on raw images

The dataset was fit on various pretrained architecture and significant improvement was observed in comparison to simple CNN networks. Test accuracies approximately lied in range of 70% to 79% with kappa score of 0.60 and ROC area of 0.87, ResNet50 performed best among all architectures (Table 5.3).

iv. Pretrained networks (transfer learning) on masked images

Pretrained networks fit well on masked images and test accuracies are in range of 90%–98% except for AlexNet. VGG19 architecture performed best among all other architectures. Pretrained models showed better results than multilayered CNN architecture and significant improvement was observed in kappa and F1 score for both the dataset masked and unmasked (Table 5.4).

v. Deep feature extraction with pretrained models followed by classification

VGG16 is simple arrangement of convolution and max pool layers consistently throughout the whole architecture followed by two fully connected network. Each convolution layers of 3×3 filter with a stride 1, uses same padding and max pool layer of 2×2 filter of stride 2.

Overfitting is observed in Random Forest, XGBoost, and ANN. Test and validation set accuracies were consistently in the range of 65–75 (Table 5.5).

The results on Masked images are quite impressive although Random Forest, XGBoost, and ANN were overfitting on training set. KNN and LSTM models showed impressive results with test accuracy of 98% (Table 5.6).

VGG19 architecture is same as that of VGG16 but it contains additional three more CNN layers than VGG16. ANN and XGBoost performed best with test accuracy of approx. 76% but they overfit on training data (Table 5.7).

Significant improvement is observed on training same network on masked images and SVM performing best with test accuracy of 97% (Table 5.8).

TABLE 5.1 Performance of the classifiers for multilayered CNN networks on images.

Model	Training accuracy	Validation accuracy	Test accuracy	F1 measure	Kappa	ROC area
CNN 2 Layer	0.979866	0.701493	0.649573	0.637768	0.402913	0.828568
CNN 3 Layer	0.776846	0.611940	0.666667	0.642752	0.431614	0.808549
CNN 4 Layer	0.651007	0.656716	0.606838	0.524100	0.267157	0.765958
CNN 5 Layer	0.744966	0.641791	0.683761	0.663500	0.459753	0.809080
CNN 6 Layer	0.706376	0.686567	0.658120	0.580778	0.369952	0.807845
CNN 7 Layer	0.911074	0.686567	0.726496	0.718569	0.539143	0.849139
CNN 8 Layer	0.652685	0.641791	0.555556	0.473127	0.161983	0.738786

TABLE 5.2 Performance of the classifiers for multilayered CNN networks on masked images.

Model	Training accuracy	Validation accuracy	Test accuracy	F1 measure	Kappa	ROC area
CNN 2 Layer	0.968121	0.791045	0.760684	0.751484	0.603486	0.838874
CNN 3 Layer	0.996644	0.850746	0.837607	0.835283	0.717427	0.891374
CNN 4 Layer	0.956376	0.865672	0.837607	0.838276	0.722368	0.919463
CNN 5 Layer	0.978188	0.880597	0.905983	0.907059	0.840659	0.956980
CNN 6 Layer	1.000000	0.850746	0.871795	0.872323	0.780817	0.956756
CNN 7 Layer	0.939597	0.835821	0.888889	0.891866	0.816460	0.966787
CNN 8 Layer	0.986577	0.925373	0.914530	0.915208	0.854514	0.983993

TABLE 5.3 Performance of the classifiers on pretrained networks on images.

Model	Training accuracy	Validation accuracy	Test accuracy	F1 measure	Kappa	ROC area
ResNet50	0.953020	0.716418	0.794872	0.791786	0.661483	0.895108
VGG16	0.909396	0.761194	0.743590	0.741204	0.580194	0.864333
VGG19	0.909396	0.701493	0.700855	0.697843	0.505435	0.880529
Inception_v3	0.931208	0.731343	0.726496	0.720230	0.540839	0.879365
MobileNet	0.867450	0.805970	0.726496	0.718394	0.536748	0.858062
DenseNet169	0.892617	0.731343	0.777778	0.775716	0.633803	0.879575
DenseNet121	0.917785	0.761194	0.786325	0.786257	0.655355	0.884861
InceptionResNetV2	0.932886	0.716418	0.769231	0.767404	0.619443	0.889081
MobileNetV2	0.773490	0.716418	0.709402	0.693798	0.502626	0.823654
ResNet101	0.832215	0.701493	0.709402	0.708545	0.532110	0.840034
AlexNet	0.994966	0.761194	0.752137	0.749579	0.588727	0.870327

ResNet (Residual network) is deep network with skip connection that helps in vanishing gradient problem and decrease error on higher layered model compared to conventional models. Feature extracted from ResNet50 followed by XGBoost resulted in 78% of test accuracy (Table 5.9).

ANN model on masked image resulted in 94% of accuracy with good kappa and F1 score (Table 5.10).

ResNet101 is 101-layered deep neural network with residual networks, the features extracted are trained to ML models; ANN and XGBoost performed best among all other models (Table 5.11).

Bagging model performed quiet impressively with test accuracy of 91.45% on ResNet101 feature vectors (Table 5.12).

Inverted residual blocks are introduced in this architecture which helps to reduce the computation cost and model size. SVM

TABLE 5.4 Performance of the classifiers on pretrained networks on masked images.

Model	Training accuracy	Validation accuracy	Test accuracy	F1 measure	Kappa	ROC area
ResNet50	0.971476	0.865672	0.923077	0.922888	0.868670	0.986087
VGG16	0.989933	0.940298	0.974359	0.974245	0.955777	0.998880
VGG19	0.988255	0.910448	0.982906	0.982749	0.970387	0.999677
Inception_v3	0.963087	0.880597	0.923077	0.924044	0.872022	0.992345
MobileNet	0.971476	0.910448	0.948718	0.949125	0.912708	0.990025
DenseNet169	0.988255	0.940298	0.965812	0.966083	0.941806	0.989245
DenseNet121	0.976510	0.910448	0.965812	0.965812	0.941295	0.998208
InceptionResNetV2	0.973154	0.925373	0.905983	0.907244	0.843545	0.975680
MobileNetV2	0.969799	0.910448	0.957265	0.957075	0.926295	0.995967
ResNet101	0.963087	0.925373	0.897436	0.896492	0.822323	0.972852
AlexNet	0.802013	0.776119	0.794872	0.753114	0.610108	0.972852

TABLE 5.5 Performance of the classifiers using VGG16 pretrained model for feature extraction.

Model	Training accuracy	Validation accuracy	Test accuracy	F1 measure	Kappa	Recall	Precision
SVM	0.7097	0.6866	0.6325	0.559	0.3188	0.6325	0.5513
Random Forest	1	0.7463	0.7436	0.7271	0.5512	0.7436	0.7904
AdaBoost	0.8456	0.7313	0.6752	0.6627	0.4397	0.6752	0.695
KNN	0.8339	0.6567	0.6667	0.6674	0.4654	0.6667	0.6693
XGBoost	1	0.7463	0.7692	0.7658	0.6144	0.7692	0.7744
Bagging	0.9849	0.6866	0.6667	0.6493	0.421	0.6667	0.68
ANN	1	0.7313	0.7607	0.7586	0.608	0.7607	0.7583
LSTM	0.802	0.7313	0.6923	0.6819	0.4769	0.6923	0.6955
Bi-LSTM	0.7399	0.6866	0.6838	0.6686	0.4658	0.6838	0.6849

TABLE 5.6 Performance of the classifiers using VGG16 pretrained model for feature extraction on masked images.

Model	Training accuracy	Validation accuracy	Test accuracy	F1 Measure	Kappa	Recall	Precision
SVM	0.9631	0.8955	0.9573	0.957	0.9264	0.9573	0.9575
Random Forest	1	0.9254	0.9658	0.9658	0.9413	0.9658	0.9658
AdaBoost	0.9362	0.8209	0.9573	0.9578	0.9276	0.9573	0.9599
KNN	0.9748	0.9254	0.9744	0.9742	0.9558	0.9744	0.9743
XGBoost	1	0.9403	0.9744	0.9745	0.9562	0.9744	0.9748
Bagging	0.9983	0.9104	0.9744	0.9745	0.9562	0.9744	0.9748
ANN	1	0.9403	0.9744	0.9742	0.9558	0.9744	0.9743
LSTM	0.9899	0.9552	0.9829	0.983	0.9709	0.9829	0.9839
Bi-LSTM	0.9883	0.9701	0.9744	0.9747	0.9565	0.9744	0.9766

TABLE 5.7 Performance of the classifiers using VGG19 pretrained model for feature extraction.

Model	Training accuracy	Validation accuracy	Test accuracy	F1 Measure	Kappa	Recall	Precision
SVM	0.6544	0.6866	0.5556	0.4504	0.1421	0.5556	0.5427
Random Forest	1	0.7164	0.7094	0.6929	0.4868	0.7094	0.7635
AdaBoost	0.8591	0.7015	0.7094	0.7021	0.5006	0.7094	0.7348
KNN	0.8322	0.7015	0.6838	0.6834	0.4881	0.6838	0.6832
XGBoost	1	0.7313	0.7692	0.7633	0.6093	0.7692	0.7837
Bagging	0.9899	0.7164	0.7521	0.7426	0.5706	0.7521	0.7849
ANN	1	0.7761	0.7607	0.7594	0.6076	0.7607	0.7614
LSTM	0.8054	0.6716	0.6581	0.6576	0.4452	0.6581	0.6572
Bi-LSTM	0.8322	0.6866	0.7009	0.7002	0.5146	0.7009	0.6998

TABLE 5.8 Performance of the classifiers using VGG19 pretrained model for feature extraction on masked images.

Model	Training accuracy	Validation accuracy	Test accuracy	F1 measure	Kappa	Recall	Precision
SVM	0.9715	0.9254	0.9744	0.974	0.9554	0.9744	0.9755
Random Forest	1	0.9254	0.9487	0.9491	0.9127	0.9487	0.9503
AdaBoost	0.9446	0.8657	0.9145	0.9152	0.8545	0.9145	0.9166
KNN	0.9681	0.9254	0.9744	0.9742	0.9558	0.9744	0.9743
XGBoost	1	0.9254	0.9658	0.9658	0.9413	0.9658	0.9658
Bagging	1	0.9104	0.9658	0.9658	0.9413	0.9658	0.9658
ANN	1	0.9403	1	1	1	1	1
LSTM	0.9815	1	0.9231	0.9251	0.8729	0.9231	0.9404
Bi-LSTM	0.9849	0.9851	0.9316	0.9333	0.8866	0.9316	0.9456

TABLE 5.9 Performance of the classifiers using ResNet50 pretrained model for feature extraction.

	Training accuracy	Validation accuracy	Test accuracy	F1 Measure	Kappa	Recall	Precision
Model	0.5721	0.5672	0.4957	0.3286	0	0.4957	0.2457
Random Forest	1	0.7015	0.6667	0.6511	0.413	0.6667	0.7153
AdaBoost	0.8238	0.6866	0.641	0.6282	0.3777	0.641	0.6609
KNN	0.8054	0.7313	0.6752	0.6752	0.4752	0.6752	0.6752
XGBoost	1	0.7463	0.7778	0.7671	0.6201	0.7778	0.7929
Bagging	0.9832	0.6866	0.7009	0.689	0.4859	0.7009	0.7152
ANN	0.8775	0.7761	0.735	0.7169	0.54	0.735	0.7617
LSTM	0.6409	0.6567	0.5726	0.4851	0.1988	0.5726	0.4779
Bi-LSTM	0.6426	0.6866	0.5726	0.4813	0.1872	0.5726	0.518

TABLE 5.10 Performance of the classifiers using ResNet50 pretrained model for feature extraction on masked images.

Model	Training accuracy	Validation accuracy	Test accuracy	F1 measure	Kappa	Recall	Precision
SVM	0.854	0.8358	0.9316	0.9285	0.8783	0.9316	0.939
Random Forest	1	0.8507	0.9231	0.9227	0.8673	0.9231	0.9226
AdaBoost	0.8339	0.7164	0.8205	0.8231	0.7014	0.8205	0.8318
KNN	0.9262	0.806	0.8889	0.8855	0.8052	0.8889	0.8886
XGBoost	1	0.8358	0.9145	0.9155	0.8572	0.9145	0.9215
Bagging	0.9933	0.8657	0.906	0.9063	0.8409	0.906	0.9079
ANN	1	0.8806	0.9402	0.9399	0.8968	0.9402	0.9398
LSTM	0.9211	0.8507	0.9316	0.931	0.8815	0.9316	0.9311
Bi-LSTM	0.9295	0.8507	0.9402	0.9399	0.8968	0.9402	0.9398

TABLE 5.11 Performance of the classifiers using ResNet101 pretrained model for feature extraction.

Model	Training accuracy	Validation accuracy	Test accuracy	F1 Measure	Kappa	Recall	Precision
SVM	0.5721	0.5672	0.4957	0.3286	0	0.4957	0.2457
Random Forest	1	0.6866	0.6838	0.6624	0.4454	0.6838	0.7286
AdaBoost	0.8154	0.597	0.6496	0.6373	0.4082	0.6496	0.6431
KNN	0.7718	0.7015	0.6838	0.6728	0.4693	0.6838	0.6734
XGBoost	1	0.6866	0.7607	0.7549	0.5994	0.7607	0.7627
Bagging	0.9765	0.6866	0.6325	0.6129	0.3665	0.6325	0.6355
ANN	0.9295	0.6866	0.7521	0.7469	0.5834	0.7521	0.758
LSTM	0.6258	0.6567	0.5897	0.5101	0.2418	0.5897	0.4855
Bi-LSTM	0.6275	0.6418	0.5812	0.518	0.2616	0.5812	0.4685

performed best with test accuracy if 79.5% and F1 score of 0.7871 (Table 5.13).

Feature vector extracted from MobileNetV2 for masked image shows test accuracy in the range 95.5% to 98.3% (Table 5.14).

Feature extracted from MobileNet architecture obtained test accuracies in range of 67%–80%, where ANN performance best with F1 score of 0.7992 (Table 5.15).

Feature extracted from MobileNet architecture for masked images obtained test accuracies in range of 92%–98.2%, Here again ANN performs best among all other models with F1 score of 0.9706 (Table 5.16).

Random Forest and XGBoost were overfitting on training set, the accuracies were in range of 60%–73.5%, SVM performed best among all other models with F1 score of 0.7812 (Table 5.17).

Training accuracies for Random Forest, XGBoost, and ANN were 100% and clearly it is overfitting. KNN performed best among all

TABLE 5.12 Performance of the classifiers using ResNet101 pretrained model for feature extraction on masked images.

Model	Training accuracy	Validation accuracy	Test accuracy	F1 Measure	Kappa	Recall	Precision
SVM	0.8641	0.806	0.8632	0.8561	0.7573	0.8632	0.8677
Random Forest	1	0.8209	0.906	0.9056	0.8378	0.906	0.9053
AdaBoost	0.8372	0.7313	0.8205	0.8259	0.706	0.8205	0.8478
KNN	0.9178	0.806	0.8547	0.8526	0.7472	0.8547	0.852
XGBoost	1	0.8806	0.8889	0.8891	0.8111	0.8889	0.8898
Bagging	0.9916	0.8507	0.9145	0.9128	0.8506	0.9145	0.9142
ANN	1	0.8806	0.9231	0.9227	0.8673	0.9231	0.9226
LSTM	0.9346	0.806	0.906	0.9071	0.8407	0.906	0.9096
Bi-LSTM	0.9211	0.8358	0.8803	0.8792	0.7927	0.8803	0.8788

TABLE 5.13 Performance of the classifiers using MobileNetV2 pretrained model for feature extraction.

Model	Training accuracy	Validation accuracy	Test accuracy	F1 Measure	Kappa	Recall	Precision
SVM	0.9279	0.791	0.7949	0.7871	0.6506	0.7949	0.8087
Random Forest	1	0.7612	0.6923	0.6674	0.4451	0.6923	0.7794
AdaBoost	0.8322	0.7761	0.6239	0.5883	0.3257	0.6239	0.6605
KNN	0.8154	0.6716	0.7179	0.7145	0.5413	0.7179	0.7155
XGBoost	1	0.7463	0.7949	0.7879	0.6479	0.7949	0.8227
Bagging	0.9866	0.7313	0.6838	0.6631	0.4426	0.6838	0.7214
ANN	1	0.8358	0.8034	0.7986	0.6681	0.8034	0.8144
LSTM	0.8322	0.7463	0.735	0.7259	0.5532	0.735	0.7385
Bi-LSTM	0.8674	0.7612	0.7863	0.7789	0.6392	0.7863	0.7965

TABLE 5.14 Performance of the classifiers using MobileNetV2 pretrained model for feature extraction on masked images.

Model	Training accuracy	Validation accuracy	Test accuracy	F1 measure	Kappa	Recall	Precision
SVM	0.9916	0.9104	0.9829	0.9829	0.9706	0.9829	0.9829
Random Forest	1	0.9104	0.9744	0.9742	0.9558	0.9744	0.9743
AdaBoost	0.948	0.8955	0.906	0.9076	0.842	0.906	0.9134
KNN	0.9799	0.9104	0.9829	0.9827	0.9704	0.9829	0.9834
XGBoost	1	0.9104	0.9573	0.9574	0.9269	0.9573	0.9578
Bagging	0.9983	0.9104	0.9658	0.9658	0.9413	0.9658	0.9658
ANN	1	0.9104	0.9744	0.9742	0.9558	0.9744	0.9743
LSTM	0.9883	0.9403	0.9744	0.9742	0.9558	0.9744	0.9743
Bi-LSTM	0.9983	0.9552	0.9744	0.9742	0.9558	0.9744	0.9743

TABLE 5.15 Performance of the classifiers using MobileNet pretrained model for feature extraction.

Model	Training accuracy	Validation accuracy	Test accuracy	F1 measure	Kappa	Recall	Precision
SVM	0.948	0.7463	0.7692	0.7598	0.6046	0.7692	0.7889
Random Forest	1	0.7164	0.7607	0.7417	0.5782	0.7607	0.8055
AdaBoost	0.8272	0.6567	0.6923	0.6816	0.4666	0.6923	0.7213
KNN	0.8205	0.6269	0.7094	0.7135	0.542	0.7094	0.7314
XGBoost	1	0.7463	0.7607	0.7524	0.5899	0.7607	0.7824
Bagging	0.9916	0.6567	0.6752	0.6542	0.4302	0.6752	0.6983
ANN	1	0.7463	0.8034	0.7992	0.6688	0.8034	0.8154
LSTM	0.9178	0.7463	0.7607	0.7536	0.5954	0.7607	0.7737
Bi-LSTM	0.9128	0.7164	0.7692	0.7633	0.6103	0.7692	0.7821

TABLE 5.16 Performance of the classifiers using MobileNet pretrained model for feature extraction on masked images.

Model	Training accuracy	Validation accuracy	Test accuracy	F1 measure	Kappa	Recall	Precision
SVM	0.9899	0.9104	0.9744	0.9742	0.9558	0.9744	0.9743
Random Forest	1	0.8955	0.9573	0.9574	0.9269	0.9573	0.9578
AdaBoost	0.9933	0.9254	0.9573	0.9574	0.9269	0.9573	0.9578
KNN	0.9748	0.9104	0.9658	0.9655	0.9408	0.9658	0.966
XGBoost	1	0.9403	0.9487	0.949	0.9136	0.9487	0.9516
Bagging	0.9983	0.9254	0.9231	0.9244	0.8707	0.9231	0.93
ANN	1	0.9104	0.9829	0.9829	0.9706	0.9829	0.9829
LSTM	0.9933	0.9701	0.9573	0.9578	0.9276	0.9573	0.9599
Bi-LSTM	0.9966	0.9552	0.9744	0.9747	0.9565	0.9744	0.9766

other models with test accuracy of 94.87% and F1 score of 0.9491 and kappa score 0.9127 (Table 5.18).

InceptionResNetV2 combines Inception network with residual linkage. Bi-LSTM showed best results with test accuracy 76% and 0.7574 and 0.5981 F1 score and kappa scores, respectively (Table 5.19).

Test accuracies were consistently in range 91%–95%, LSTM and Bi-LSTM performed best with test accuracy of 95.73% (Table 5.20).

Model was overfitting on Random Forest, XGBoost, and ANN model. Highest test accuracy 79.49% was observed in ANN with F1 score of 0.7897 and kappa score of 0.6558 (Table 5.21).

Approximately same performance was observed by SVM, KNN, and bagging model with highest accuracy of 97.44% (Table 5.22).

Highest test accuracy of 80.34% was obtained in XGBoost. Bi-LSTM model showed good results with test accuracy approx. 80%

TABLE 5.17 Performance of the classifiers using InceptionV3 pretrained model for feature extraction.

Model	Training accuracy	Validation accuracy	Test accuracy	F1 measure	Kappa	Recall	Precision
SVM	0.9228	0.7015	0.7265	0.7182	0.5331	0.7265	0.7486
Random Forest	1	0.6716	0.6325	0.5961	0.3411	0.6325	0.6714
AdaBoost	0.7617	0.5522	0.6068	0.5959	0.323	0.6068	0.6214
KNN	0.7752	0.5672	0.641	0.6418	0.4234	0.641	0.6431
XGBoost	1	0.6716	0.735	0.725	0.5451	0.735	0.7528
Bagging	0.9832	0.7015	0.6325	0.6131	0.3535	0.6325	0.6621
ANN	0.9983	0.6716	0.7094	0.7082	0.5194	0.7094	0.7149
LSTM	0.9178	0.7313	0.7009	0.6967	0.5007	0.7009	0.7029
Bi-LSTM	0.8674	0.7761	0.7179	0.7142	0.532	0.7179	0.7171

TABLE 5.18 Performance of the classifiers using InceptionV3 pretrained model for feature extraction on masked images.

Model	Training accuracy	Validation accuracy	Test accuracy	F1 measure	Kappa	Recall	Precision
SVM	0.9849	0.9104	0.9487	0.9491	0.9127	0.9487	0.9503
Random Forest	1	0.8806	0.9573	0.9574	0.9269	0.9573	0.9578
AdaBoost	0.8893	0.7164	0.8205	0.8263	0.7084	0.8205	0.8551
KNN	0.9681	0.8955	0.9487	0.9491	0.9127	0.9487	0.9503
XGBoost	1	0.9254	0.9402	0.9409	0.8986	0.9402	0.9431
Bagging	0.9983	0.8806	0.9145	0.9137	0.8519	0.9145	0.9137
ANN	1	0.9254	0.9487	0.9495	0.9135	0.9487	0.9531
LSTM	0.9966	0.9403	0.9487	0.9495	0.9135	0.9487	0.9531
Bi-LSTM	0.9916	0.9254	0.9316	0.9333	0.8866	0.9316	0.9456

TABLE 5.19 Performance of the classifiers using InceptionResNetV2 pretrained model for feature extraction.

Model	Training accuracy	Validation accuracy	Test accuracy	F1 measure	Kappa	Recall	Precision
SVM	0.8893	0.7313	0.7436	0.7291	0.549	0.7436	0.7941
Random Forest	1	0.6716	0.6923	0.6659	0.4481	0.6923	0.7595
AdaBoost	0.7819	0.6866	0.5983	0.5848	0.302	0.5983	0.6233
KNN	0.8188	0.6716	0.7607	0.7565	0.5985	0.7607	0.7687
XGBoost	1	0.7313	0.735	0.7275	0.5451	0.735	0.758
Bagging	0.9866	0.7313	0.6752	0.6469	0.4245	0.6752	0.7012
ANN	1	0.7761	0.7436	0.7347	0.5613	0.7436	0.7605
LSTM	0.9279	0.7612	0.7521	0.7448	0.5751	0.7521	0.7757
Bi-LSTM	0.9195	0.7612	0.7607	0.7574	0.5981	0.7607	0.77

TABLE 5.20 Performance of the classifiers using InceptionResNetV2 pretrained model for feature extraction on masked images.

Model	Training accuracy	Validation accuracy	Test accuracy	F1 measure	Kappa	Recall	Precision
SVM	0.9815	0.9254	0.9402	0.9404	0.8977	0.9402	0.9408
Random Forest	1	0.9254	0.9487	0.9495	0.9135	0.9487	0.9531
AdaBoost	0.9681	0.8806	0.9402	0.9415	0.9003	0.9402	0.9512
KNN	0.9631	0.9104	0.9145	0.9151	0.8553	0.9145	0.9168
XGBoost	1	0.9104	0.9316	0.9326	0.8853	0.9316	0.9372
Bagging	0.9983	0.9403	0.9402	0.9412	0.8995	0.9402	0.9466
ANN	1	0.9254	0.9487	0.9491	0.9127	0.9487	0.9503
LSTM	0.9966	0.9254	0.9573	0.958	0.9282	0.9573	0.9632
Bi-LSTM	0.9983	0.9254	0.9573	0.9578	0.9276	0.9573	0.9599

TABLE 5.21 Performance of the classifiers using DenseNet169 pretrained model for feature extraction.

Model	Training accuracy	Validation accuracy	Test accuracy	F1 measure	Kappa	Recall	Precision
SVM	0.8809	0.791	0.735	0.7254	0.5419	0.735	0.7635
Random Forest	1	0.7761	0.7009	0.6803	0.466	0.7009	0.7667
AdaBoost	0.8523	0.6716	0.641	0.6193	0.3626	0.641	0.6846
KNN	0.797	0.7164	0.6667	0.6612	0.4443	0.6667	0.6692
XGBoost	1	0.806	0.735	0.7218	0.5414	0.735	0.7568
Bagging	0.995	0.7313	0.6838	0.6629	0.4381	0.6838	0.7365
ANN	1	0.7313	0.7949	0.7897	0.6558	0.7949	0.8005
LSTM	0.8691	0.7612	0.6923	0.6741	0.4641	0.6923	0.7098
Bi-LSTM	0.9446	0.8358	0.7607	0.7531	0.5968	0.7607	0.7647

TABLE 5.22 Performance of the classifiers using DenseNet169 pretrained model for feature extraction on masked images.

Model	Training accuracy	Validation accuracy	Test accuracy	F1 measure	Kappa	Recall	Precision
SVM	0.9883	0.9254	0.9744	0.9745	0.9562	0.9744	0.9748
Random Forest	1	0.9104	0.9744	0.9745	0.9562	0.9744	0.9748
AdaBoost	0.9597	0.8806	0.9658	0.9655	0.9408	0.9658	0.966
KNN	0.9782	0.9104	0.9744	0.9745	0.9562	0.9744	0.9748
XGBoost	1	0.9403	0.9402	0.9409	0.8986	0.9402	0.9431
Bagging	0.9933	0.9104	0.9744	0.9742	0.9558	0.9744	0.9743
ANN	1	0.9552	0.9487	0.9495	0.9135	0.9487	0.9531
LSTM	0.9933	0.9701	0.9487	0.9495	0.9135	0.9487	0.9531
Bi-LSTM	0.9966	0.9701	0.9231	0.9248	0.8719	0.9231	0.9347

TABLE 5.23 Performance of the classifiers using DenseNet121 pretrained model for feature extraction.

Model	Training accuracy	Validation accuracy	Test accuracy	F1 measure	Kappa	Recall	Precision
SVM	0.8909	0.791	0.7521	0.7373	0.5678	0.7521	0.7854
Random Forest	1	0.7164	0.7521	0.7361	0.5655	0.7521	0.7917
AdaBoost	0.854	0.6866	0.641	0.6218	0.3712	0.641	0.6645
KNN	0.8087	0.6269	0.6496	0.6451	0.4227	0.6496	0.6458
XGBoost	1	0.7164	0.8034	0.7969	0.6637	0.8034	0.8266
Bagging	0.9933	0.6567	0.7009	0.6898	0.4851	0.7009	0.7181
ANN	1	0.7015	0.7778	0.7717	0.6271	0.7778	0.7806
LSTM	0.8893	0.7612	0.7521	0.7462	0.5819	0.7521	0.7587
Bi-LSTM	0.9581	0.7313	0.7949	0.7932	0.6626	0.7949	0.7949

and F1 score of 0.7932 and kappa score of 0.6626 (Table 5.23).

All the model performed extremely good with accuracies in range 96%–99% (Table 5.24).

Xception is a convolutional neural network that is 71-layer deep. Random Forest, XGBoost, and ANN were overfitting on training set, accuracies are in range of61%–76% with ANN performing best of all other models (Table 5.25).

Xception architecture on masked images yield test accuracies in range 93%–97%. Bi-LSTM performed best with accuracy of 97.44% and F1 score of 0.9747 and kappa score of 0.9565 (Table 5.26).

5.5 Discussion

In this chapter, we have trained various DL architectures on unmasked and masked images, and the performance of each model is quantitatively compared. The multilayered

TABLE 5.24 Performance of the classifiers using DenseNet121 pretrained model for feature extraction on masked images.

Model	Training accuracy	Validation accuracy	Test accuracy	F1 measure	Kappa	Recall	Precision
SVM	0.9832	0.9104	0.9829	0.9829	0.9706	0.9829	0.9829
Random Forest	1	0.9104	0.9915	0.9914	0.9853	0.9915	0.9916
AdaBoost	0.9513	0.791	0.9402	0.9412	0.8995	0.9402	0.9466
KNN	0.9715	0.8806	0.9744	0.9742	0.9558	0.9744	0.9743
XGBoost	1	0.9104	1	1	1	1	1
Bagging	0.9983	0.8955	0.9829	0.9827	0.9704	0.9829	0.9834
ANN	1	0.9403	0.9573	0.9578	0.9276	0.9573	0.9599
LSTM	0.9966	0.9701	0.9658	0.9663	0.9423	0.9658	0.9697
Bi-LSTM	0.9983	0.9701	0.9658	0.9663	0.9423	0.9658	0.9697

TABLE 5.25 Performance of the classifiers using Xception pretrained model for feature extraction.

Model	Training accuracy	Validation accuracy	Test accuracy	F1 measure	Kappa	Recall	Precision
SVM	0.8977	0.7313	0.6667	0.6482	0.4172	0.6667	0.689
Random Forest	1	0.7164	0.7009	0.6851	0.4742	0.7009	0.7426
AdaBoost	0.8523	0.6866	0.641	0.6036	0.3546	0.641	0.7103
KNN	0.8003	0.6866	0.6154	0.6141	0.3721	0.6154	0.6142
XGBoost	1	0.7164	0.7436	0.7356	0.5578	0.7436	0.7725
Bagging	0.9899	0.6866	0.6496	0.6225	0.3795	0.6496	0.6694
ANN	1	0.7015	0.7692	0.7657	0.6161	0.7692	0.7705
LSTM	0.807	0.6716	0.641	0.6242	0.3777	0.641	0.6509
Bi-LSTM	0.745	0.6866	0.6154	0.5693	0.2977	0.6154	0.6592

TABLE 5.26 Performance of the classifiers using Xception pretrained model for feature extraction on masked images.

Model	Training accuracy	Validation accuracy	Test accuracy	F1 measure	Kappa	Recall	Precision
SVM	0.9782	0.8955	0.9487	0.9486	0.9126	0.9487	0.9494
Random Forest	1	0.9254	0.9744	0.9742	0.9558	0.9744	0.9743
AdaBoost	0.9027	0.8507	0.8974	0.8995	0.8284	0.8974	0.9073
KNN	0.9715	0.8806	0.9316	0.9315	0.8833	0.9316	0.9316
XGBoost	1	0.9254	0.9658	0.9661	0.9418	0.9658	0.9671
Bagging	0.995	0.9104	0.9658	0.9658	0.9413	0.9658	0.9658
ANN	1	0.9254	0.9658	0.9661	0.9418	0.9658	0.9671
LSTM	0.9681	0.9254	0.9316	0.9333	0.8866	0.9316	0.9456
Bi-LSTM	0.9883	0.9552	0.9744	0.9747	0.9565	0.9744	0.9766

CNN architecture followed by the dense neural network to predict each image as benign, normal, or malignant on unmasked images showed average results with validation and test accuracies in the range of 60%–70%. Overfitting was observed on the training set due to a small dataset, so batch normalization and dropout layer were introduced. No significant improvement was observed on stacking many CNN layers. On the other hand, CNN architecture on masked images showed good results and increasing CNN layers, the model learned better and performance metrics were quite impressive.

To overcome the problem of insufficient data, pretraining models and training a classifier on top of it were implemented. For unmasked data, training accuracies were in the range of 90%–95%, while validation and test accuracies were in the range of 70%–80%. For masked data, the training accuracies were in the range of 97%–98%, while validation and test accuracies were in the range of 90%–95%. There were significant improvements on F1, kappa score, and ROC area as well.

Further ML classifiers such as SVM, AdaBoost, Random Forest, XGBoost, KNN, ANN, LSTM, and Bi-LSTM were trained on deep features extracted from pretrained models. These showed satisfactory results. Bi-LSTM and LSTM models outperformed other models.

The number of hyperparameters such as the total number of hidden layers, number of neurons in each hidden layer, number of epochs to train, learning rate, and choice of optimizer were fine-tuned on validation accuracies. This plays important role in determining the performance of the model.

5.6 Conclusion

In this chapter, we classified breast cancer as benign, malignant, or normal using ultrasound images by training various CNN models and pretrained architecture, optimizing, fine-tuning, and evaluating them with various performance metrics. The performance on masked imaged with such a small dataset was impressive and test accuracies up to 98% were observed. Training and transfer learning made it possible to give satisfactory results on unmasked data and test accuracy of 80% was achieved.

References

[1] H. Sung, et al., Global Cancer Statistics 2020: GLOBOCAN Estimates of Incidence and Mortality Worldwide for 36 Cancers in 185 Countries, CA. Cancer J. Clin. 71 (3) (2021) 209–249. Available from: https://doi.org/10.3322/caac.21660.

[2] Cancer.Net, Breast cancer - diagnosis, https://www.cancer.net/cancer-types/breast-cancer/diagnosis, Jun. 25, 2012 (accessed 26.11.21).

[3] Radiologyinfo.org., R. S. of N. A. (RSNA) and A. C. of Radiology (ACR), Ultrasound - breast, https://www.radiologyinfo.org/en/info/breastus, June 15, 2020 (accessed 26.11.21).

[4] M.H.-M. Khan, et al., Multi- class classification of breast cancer abnormalities using deep convolutional neural network (CNN), PLOS ONE 16 (8) (2021) e0256500. Available from: https://doi.org/10.1371/journal.pone.0256500.

[5] M. Talo, Automated classification of histopathology images using transfer learning, Artif. Intell. Med. 101 (2019) 101743. Available from: https://doi.org/10.1016/j.artmed.2019.101743.

[6] S.M. Shah, R.A. Khan, S. Arif, U. Sajid, Artificial intelligence for breast cancer detection: trends & directions, ArXiv211000942 Cs Eess, http://arxiv.org/abs/2110.00942, 2021 (accessed 18.12.21)

[7] S.S. Yadav, S.M. Jadhav, Deep convolutional neural network based medical image classification for disease diagnosis, J. Big Data 6 (1) (2019) 113. Available from: https://doi.org/10.1186/s40537-019-0276-2.

[8] G.E. Hinton, S. Osindero, Y.-W. Teh, A fast learning algorithm for deep belief nets, Neural Comput 18 (7) (2006) 1527–1554. Available from: https://doi.org/10.1162/neco.2006.18.7.1527.

[9] S. Bhise, S. Gadekar, A.S. Gaur, S. Bepari, D. Kale, D.S. Aswale, Breast cancer detection using machine learning techniques, Int. J. Eng. Res. Technol. 10 (7) (2021) Accessed: Dec. 18, 2021. [Online]. Available from: https://www.ijert.org/research/breast-cancer-detection-using-machine-learning-techniques-IJERTV10IS070064.pdf. Available from: https://www.ijert.org/breast-cancer-detection-using-machine-learning-techniques.

[10] A.R. Vaka, B. Soni, K. Sudheer Reddy, Breast cancer detection by leveraging machine learning, ICT Express 6 (4) (2020) 320–324. Available from: https://doi.org/10.1016/j.icte.2020.04.009.

[11] C. Muñoz-Meza, W. Gómez, "A feature selection methodology for breast ultrasound classification," in 2013 10th International Conference on Electrical Engineering, Computing Science and Automatic Control (CCE), Sep. 2013, pp. 245–249. Available from: https://doi.org/10.1109/ICEEE.2013.6676056.

[12] S. Kwok, Multiclass classification of breast cancer in whole-slide images, Image Analysis and Recognition, Cham, 2018, pp. 931–940. Available from: https://doi.org/10.1007/978-3-319-93000-8_106.

[13] W. Nawaz, S. Ahmed, A. Tahir, H.A. Khan, Classification of Breast Cancer Histology Images Using ALEXNET, SpringerLink, 2018. https://link.springer.com/chapter/10.1007/978-3-319-93000-8_99 (accessed 23.03.22).

[14] N. Dhungel, G. Carneiro, A.P. Bradley, Deep learning and structured prediction for the segmentation of mass in mammograms, Medical Image Computing and Computer-Assisted Intervention, Cham, 2015, pp. 605–612. Available from: https://doi.org/10.1007/978-3-319-24553-9_74.

[15] J.-J. Mordang, T. Janssen, A. Bria, T. Kooi, A. Gubern-Mérida, N. Karssemeijer, Automatic microcalcification detection in multi-vendor mammography using convolutional neural networks, Breast Imaging, Cham, 2016, pp. 35–42. Available from: https://doi.org/10.1007/978-3-319-41546-8_5.

[16] C.K. Ahn, C. Heo, H. Jin, J.H. Kim. A novel deep learning-based approach to high accuracy breast density estimation in digital mammography. Proc. SPIE 10134, Medical Imaging 2017: Computer-Aided Diagnosis, 101342O, March 3, 2017. Available from: https://doi.org/10.1117/12.2254264.

[17] B. Huynh, K. Drukker, M. Giger, MO-DE-207B-06: computer-aided diagnosis of breast ultrasound images using transfer learning from deep convolutional neural networks - Huynh - 2016 - Medical Physics - Wiley Online Library. <https://aapm.onlinelibrary.wiley.com/doi/abs/10.1118/1.4957255> (accessed Mar. 23, 2022).

[18] M. Kim, et al., Deep learning in medical imaging, Neurospine 16 (4) (2019) 657–668. Available from: https://doi.org/10.14245/ns.1938396.198.

[19] S.-C.B. Lo, H.-P. Chan, J.-S. Lin, H. Li, M.T. Freedman, S.K. Mun, Artificial convolution neural network for medical image pattern recognition, Neural Netw. 8 (7) (1995) 1201–1214. Available from: https://doi.org/10.1016/0893-6080(95)00061-5.

[20] W.S. Gan, Application of neural networks to the processing of medical images, in: Proc. 1991 IEEE International Joint Conference on Neural Networks, 1991, pp. 300–306 vol. 1. Available from: https://doi.org/10.1109/IJCNN.1991.170419.

[21] M.M. Mehdy, P.Y. Ng, E.F. Shair, N.I.M. Saleh, C. Gomes, Artificial neural networks in image processing for early detection of breast cancer, Comput. Math. Methods Med. 2017 (2017) e2610628. Available from: https://doi.org/10.1155/2017/2610628.

[22] ScienceDirect Topics, Feature extraction network - an overview. 2019. https://www.sciencedirect.com/topics/computer-science/feature-extraction-network (accessed 06.01.22).

[23] W. Al-Dhabyani, M. Gomaa, H. Khaled, A. Fahmy, Dataset of breast ultrasound images, Data Brief. 28 (2020) 104863. Available from: https://doi.org/10.1016/j.dib.2019.104863.

[24] M.L. McHugh, Interrater reliability: the kappa statistic, Biochem. Medica 22 (3) (Oct. 2012) 276–282.

[25] Analytics Vidhya, AUC-ROC curve in machine learning clearly explained. https://www.analyticsvidhya.com/blog/2020/06/auc-roc-curve-machine-learning/, Jun. 15, 2020 (accessed 08.01.22).

[26] M. Hossin, M.N. Sulaiman, A review on evaluation metrics for data classification evaluations, Int. J. Data Min. Knowl. Manag. Process. 5 (2015) 01–11. Available from: https://doi.org/10.5121/ijdkp.2015.5201.

CHAPTER

6

Artificial intelligence-based skin cancer diagnosis

Abdulhamit Subasi[1,2] *and Saqib Ahmed Qureshi*[3]

[1]Institute of Biomedicine, Faculty of Medicine, University of Turku, Turku, Finland

[2]Department of Computer Science, College of Engineering, Effat University, Jeddah, Saudi Arabia

[3]Indian Institute of Technology Kharagpur, Kharagpur, West Bengal, India

OUTLINE

6.1 Introduction

Skin cancer is associated with abnormal or uncontrolled growth of skin cells. Like other skin cancers, melanoma typically named as malignant melanoma is a type of skin cancer. It is less common but the most serious type of skin cancer because it spreads to other parts of the body [1]. Its seriousness depends upon how early you were aware about it and start the

Applications of Artificial Intelligence in Medical Imaging
DOI: https://doi.org/10.1016/B978-0-443-18450-5.00006-2

diagnosis [2]. Melanoma as a disease was first discovered by Rene Laennec in 1804 by the name melanose. The heterogeneous nature of some melanoma tumors was first observed by William Norris in 1820. Sir Robert Carswell introduced the term melanoma in 1838 [3]. In the year 2020 in the United States nearly 100,350 new cases of melanoma were diagnosed. It affected 60,190 men and 40,160 women. And in 2020 in the United States, an estimated 6850 deaths from melanoma were expected, which include 4610 men and 2240 women [4].

Melanoma cancer developed through the pigment producing cells known as melanocytes. It is also caused due to exposure of UV (ultraviolet rays). It is like moles in the body, have black/pink color, and can develop anywhere on the skin. In men, these are more likely to begin on the chest and back, while in case of women, it generally appears on the legs. The neck and face are other common sites where melanoma would appear. It is more likely to be found among the fair-skinned people, whereas people with dark skin have very less possibility to develop melanoma, but it can develop between the gaps of fingers, under the nails, and also sometimes inside the mouth and eyes [5]. Melanoma can affect people of all ages. It is often considered serious for older people with average age of 65 [6], while the incidence of melanoma is rapidly rising in young adults. It has been observed that melanoma is now the most common form of cancer in men and women aged 20–39 [7]. The risk of melanoma seems to be increasing under the age of 40 especially in women. It is also the most common cancers in young adults [8].

There are several types of melanoma cancer but most common are cutaneous. Other types include superficial spreading melanoma, acral lentiginous melanoma, nodular melanoma, amelanotic and desmoplastic melanomas, lentigo maligna melanoma, metastatic melanoma, and ocular melanoma [9]. First important symptom of melanoma that defines whether you are infected is a new spot on the skin (likely moles) or a spot that is changing in size, shape, or color. However, you can also determine the symptoms with the help of ABCDE rule [10]:

- Asymmetry: Shape of the mole is irregular.
- Border: It has irregular edges instead of smooth.
- Color: The mole has dark spots or uneven shading.
- Diameter: The spot is larger than the size of a pencil eraser.
- Evolving or elevation: It is continuously changing in size, shape, or texture.

Other symptoms are such as a sore that does not heal, redness or a new swelling, changes in sensation such as itchiness, pain, or tenderness, change in size of the mole, bleeding, oozing, or the development of lump or bump. Due to changing signs of cancer, it is better to take consultation of skin specialist as soon as possible to diagnose melanoma. Detection of cancer early often allows for more treatment options. Like other cancers, melanoma also has four stages. As a rule, the lower the number, the less the cancer has spread, so melanoma can be diagnosed successfully after detecting the early stage [11].

Prevention of melanoma cancer can be done through limiting your exposure to UV rays, by avoiding the use of tanning beds and sunlamp, by wearing protective clothing and putting moisturizers on skin, and by boosting the immune system, as having a weakened immune system increases not only risk of getting melanoma but also other types of skin cancer also. Melanoma treatment includes chemotherapy, radiotherapy, surgery, radiation therapy, immunotherapy, or targeted therapy, either alone or in combination. Treatment may include a combination of procedures or not depends upon how early you have detected the cancer [12].

Chemotherapy is recommended for patients who have metastatic melanoma which has spread over distant parts of the body. It is used to find the location of a tumor [13].

Immunotherapy drugs may be used as a first-line treatment for melanoma and some other cancers [14]. Several types of immunotherapies may be treated as options to treat melanoma, including cytokines and oncolytic virus therapy.

Radiation therapy may be used for patients with advanced melanoma. These therapies include external beam radiation therapy, intensity-modulated radiation therapy, and Tomo Therapy [15].

Surgery is the primary treatment for melanoma, and it can be considered as a treatment option for melanoma that has metastasized. Surgical procedures for localized melanoma include excision, reconstructive surgery, sentinel lymph node biopsy, lymph node biopsy removal, and surgery for metastatic melanoma [16].

Artificial intelligence (AI) has transformed the world in various domains and gave us new ways for solving real-life problems. Specifically, in medical science it has shown great development, as a result of which many diseases are detected at much ease and pace. Due to advancement in deep learning, reinforcement learning, natural language processing, computer vision, etc., it has become possible to detect diseases without the help of a doctor or radiologist. Especially, when the number of doctors is less and we need to detect patients who are actually suffering from disease, this could turn out to be a great helping hand. It could help us in detecting anomalous images of skin cancer with the help of computer vision. The challenge, which AI or machine learning faces today, is lack of labeled data. Therefore, for using machine learning algorithms, we need to first collect labeled data from trusted source. Although there are algorithms, which can make use of unlabeled data, called unsupervised learning, it is always good to have labeled data.

Various experiments are done to find out whether the person is suffering from skin cancer disease or not from the image of the sore using deep learning. Deep learning algorithms are so powerful such that they can even surpass the human-level performance in various tasks. It has a very significant impact on various tasks such as object detection, face recognition, classification problems, machine translation, etc. Initially in this experiment, AI algorithms were used for prediction in an end-to-end process. This is known as single-stage detection methods. In dual-stage detection algorithms or methods, features were extracted from the images using various convolutional neural networks (CNNs) and then these features were fed into machine learning classifier for classification.

Melanoma is one of the most dangerous cancers. It occurs in the cells, which produces melanin. Melanin is a pigment that gives color to the skin and the cells which produce it are known as melanocytes. It generally occurs on the skin and sometimes in eyes but very rarely it happens inside the body, that is, nose or throat. In a normal person, new and healthy skin cells grow and push the older one upward. When these older cells die, they shed off from the skin and new cells take their place. It happens in a controlled and systematic matter. But when a person is suffering from melanoma, there is an uncontrolled growth of these cells which is due to the damage in DNA of these cells. Doctors believe that this damage of DNA primarily happens because of exposure to UV radiation. But the actual reason is still not clear, as melanoma also occurs at the part of skin which is not exposed to UV radiations. Hence, it can be due to the combination of many factors, such as environmental, genetical, etc. As this disease primarily happens on the outer skin and produces moles, which results in the change of skin color, therefore these moles can be distinguished from normal moles using computer vision with the help of images.

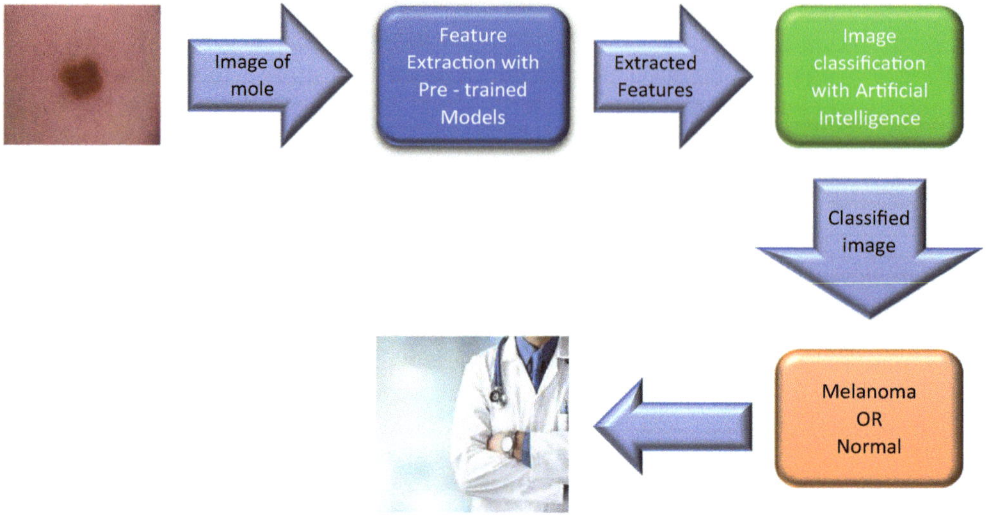

FIGURE 6.1 Skin cancer detection framework.

The main method or test for detecting melanoma skin cancer is biopsy. Biopsy is a process of removal of skin tissues for testing [17]. What type of biopsy is to be used depends on doctor and the condition of skin? It is costlier and time-consuming process as compare to deep learning techniques. Hence, it is better to use deep learning techniques first and then these biopsy testing methods can be done as subsequent steps. It will save money and time as well. As there are several Deep Learning models, they can be deployed on web or made available in the form of apps for maximizing the reach.

In this chapter, several deep learning methods are used for the detection of skin cancer using images (Fig. 6.1). Apart from classical methods, transfer learning is also used which uses pretrained models. The below image shows the pipeline of transfer learning used.

6.2 Literature review

Li and Shen [18] developed a framework using deep learning which can simultaneously produce the segmentation and coarse classification results. It consisted of two fully convolutional residual networks. The classification was done in two steps: a simple CNN was used for feature extraction and a lesion index calculation unit is developed to refine the coarse classification results by calculating the distance heat map. Accuracies of this framework showed promising results, that is, for task 1, task 2, and task 3, 0.753, 0.848, and 0.912 were obtained respectively.

Castro et al. [19] developed an accessible mobile app which contains a CNN model trained on images collected from smartphones and lesion clinical information. The dataset used for this problem was highly imbalanced. From the proposed approach, promising results were obtained with an accuracy of 92% and a recall of 94%.

Gulati and Bhogal [20] used two pretrained models AlexNet and VGG16 in two different ways. They used these pretrained models for transfer learning and feature extraction. The results showed that transfer learning was more efficient for both CNN models as compared to feature extraction method. In transfer learning method, AlexNet gave 95% of accuracy and VGG16 gave 97.5% accuracy; while in feature extraction method AlexNet gave 90% of accuracy and VGG16 gave 95% accuracy.

Adegun and Viriri [21] proposed a system that makes use of a softmax classifier for pixel-wise classification of melanoma wounds. It was a multistage and multiscale approach. They performed the experiment on two different publicly available datasets, Hospital Pedro Hispano (PH2) and International Symposium on Biomedical Imaging (ISBI) 2017. They obtained an accuracy of 95% on both the International Skin Imaging Collaboration (ISIC) 2017 dataset and PH2 datasets.

Codella [22] with his team proposed a system that uses ensemble learning which combines deep learning along with machine learning approaches and is capable of separating skin lesions, as well as analyzing the affected area. There was a significant improvement in all the performance-related characteristics. An increment of 4% in average precision, 7.5% in the area under the receiver operating characteristic (ROC) curve (AUC) was observed compared to the previous state of the art. The proposed system's accuracy was also compared with the average of eight expert dermatologists on a subset of 100 test images, it was observed that the system obtained higher accuracy (76% vs 70.5%), and specificity (62% vs 59%) evaluated at an equivalent sensitivity (82%).

Lopez et al. [23] proposed a method for solving the problem of distinguishing a dermoscopic image containing a skin lesion as malignant or benign. The developed method used transfer learning with the help of VGGNet. Experimental results were promising as it was observed that the proposed method achieves a sensitivity value of 78.66% on the ISIC archive dataset.

Astorino et al. [24] presented a multiple instance learning (MIL) approach, the objective of which, in binary case, is to differentiate between positive and negative sets of items. According to MIL technology, these sets are called bags and the items inside the bags are called instances. The hypothesis of this technology states that a bag is negative if all its instances are negative and it is positive if at least one of its instances is positive. In case of image classification, this approach fits well, since an image (bag) is basically distinguished on the basis of some of its subregions (instances). This method performed better both in accuracy and sensitivity than the other classical machine learning approaches such as SVM. The results were promising as follows: accuracy = 92.50%, sensitivity = 97.50%, and specificity = 87.50%.

Ali et al. [25] proposed an ensemble technique which includes VGG19-UNet, DeeplabV3 + , and other preprocessing methodologies. For evaluation purpose, ISIC 2018 datasets which contain 2594 dermoscopy images with their ground-truth segmentation tasks were utilized. The proposed model showed promising results on the testing dataset: accuracy = 93.6%, average Jaccard index = 0.815, and dice coefficient = 0.887.

6.3 Machine learning techniques

6.3.1 Artificial neural network

The idea of artificial neural network (ANN) comes from the neurons which transmit information from one cell to another cell. These ANNs exhibit same behavior as biological neurons and send information from one layer to another. There are various types of ANNs, among them the most common one is multilayer perceptron (MLP) which is used in this experiment as well [26]. In MLP, there are three types of layers: input, hidden, and output. Input layer takes the input in the form of vectors, hidden layer process them to extract out the features, and then finally output layer displays the final output. In this algorithm, inputs are multiplied with weights and a bias term is added to them and then it is passed through a nonlinear function. This process continues till output layer is reached [27]. During this whole process, neurons of only one layer remain active at a time; the output of one hidden layer acts as an input for the next hidden layer. As the number of hidden layer increases, the complexity of algorithm also increases. If nonlinear function is not used, then the expression in the output layer can be written as a linear combination of the inputs. This would eliminate the purpose of all hidden layers and the whole system will act as a single-layer network. Also, this nonlinearity function makes the model more complex and helps it learn more complicated features. The bias term is added to generalize the model and make it more flexible. The weights are updated by using an algorithm named as backpropagation algorithm [28]. During this process, differentials are taken with respect to weights and then they are multiplied in subsequent steps. This gives rise to two major problems: exploding gradient and vanishing gradient problem. In exploding gradient problem, sometimes the differentials take very big values, which leads to overflow and makes the gradient "NaN," that is, not a number. While in vanishing gradient problem, the differentials take very small values which tend toward zero after subsequent multiplication; it results in no update of weights. Sample Python code for ANN is given below.

```
import tensorflow as tf
model_ANN = tf.keras.models.Sequential([
    tf.keras.Input(shape=(last_layer_shape[1],last_layer_shape[2],last_layer_shape[3])),
    tf.keras.layers.Flatten(),
    tf.keras.layers.Dense(32, activation = 'relu'),
    tf.keras.layers.Dense(1, activation = 'sigmoid')
])

LEARNING_RATE = 1e-4
OPTIMIZER = RMSprop(lr=LEARNING_RATE,decay=1e-2)
LOSS = 'binary_crossentropy'
METRICS = [
    'accuracy',
    'AUC'
]

model_ANN.compile(
    loss=LOSS,
    metrics=METRICS,
    optimizer=OPTIMIZER,
)

print("fit model on gpu")
history_ANN = model_ANN.fit(
    train_features, train_labels,
    epochs=EPOCHS,
    verbose=VERBOSE_LEVEL,
    validation_data=(val_features,val_labels)
)

        y_pred_ANN = model_ANN.predict(X_test)
```

6.3.2 k-nearest neighbor

k-nearest neighbor (k-NN) is one of the easiest and straightforward machine learning algorithms. It can be used for both regression and classification. It does not build a model unlike other machine learning algorithms; it does not have any trainable parameters. For every new test sample, it computes distances between this test sample and all training samples. Among all these distances, it chooses the "k" nearest training samples and then checks which class has maximum elements in the "k" closest set; it labels the test sample with the class having the maximum elements in the "k" closest set. The value of "k" is chosen empirically, it shouldn't be too large or too small. The selection of distance function is very important; it depends on the application. The disadvantage is that for every new sample, we have to compute distances with every training sample present in the dataset. Thus it is very expensive computationally. Also, some preprocessing such as normalization should be done to ensure that algorithm does not favor any single distance just because it is closer specifically [29]. Sample Python code for k-NN is given below.

```
from sklearn.neighbors import KNeighborsClassifier
classifier_kNN = KNeighborsClassifier(n_neighbors = 5, algorithm='ball_tree', leaf_size=30)
classifier_kNN.fit(X_train, y_train)

    test_acc_kNN = classifier_kNN.score(X_test, y_test)

    y_pred_kNN = classifier_kNN.predict(X_test)
```

6.3.3 Support vector machine

Support vector machine (SVM) [30] is a bunch of supervised learning algorithms which is used for classification, regression, and anomaly detection. It takes all the training points and produces a decision boundary which is called as maximum margin hyperplane which separates different classes. This hyperplane is the most optimal one as it maximizes the distance between two nearest points of different classes. In case of two-dimensional data and linear classifier, this hyperplane becomes a straight line which divides the space into two halves belonging to two different classes. When data is nonlinear, it cannot be separated using a straight line, then to solve this problem, SVM uses different kernels which transforms the data into higher dimensional space and makes a nonlinear decision boundary. Compared to neural networks they have better speed and also, they do not have vanishing and exploding gradient problems. Sample Python code for SVM is given below.

```
from sklearn.svm import SVC
classifier_SVM = SVC(kernel = 'rbf', random_state = 0)
classifier_SVM.fit(X_train, y_train)

test_acc_SVM = classifier_SVM.score(X_test, y_test)

    y_pred_SVM = classifier_SVM.predict(X_test)
```

6.3.4 Random Forest

Random Forest [31] is one of the mostly used ensemble learning algorithm which produces promising results most of the times. It is used at many places due to its simplicity and diversity. It can be used for both regression and classification problems. The word "forest" means collection of trees; here trees refer to decision tree classifiers. Hence, it makes a forest of decision tree classifiers. They follow "Bagging" approach which says that the collection of learning algorithms increases the overall performance of the model. Decision trees are very interpretable and straightforwardly deterministic. For a given feature set, they always produce the same regression or classification model structure. To add some amount of randomness or fuzziness, Random Forest classifier is used. There can be some cases where one decision tree classifier is overfitting, while the other one is highly biased; they cancel errors by supporting each other and as a whole become more robust model. During model building, whole dataset is not taken at once; instead, some randomly selected data points were given to individual decision tree classifier. Another important function of Random Forest is that it can be used to determine the features which are more important. At first, it creates the model on the whole dataset and calculates the score. Then it shuffles one of the features randomly and calculates the score on the remaining features. By observing the magnitude of increment or decrement of the score, the importance of feature can be easily determined. Sample Python code for Random Forest is given below.

```
from sklearn.ensemble import RandomForestClassifier
classifier_RF = RandomForestClassifier(n_estimators = 800, criterion = 'entropy', random_state=0)
classifier_RF.fit(X_train,y_train)

test_acc_RF = classifier_RF.score(X_test, y_test)

y_pred_RF = classifier_RF.predict(X_test)
```

6.3.5 XGBoost

XGBoost [32] stands for extreme gradient boosting. It is an ensemble machine learning algorithm which uses Decision Tree classifier as a basic model and then builds the whole model using "Boosting" approach. In this technique, new decision tree models are created to rectify the mistakes made by existing models. These models are created sequentially on top of each other following an iterative approach. This process continues till stopping criteria is achieved. Gradient descent algorithm is used to minimize the loss while making new models, therefore it is called gradient boosting. Execution speed of XGBoost is very fast as it uses parallelization to make use of all cores of CPU. During preprocessing of data, if there exist missing values or variables need to be one-hot encoded it makes the data sparse. XGBoost has an inbuilt algorithm which handles different types of patterns in sparse data. In order to tackle the problem of overfitting, it uses L1 and L2 regularization. Sample Python code for XGBoost is given below.

```
import xgboost as xgb
classifier_xgb = xgb.XGBClassifier(n_estimators = 300)
classifier_xgb.fit(X_train, y_train)

    test_acc_xgb = classifier_xgb.score(X_test, y_test)

    y_pred_xgb = classifier_xgb.predict(X_test)
```

6.3.6 AdaBoost

AdaBoost [33] stands for Adaptive Boosting. It is an ensemble machine learning algorithm which follows "Boosting" technique. It assigns certain weights to the training samples. The choice of base learner depends upon whether it accepts training data with weights or not; in many cases decision tree classifier is used as base learner. It builds model iteratively on top of one another and reassign weights in each iteration. It gives higher weights to the misclassified examples so that in next iteration these examples are given more priority. This process continues until there is no error and all the examples are classified properly, or maximum number of base estimators is achieved. Sample Python code for AdaBoost is given below.

```
from sklearn.ensemble import AdaBoostClassifier
classifier_AdaBoost = AdaBoostClassifier(n_estimators = 100)
classifier_AdaBoost.fit(X_train, y_train)

test_acc_AdaBoost = classifier_AdaBoost.score(X_test, y_test)

y_pred_AdaBoost = classifier_AdaBoost.predict(X_test)
```

6.3.7 Bagging

Bagging [34] classifier is based upon ensemble machine learning algorithm. This algorithm fits base classifiers on a random choice of the portion of original training data. This random allotment of training data is independent of each other. Sometimes because of this random allotment, some portion of data is used repeatedly while some portion is left out. There can be some base learners which were having the problem of overfitting while some of the learners dealing with underfitting. Base learners try to cancel errors of one another and give the final output by aggregating the predictions from all the learners either by voting or averaging. Sample Python code for Bagging is given below.

```
from sklearn.ensemble import BaggingClassifier
classifier_Bagging = BaggingClassifier(n_estimators=100)
classifier_Bagging.fit(X_train,y_train)

test_acc_Bagging = classifier_Bagging.score(X_test, y_test)

y_pred_Bagging = classifier_Bagging.predict(X_test)
```

6.3.8 Long short-term memory

LSTM [35] stands for long short-term memory. Recurrent neural network (RNN) is a type of neural network that has an internal memory to store the previous states. Unlike other neural networks, inputs are not independent to each other because RNN makes decision by considering both input and the output of the previous states. This property of RNN makes them applicable for many tasks such as speech recognition, music generation, etc. There is a problem which arises while using RNN architecture. During backpropagation in time, if the sequence is very large then vanishing and exploding gradient problem come into the picture. If one of derivative takes very large value, then it may cause overflow problem in subsequent derivatives. On the other hand, if a derivative takes very small value, then the value of subsequent derivatives in time become negligible; information does not transfer properly and the effect of previous time states on the current state becomes negligible. Exploding gradient problem can be solved by clipping them in the desired range. To solve the problem of vanishing gradients, LSTM architecture is developed, which makes it easier to remember the previous hidden states. They have the ability to control which information they want to remember and which one to forget. It has three gates, which control the flow of information: input gate, forget gate, and output gate. Input gate decides which value from the input should be used to update the memory. Forget gate decides which information to be discarded or which should be retained. And finally, output gate decides the flow of output information by making use of input and memory of the cell. Sample Python code for LSTM is given below.

```
import tensorflow as tf
model_LSTM = tf.keras.models.Sequential([
    tf.keras.Input(shape=(last_layer_shape[1], last_layer_shape[2]*last_layer_shape[3])),
    tf.keras.layers.LSTM(100, return_sequences=True),
    tf.keras.layers.LSTM(32),
    tf.keras.layers.Dense(1, activation = 'sigmoid')
])

LEARNING_RATE = 1e-4
OPTIMIZER = Adam(lr=LEARNING_RATE,decay=1e-2)
LOSS = 'binary_crossentropy'
METRICS = [
    'accuracy',
    'AUC'
]

model_LSTM.compile(
    loss=LOSS,
    metrics=METRICS,
    optimizer=OPTIMIZER,
)

print("fit model on gpu")
history_LSTM = model_LSTM.fit(
    train_features_2d, train_labels,
    epochs=EPOCHS,
    verbose=VERBOSE_LEVEL,
    validation_data=(val_features_2d, val_labels)
)

y_pred_LSTM = model_LSTM.predict(test_features_2d)
```

6.3.9 Bidirectional long short-term memory

Bidirectional LSTM (Bi-LSTM) [36] is the extended version of traditional LSTMs. This structure considers information not only from past states but also from future states. It enables the current state to take decision by considering both past and present information. It uses two LSTMs and makes the inputs to run in both forward and backward direction. This makes model more robust and efficient as for taking the decision both future and past hidden states are utilized. All other properties of Bi-LSTM are similar to unidirectional LSTM. Sample Python code for Bi-LSTM is given below.

```
model_Bi_LSTM.compile(
    loss=LOSS,
    metrics=METRICS,
    optimizer=OPTIMIZER,
)

print("fit model on gpu")
history_Bi_LSTM = model_Bi_LSTM.fit(
    train_features_2d, train_labels,
    epochs=EPOCHS,
    verbose=VERBOSE_LEVEL,
    validation_data=(val_features_2d, val_labels)
)

y_pred_Bi_LSTM = model_Bi_LSTM.predict(test_features_2d)
```

```
import tensorflow as tf
model_Bi_LSTM = tf.keras.models.Sequential([
    tf.keras.Input(shape=(last_layer_shape[1], last_layer_shape[2]*last_layer_shape[3])),
    tf.keras.layers.Bidirectional(tf.keras.layers.LSTM(100, return_sequences=True)),
    tf.keras.layers.Bidirectional(tf.keras.layers.LSTM(100)),
    tf.keras.layers.Dense(1, activation = 'sigmoid')
])

LEARNING_RATE = 1e-4
OPTIMIZER = Adam(lr=LEARNING_RATE,decay=1e-2)
LOSS = 'binary_crossentropy'
METRICS = [
    'accuracy',
    'AUC'
]
```

6.3.10 Convolutional neural network

CNN [37] is one of the special types of deep neural networks which are specialized in analyzing visual imagery. The term "convolution" comes from the linear operation in mathematics which is performed between matrices. CNN has performed exceptionally well in computer vision, natural language processing, voice recognition, etc. The features should not be spatially dependent; it is very important assumption, which is considered for all problems solved by CNNs. CNNs consist of filters in the form of matrix which extract different features from the image. For example, if a boundary detection filter is passed through the image, it extracts all the boundaries present in the image. It actually creates an activation map of that particular feature so that it can be easily seen which part of the image is activated or what features are extracted by this filter. A CNN architecture has many layers: convolutional layer, pooling layer, nonlinearity layer, and fully connected layer. In convolutional layer, instead of looking for the full image, local regions are focused so that trainable parameters can be reduced. For example, if an image has resolution $32 \times 32 \times 3$ raw pixels are used as an input. To connect this input layer with one neuron only $32 \times 32 \times 3$ weights, that is, 3072 weights are needed. Now if one more neuron is added, it becomes $32 \times 32 \times 3 \times 2$, that is, more than 6000 weights. For such a small resolution image only, there are so many weights need to be trained. Hence, it is more fruitful and practical approach to look for the local regions in an image rather than full image. Parameters or weights can be decreased more rapidly by increasing the strides. In convolutional layer, there is a problem of loss of information on borders of the image, which can be easily overcome by using zero padding. Pooling layer is used to downsample the image for reducing the complexity for further layers. Also, it does not contain any trainable parameters or weights. Nonlinearity layer is used to adjust or saturate the output. The most common function which is used for nonlinearity is Rectified Linear Unit also known as "ReLU"; other functions are "sigmoid" and "tanh." After the processing of image is done by going through required number of layers, global pooling is done and the matrix is flattened into a vector to extract out final features. These features are fed into the fully connected layer. The fully connected layer is the same as traditional neural networks. It is used to give the final output from the model by using "sigmoid" or "softmax" function depending upon the number of classes. Sample Python code for CNN is given below.

```
from tensorflow.keras.models import Sequential
from keras.layers import Conv2D, MaxPooling2D, Flatten, Dense, Dropout
from keras import optimizers

def create_model():
    print("create model")

    model = Sequential()

    model.add(Conv2D(16, kernel_size=(3, 3), activation='relu',
                 input_shape=INPUT_SHAPE))
    model.add(MaxPooling2D(pool_size=(2, 2)))

    model.add(Conv2D(32, kernel_size=(3, 3), activation='relu'))
    model.add(MaxPooling2D(pool_size=(2, 2)))

    model.add(Conv2D(64, kernel_size=(3, 3), activation='relu'))
    model.add(MaxPooling2D(pool_size=(2, 2)))

    model.add(Conv2D(128, kernel_size=(3, 3), activation='relu'))
    model.add(MaxPooling2D(pool_size=(2, 2)))

    model.add(Conv2D(128, kernel_size=(3, 3), activation='relu'))
    model.add(MaxPooling2D(pool_size=(2, 2)))

    model.add(Conv2D(256, kernel_size=(3, 3), activation='relu'))
    model.add(MaxPooling2D(pool_size=(2, 2)))

    model.add(Flatten())
    model.add(Dense(512, activation='relu'))
    model.add(Dropout(0.2))
    model.add(Dense(512, activation='relu'))
    model.add(Dropout(0.2))
    model.add(Dense(1, activation='sigmoid'))

    return model

LEARNING_RATE = 1e-4
OPTIMIZER = Adam(lr=LEARNING_RATE,clipvalue=0.45)
LOSS = 'binary_crossentropy'
METRICS = [
    'accuracy',
    'AUC'
]

model.compile(
    loss=LOSS,
    metrics=METRICS,
    optimizer=OPTIMIZER,
)

print("fit model on gpu")
history = model.fit(
    train_gen,
    epochs=EPOCHS,
    verbose=VERBOSE_LEVEL,
    validation_data=(valX, valY),
)

y_p = model.predict_generator(test_gen)
```

6.3.11 Transfer learning

Transfer learning [38] is a method in AI which uses the learning or feature extraction done in one task for solving the problem in another task. The model made in one task behaves as a starting point in another task. It tries to store the knowledge gained in a task, so that again training from scratch is not required for similar kind of tasks; saved model can be reused. In deep learning tasks, this technique is widely used as promising results are obtained from it. As it is known that in CNN architecture different layers learn different features; the layers present in the initial stage learn very basic features; for example, edge detection or contour detection as compared to layers present in the later stage. Hence, these basic features such as edge detection and contour detection can be used in the similar tasks using transfer learning, without doing training from scratch in the other task. For example, if weights are saved for anomaly detection task previously, then these weights can be used again in detecting a benign tumor or melanoma. Also, when there is lack of data, a big CNN architecture cannot be trained from scratch. In these cases, use of transfer learning becomes the best option. The weights of many trained networks are available on internet in various frameworks. The CNN architectures used in this chapter are VGG16, VGG19, ResNet, ResNet50, InceptionV3, InceptionResNetV2, DenseNet121, DenseNet169, MobileNet, MobileNetV2, and Xception. Sample Python code for transfer learning (MobileNet) is given below.

```
from keras.applications.mobilenet import MobileNet
from tensorflow.keras import layers
from tensorflow.keras.models import Sequential

def load_pretrained_model():
    base_model = MobileNet(
        input_shape=INPUT_SHAPE,
        include_top=False,
        weights='imagenet'
    )

    # freeze the first 75 layers of the base model. All other layers are trainable.
    for layer in base_model.layers[0:75]:
        layer.trainable = False

    return base_model


def create_model():
    print("create model")

    model = Sequential()

    model.add(load_pretrained_model())
    model.add(layers.Flatten())

    model.add(layers.Dense(128, activation='relu'))
    model.add(layers.Dropout(0.3))

    model.add(layers.Dense(64, activation='relu'))
    model.add(layers.Dropout(0.3))

    model.add(layers.Dense(32, activation='relu'))

    model.add(layers.Dense(1, activation='sigmoid'))

    return model


model = create_model()
```

6.4 Results and discussions

6.4.1 Dataset

The dataset used in this experiment is taken from Kaggle website [39]. It was generated and available by the ISIC and images are from the following sources: Memorial Sloan Kettering Cancer Center, Melanoma Institute Australia, Hospital Clinic de Barcelona, Medical University of Vienna, The University of Athens Medical School, and the University of Queensland. The dataset was highly imbalanced as the number of melanoma cases was only 575, while on the other hand the number of benign cases was 31,956. For balancing the dataset, equal number of benign and melanoma cases was taken, that is, 575. The height and the width of images varied a lot. To make all the images have uniform size, they were converted to the size 224 × 224.

6.4.2 Experimental setup

In the first stage, CNNs were used for classification of benign and melanoma images. A total of 11 different pretrained models (VGG16, VGG19, MobileNet, MobileNetV2, InceptionV3, InceptionResNetV2, DenseNet121, DenseNet169, Xception, ResNet, and ResNet50) with some fine-tunings were used for classification and their different accuracy parameters were noted down. After this step, different convolutional layer models were built ranging from two to eight convolutional layers. This whole process completes the first stage of the experiment.

In the second stage, features of the image were extracted from different pretrained models (mentioned above). These features were passed through global average pooling layer and a classifier was put on the top of it. Several machine learning techniques were used for final classification such as XGBoost, Random Forest, SVM, k-NN, AdaBoost, ANN, LSTM, Bi-LSTM, and Bagging classifier.

6.4.3 Performance metrics

The dataset was initially imbalanced, for making it balanced equal number of positive and negative cases was taken. It was then further divided into three sets, that is, train, validation, and test set and accuracies were recorded separately for them. Training was done on training set, hyperparameter tuning was done on validation set and finally model was tested on test set to see how the model is generalized. After these steps, F1 score was calculated on test set as it includes both precision and recall which gives us better idea whether the model is biased for one particular class or not. Finally, AUC is calculated for test set to check the model robustness further.

```
from sklearn import metrics
import numpy as np
def print_performance_metrics(test_labels,predict):
    print('Accuracy:', np.round(metrics.accuracy_score(test_labels, predict),4))
    print('ROC Area:', np.round(metrics.roc_auc_score(test_labels, predict),4))
    print('Precision:', np.round(metrics.precision_score(test_labels, predict,
                                                  average='weighted'),4))
    print('Recall:', np.round(metrics.recall_score(test_labels, predict,
                                                  average='weighted'),4))
    print('F1 Score:', np.round(metrics.f1_score(test_labels, predict,
                                                  average='weighted'),4))
    print('Cohen Kappa Score:', np.round(metrics.cohen_kappa_score(test_labels, predict),4))
    print('Matthews Corrcoef:', np.round(metrics.matthews_corrcoef(test_labels, predict),4))
    print('\t\tClassification Report:\n', metrics.classification_report(test_labels, predict))

print_performance_metrics(y_test,y_pred)
```

6.4.4 Experimental results

Different pretrained architectures were used for classification which contains pretrained weights. Apart from these pretrained architectures, several CNNs with different number of layers were also used.

From Table 6.1, it can be observed that in pretrained models "MobileNet" architecture performed the best with an F1 score of 0.8014. On the other hand, "ResNet50" performed the worst with an F1 score of 0.371. Among the CNN layers, the model with eight layers performed the best with an F1 score of 0.7675, while the model with four layers performed the worst with an F1 score of 0.657.

From Table 6.2 to Table 6.12, deep feature extraction techniques were used. Pretrained model architectures were used to extract the features and then several machine learning classifiers were put on top of that to give the final output.

In Table 6.2, VGG16 architecture was used for feature extraction. It can be clearly seen that "LSTM" classifier performed the best with an F1 score of 0.802. The worst performance was given by the "k-NN" classifier with an F1 score of 0.6446.

TABLE 6.1 Pretrained and convolutional neural network (CNN) models.

Model	Training accuracy	Validation accuracy	Test accuracy	F1 score	ROC area
VGG16	0.79	0.7969	0.7647	0.7603	0.7654
VGG19	0.7747	0.7344	0.7594	0.7593	0.7593
MobileNet	0.8359	0.7812	0.8021	0.8014	0.8018
MobileNetV2	0.7941	0.8438	0.7754	0.7752	0.7756
InceptionV3	0.8451	0.8438	0.7647	0.7644	0.7645
InceptionResNetV2	0.8889	0.8281	0.7914	0.7913	0.7913
DenseNet121	0.844	0.9062	0.754	0.7531	0.7543
DenseNet169	0.8756	0.8438	0.7594	0.7591	0.7596
Xception	0.8328	0.7344	0.6845	0.6828	0.6841
ResNet	0.5015	0.6719	0.5936	0.5911	0.5932
ResNet50	0.5413	0.4375	0.5187	0.371	0.5161
CNN 2 Layer	0.7574	0.75	0.7326	0.7316	0.733
CNN 3 Layer	0.7472	0.7656	0.7594	0.7587	0.7591
CNN 4 Layer	0.7584	0.7344	0.6578	0.657	0.6575
CNN 5 Layer	0.7299	0.7344	0.7433	0.7431	0.7435
CNN 6 Layer	0.7666	0.7812	0.7273	0.7272	0.7274
CNN 7 Layer	0.7248	0.7344	0.7273	0.7266	0.727
CNN 8 Layer	0.7798	0.7812	0.7701	0.7675	0.7695

TABLE 6.2 VGG16 architecture was used for feature extraction.

Model	Training accuracy	Validation accuracy	Test accuracy	F1 score	ROC area
SVM	0.9195	0.7991	0.7701	0.7693	0.7704
Random Forest	1	0.829	0.7807	0.7803	0.781
AdaBoost	1	0.7778	0.7166	0.7153	0.7169
K-NN	0.7543	0.6624	0.6471	0.6446	0.6475
XGBoost	1	0.8205	0.7701	0.77	0.77
Bagging	1	0.7991	0.7433	0.7427	0.7436
ANN	1	0.8034	0.7487	0.7481	0.7489
LSTM	1	0.8419	0.8021	0.802	0.8023
Bi-LSTM	1	0.8333	0.7861	0.7859	0.7863

TABLE 6.3 VGG19 architecture was used for feature extraction.

Model	Training accuracy	Validation accuracy	Test accuracy	F1 score	ROC area
SVM	0.9256	0.7735	0.738	0.738	0.7379
Random Forest	1	0.7991	0.7433	0.7428	0.7431
AdaBoost	0.999	0.7607	0.7005	0.7005	0.7006
K-NN	0.7533	0.6538	0.6203	0.607	0.6213
XGBoost	1	0.8077	0.7487	0.7477	0.7483
Bagging	1	0.829	0.7861	0.7861	0.7861
ANN	1	0.7863	0.7273	0.7272	0.7272
LSTM	1	0.8077	0.7487	0.7469	0.7482
Bi-LSTM	1	0.8077	0.754	0.7525	0.7536

In Table 6.3, VGG19 architecture was used for feature extraction. Among the classifiers, "Bagging" classifier achieved the highest F1 score, that is, 0.7861, while "k-NN" classifier achieved the lowest F1 score, that is, 0.607.

In Table 6.4, MobileNet architecture was used for feature extraction. The classifiers which achieved the highest and the lowest F1 score were "XGBoost" with a score of 0.818 and "k-NN" with a score of 0.6608, respectively.

In Table 6.5, MobileNetV2 architecture was used for feature extraction. Among the classifiers, "XGBoost" performed the best with an F1 score of 0.8128; on the other hand, "k-NN" performed poorly with an F1 score of 0.6575.

TABLE 6.4 MobileNet architecture was used for feature extraction.

Model	Training accuracy	Validation accuracy	Test accuracy	F1 score	ROC area
SVM	0.949	0.8376	0.8075	0.8074	0.8074
Random Forest	1	0.829	0.7807	0.7807	0.7808
AdaBoost	1	0.8034	0.7487	0.7486	0.7486
K-NN	0.738	0.7008	0.6791	0.6608	0.6804
XGBoost	1	0.859	0.8182	0.818	0.818
Bagging	1	0.8376	0.7914	0.7914	0.7914
ANN	1	0.8547	0.8182	0.818	0.818
LSTM	1	0.8333	0.7861	0.7854	0.7858
Bi-LSTM	1	0.8547	0.8128	0.8124	0.8126

TABLE 6.5 MobileNetV2 architecture was used for feature extraction.

Model	Training accuracy	Validation accuracy	Test accuracy	F1 score	ROC area
SVM	0.9511	0.812	0.7701	0.7697	0.7698
Random Forest	1	0.8419	0.7914	0.7914	0.7915
AdaBoost	1	0.812	0.754	0.7538	0.7542
K-NN	0.7727	0.6752	0.6684	0.6575	0.6694
XGBoost	1	0.859	0.8128	0.8128	0.8129
Bagging	1	0.812	0.754	0.7539	0.7541
ANN	1	0.8291	0.7701	0.7701	0.7701
LSTM	1	0.812	0.7487	0.7484	0.7485
Bi-LSTM	1	0.8291	0.7701	0.7695	0.7698

In Table 6.6, InceptionV3 architecture was used for feature extraction. From the table, it can be observed that "Bi-LSTM" performed the best with an F1 score of 0.7693, while "k-NN" performed the worst with an F1 score of 0.6763.

In Table 6.7, InceptionResNetV2 architecture was used for feature extraction. Among the classifiers, "ANN" classifier performed the best with an F1 score of 0.7857, while "k-NN" performed the worst with an F1 score of 0.6819.

TABLE 6.6 InceptionV3 architecture was used for feature extraction.

Model	Training accuracy	Validation accuracy	Test accuracy	F1 score	ROC area
SVM	0.946	0.7906	0.7487	0.7484	0.7489
Random Forest	1	0.7949	0.738	0.738	0.738
AdaBoost	1	0.7906	0.738	0.7378	0.7381
K-NN	0.788	0.7094	0.6791	0.6763	0.6796
XGBoost	1	0.7521	0.6845	0.6842	0.6843
Bagging	1	0.7906	0.7326	0.7325	0.7325
ANN	1	0.812	0.7594	0.7586	0.7597
LSTM	1	0.7906	0.7326	0.7322	0.7328
Bi-LSTM	1	0.8205	0.7701	0.7693	0.7704

TABLE 6.7 InceptionResNetV2 architecture was used for feature extraction.

Model	Training accuracy	Validation accuracy	Test accuracy	F1 score	ROC area
SVM	0.9266	0.7863	0.7326	0.7326	0.7326
Random Forest	1	0.8205	0.7594	0.7593	0.7593
AdaBoost	1	0.7906	0.7273	0.7272	0.7272
K-NN	0.7788	0.7179	0.6898	0.6819	0.6907
XGBoost	1	0.812	0.7487	0.748	0.7484
Bagging	1	0.8291	0.7701	0.7695	0.7698
ANN	1	0.8419	0.7861	0.7857	0.7859
LSTM	1	0.8205	0.7594	0.7593	0.7593
Bi-LSTM	1	0.8077	0.738	0.7377	0.7378

In Table 6.8, DenseNet121 architecture was used for feature extraction. The classifiers with the highest and the lowest F1 score were "ANN" with a score of 0.7968 and "k-NN" with a score of 0.6416, respectively.

In Table 6.9, DenseNet169 architecture was used for feature extraction. "Bi-LSTM" classifier achieved the highest F1 score, that is, 0.8075, while "AdaBoost" classifier achieved the lowest F1 score, that is, 0.6898.

In Table 6.10, Xception architecture was used for feature extraction. "Random Forest" classifier achieved the highest F1 score, that is, 0.7366, while "Bagging" classifier achieved the lowest F1 score, that is, 0.6969.

TABLE 6.8 DenseNet121 architecture was used for feature extraction.

Model	Training accuracy	Validation accuracy	Test accuracy	F1 score	ROC area
SVM	0.9368	0.8077	0.7807	0.7807	0.7807
Random Forest	1	0.8248	0.7754	0.7751	0.7752
AdaBoost	1	0.7906	0.7326	0.7326	0.7326
K-NN	0.7808	0.6838	0.6524	0.6416	0.6533
XGBoost	1	0.812	0.7594	0.7587	0.7591
Bagging	1	0.7991	0.7487	0.7486	0.7486
ANN	1	0.8504	0.7968	0.7968	0.7967
LSTM	0.999	0.8248	0.7754	0.7754	0.7755
Bi-LSTM	1	0.8419	0.7861	0.7861	0.7861

TABLE 6.9 DenseNet169 architecture was used for feature extraction.

Model	Training accuracy	Validation accuracy	Test accuracy	F1 score	ROC area
SVM	0.9246	0.7949	0.7701	0.77	0.77
Random Forest	1	0.8333	0.7914	0.7914	0.7914
AdaBoost	1	0.7564	0.6898	0.6898	0.6899
K-NN	0.8114	0.7222	0.7059	0.7036	0.7064
XGBoost	1	0.8205	0.7754	0.7754	0.7755
Bagging	1	0.8248	0.7807	0.7807	0.7807
ANN	1	0.8376	0.7968	0.7968	0.7968
LSTM	1	0.8162	0.7701	0.7699	0.7699
Bi-LSTM	1	0.8462	0.8075	0.8075	0.8075

In Table 6.11, ResNet architecture was used for feature extraction. The classifiers with the highest and the lowest F1 score were "Bagging" with a score of 0.8021 and "k-NN" with a score of 0.6636, respectively.

In Table 6.12, ResNet50 architecture was used for feature extraction. From the table, it can be observed that "Random Forest" performed the best with an F1 score of 0.7967, while "k-NN" performed the worst with an F1 score of 0.6769.

6.4.5 Discussion

From the results obtained, it is observed that both end-to-end learning and deep feature extraction technique is effective in classification

TABLE 6.10 Xception architecture was used for feature extraction.

Model	Training accuracy	Validation accuracy	Test accuracy	F1 score	ROC area
SVM	0.9195	0.7607	0.7273	0.7266	0.727
Random Forest	1	0.7949	0.738	0.7366	0.7376
AdaBoost	1	0.7778	0.7112	0.7099	0.7109
K-NN	0.8043	0.7051	0.7059	0.7055	0.7061
XGBoost	1	0.7906	0.7273	0.7254	0.7268
Bagging	1	0.7692	0.7005	0.6969	0.7
ANN	1	0.7906	0.7326	0.7325	0.7325
LSTM	1	0.7778	0.7166	0.7166	0.7165
Bi-LSTM	1	0.7735	0.7112	0.7112	0.7112

TABLE 6.11 ResNet architecture was used for feature extraction.

Model	Training accuracy	Validation accuracy	Test accuracy	F1 score	ROC area
SVM	0.9521	0.8333	0.7914	0.7914	0.7914
Random Forest	1	0.8248	0.7754	0.7751	0.7752
AdaBoost	1	0.8205	0.7647	0.7644	0.7645
K-NN	0.7982	0.7009	0.6738	0.6636	0.6747
XGBoost	1	0.8205	0.7701	0.77	0.77
Bagging	1	0.8419	0.8021	0.8021	0.802
ANN	1	0.8205	0.7701	0.7698	0.7702
LSTM	1	0.8376	0.7914	0.7914	0.7915
Bi-LSTM	1	0.8376	0.7914	0.7913	0.7916

of melanoma and benign cases. Especially, MobileNet architecture performed very well in both end-to-end learning and deep feature extraction technique. In end-to-end learning, MobileNet architecture achieved the highest F1 score of 0.8014. In feature extraction technique, the model in which features were extracted from MobileNet architecture and XGBoost was used as a classifier achieved the highest F1 score of 0.818. Other classifiers such as LSTM, Random Forest, and ANN also performed well. Features extracted from VGG16, ResNet, MobileNetV2, and DenseNet169 also gave promising results and secured the F1 score of more than 80.

TABLE 6.12 ResNet50 architecture was used for feature extraction.

Model	Training accuracy	Validation accuracy	Test accuracy	F1 score	ROC area
SVM	0.9358	0.8333	0.7914	0.7914	0.7915
Random Forest	1	0.8419	0.7968	0.7967	0.7967
AdaBoost	0.999	0.7949	0.738	0.7377	0.7382
K-NN	0.789	0.7222	0.6898	0.6769	0.6909
XGBoost	1	0.8333	0.7861	0.7861	0.786
Bagging	1	0.8162	0.7647	0.7642	0.7645
ANN	1	0.8419	0.7968	0.7966	0.797
LSTM	1	0.8333	0.7861	0.7861	0.786
Bi-LSTM	1	0.8376	0.7914	0.7914	0.7914

6.5 Conclusion

Skin cancer is one of the most hazardous and common diseases which world is facing right now. Therefore proper research and actions need to be taken for reducing the spread of this disease. AI has proven its importance in every field, especially in medical science. To reduce its spread, the first step and most important step is to detect it. Many times, people ignore it and hesitate in going to the doctor because they consider it as a very rare thing. This ignorance results in great disaster and the situation becomes out of control. If some kind of app or website can be made which can deploy these machine learning models, then people will hesitate less in checking this disease. More number of cases can be detected on time and people can be cured on time. Hence, these kinds of research in medical science are very crucial.

References

[1] Wikipedia, Melanoma, <https://en.wikipedia.org/w/index.php?title = Melanoma&oldid = 1002726498>, Jan. 25, 2021 (accessed 03.02.21).

[2] The Skin Cancer Foundation, How dangerous is melanoma? It's all a matter of timing <https://www.skincancer.org/blog/dangerous-melanoma-matter-timing/>, Oct. 27, 2017 (accessed 03.02.21).

[3] V.C. Gorantla, J.M. Kirkwood, State of melanoma: an historic overview of a field in transition, Hematol. Oncol. Clin. North. Am. 28 (3) (2014) 415–435. Available from: https://doi.org/10.1016/j.hoc.2014.02.010.

[4] Melanoma Research Alliance, 2020. Melanoma mortality rates decreasing despite ongoing increase in incidence, < https://www.curemelanoma.org/blog/article/2020-melanoma-mortality-rates-decreasing-despite-ongoing-increase-in-incidence-rates > (accessed 03.02.21).

[5] Mayo Clinic, Skin cancer - symptoms and causes, < https://www.mayoclinic.org/diseases-conditions/skin-cancer/symptoms-causes/syc-20377605 > (accessed 03.02.2021).

[6] SEER, Melanoma of the skin - cancer stat facts, <https://seer.cancer.gov/statfacts/html/melan.html> (accessed 03.02.21).

[7] AIM at Melanoma Foundation, Age and risk, <https://www.aimatmelanoma.org/melanoma-101/understanding-melanoma/melanoma-risk-factors/age-and-risk/> (accessed 03.02.21).

[8] Cancer young in adults. <https://www.cancer.org/cancer/cancer-in-young-adults.html> (accessed 03.02.21).

[9] Cancer Treatment Centers of America, Types of melanoma: common, rare and more varieties. <https://www.cancercenter.com/cancer-types/melanoma/types>, Oct. 05, 2018 (accessed 03.02.21).

[10] Cancer Treatment Centers of America, What are the symptoms and signs of melanoma? <https://www.cancercenter.com/cancer-types/melanoma/symptoms>, Oct. 05, 2018 (accessed 03.02.21).

[11] Stages of melanoma skin cancer. <https://www.cancer.org/cancer/melanoma-skin-cancer/detection-diagnosis-staging/melanoma-skin-cancer-stages.html> (accessed 03.02.21).

[12] Cancer Treatment Centers of America, Melanoma treatment options & advanced therapies, <https://www.cancercenter.com/cancer-types/melanoma/treatments>, Oct. 05, 2018 (accessed 03.02.21).

[13] CTCA, Chemotherapy: personalized therapies to treat cancer. <https://www.cancercenter.com/treatment-options/chemotherapy> (accessed 03.02.21).

[14] Cancer Treatment Centers of America, Immunotherapy to treat cancer: options & side effects, Oct. 17, 2018. https://www.cancercenter.com/treatment-options/precision-medicine/immunotherapy (accessed Feb. 03, 2021).

[15] Cancer Treatment Centers of America, Radiation therapy: usages, side effects & more, Oct. 17, 2018. <https://www.cancercenter.com/treatment-options/radiation-therapy>, (accessed 03.02. 21).

[16] Cancer Treatment Centers of America, What is cancer surgery? | Options & side effects, <https://www.cancercenter.com/treatment-options/surgery>, Oct. 17, 2018 (accessed 03.02.21).

[17] Mayo Clinic, Melanoma - symptoms and causes, <https://www.mayoclinic.org/diseases-conditions/melanoma/symptoms-causes/syc-20374884> (accessed 28.02.21).

[18] Y. Li, L. Shen, Skin lesion analysis towards melanoma detection using deep learning network, Sensors 18 (2) (2018). Available from: https://doi.org/10.3390/s18020556. Art. no. 2.

[19] P.B.C. Castro, B. Krohling, A.G.C. Pacheco, R.A. Krohling, "An app to detect melanoma using deep learning: an approach to handle imbalanced data based on evolutionary algorithms," in: 2020 International Joint Conference on Neural Networks (IJCNN), Jul. 2020, pp. 1–6. Available from: https://doi.org/10.1109/IJCNN48605.2020.9207552.

[20] S. Gulati and R. K. Bhogal, "Detection of malignant melanoma using deep learning," in: International Conference on Advances in Computing and Data Science, Singapore, 2019, pp. 312–325. Available from: https://doi.org/10.1007/978981-13-9939-8_28.

[21] A.A. Adegun, S. Viriri, Deep learning-based system for automatic melanoma detection, IEEE Access. 8 (2020) 7160–7172. Available from: https://doi.org/10.1109/ACCESS.2019.2962812.

[22] N.C. Codella, et al., Deep learning ensembles for melanoma recognition in dermoscopy images, IBM J. Res. Dev. 61 (4/5) (2017). 5–1.

[23] A.R. Lopez, X. Giro-i-Nieto, J. Burdick, O. Marques, Skin lesion classification from dermoscopic images using deep learning techniques, in: 2017 13th IASTED International Conference on Biomedical Engineering (BioMed), Feb. 2017, pp. 49–54. Available from: https://doi.org/10.2316/P.2017.852-053.

[24] A. Astorino, A. Fuduli, P. Veltri, E. Vocaturo, Melanoma detection by means of multiple instance learning, Interdiscip. Sci. Comput. Life Sci 12 (1) (2020) 24–31. Available from: https://doi.org/10.1007/s12539-019-00341-y.

[25] R. Ali, R.C. Hardie, B.N. Narayanan, S.D. Silva, Deep learning ensemble methods for skin lesion analysis towards melanoma detection, in: 2019 IEEE National Aerospace and Electronics Conference (NAECON), Jul. 2019, pp. 311–316. Available from: http://10.1109/NAECON46414.2019.9058245.

[26] I. Yilmaz, N. Erik, O. Kaynar, Different types of learning algorithms of artificial neural network (ANN) models for prediction of gross calorific value (GCV) of coals, Sci. Res. Essays 5 (2010) 2242–2249.

[27] S.-C. Wang, Artificial neural network, in: S.-C. Wang (Ed.), Interdisciplinary Computing in Java Programming, Springer US, Boston, MA, 2003, pp. 81–100. Available from: https://doi.org/10.1007/978-1-4615-0377-4_5.

[28] V. Skorpil, J. Stastny, Neural networks and back propagation algorithm, Sep. 2006.

[29] G. Guo, H. Wang, D. Bell, Y. Bi, "KNN model-based approach in classification," Aug. 2004.

[30] T. Evgeniou, M. Pontil, Support vector machines: theory and applications, Advanced Course on Artificial Intelligence, vol. 2049, Springer, 2001, pp. 249–257. Available from: http://doi.org/10.1007/3-540-44673-7_12.

[31] J. Ali, R. Khan, N. Ahmad, I. Maqsood, Random Forests and Decision Trees, Int. J. Computer Sci. Issues (IJCSI) 9 (2012).

[32] T. Chen, C. Guestrin, XGBoost: a scalable tree boosting system, Aug. 2016, pp. 785–794. Available from: https://doi.org/10.1145/2939672.2939785.

[33] T. Chengsheng, L. Huacheng, X. Bing, AdaBoost typical algorithm and its application research, MATEC Web of Conferences, 139, p. 00222, Jan. 2017, Available from: https://doi.org/10.1051/matecconf/201713900222.

[34] P. Bühlmann, B. Yu, Analyzing bagging, Ann. Stat. 30 (4) (2002) 927–961. Available from: https://doi.org/10.1214/aos/1031689014.

[35] S. Hochreiter, J. Schmidhuber, Long short-term memory, Neural Comput. 9 (1997) 1735–1780. Available from: https://doi.org/10.1162/neco.1997.9.8.1735.

[36] A. Graves, S. Fernández, J. Schmidhuber, Bidirectional LSTM Networks for Improved Phoneme Classification and Recognition, International Conference on Artificial Neural Networks, Jan. 2005, pp. 799–804.

[37] S. Albawi, T.A. Mohammed, S. Al-Zawi, Understanding of a convolutional neural network, in: 2017 International Conference on Engineering and Technology (ICET), Aug. 2017, pp. 1–6. Available from: https://doi.org/10.1109/ICEngTechnol.2017.8308186.

[38] M. Hussain, J. Bird, D. Faria, A study on CNN transfer learning for image classification, UKCI 2018: 18th Annual UK Workshop on Computational Intelligence, Jun. 2018.

[39] Kaggle, SIIM-ISIC melanoma classification. <https://kaggle.com/c/siim-isic-melanoma-classification> (accessed 07.03.21).

CHAPTER

7

Brain stroke detection from computed tomography images using deep learning algorithms

Aykut Diker[1], Abdullah Elen[1] and Abdulhamit Subasi[2,3]

[1]Department of Software Engineering, Faculty of Engineering and Natural Sciences, Bandirma Onyedi Eylul University, Bandirma, Balikesir, Turkey [2]Institute of Biomedicine, Faculty of Medicine, University of Turku, Turku, Finland [3]Department of Computer Science, College of Engineering, Effat University, Jeddah, Saudi Arabia

OUTLINE

7.1 Introduction

The World Health Organization reported that in 2016, brain stroke is the second leading cause of mortality worldwide. Hereby, the generality of survivors are compelled to live with a constant or long-term injury [1,2]. It is a medical disorder in which the brain tissues lose their capability to get oxygen as a result of a diminished or completely cut blood supply. The loss of brain cells occurs quickly as a result of this. Ischemic and hemorrhagic strokes are the two forms of stroke. Ischemic stroke is the most prevalent form of stroke, and it occurs

DOI: https://doi.org/10.1016/B978-0-443-18450-5.00013-X

when the blood supply to the brain tissues is decreased; other stroke is hemorrhagic, and it occurs when a vessel inside the brain ruptures. Stroke, with the simplest definition, is a "brain attack" caused by cessation of blood flow. Although stroke is used synonymously with the term hemiplegia, it is colloquially known as paralysis. Its distribution is age-related and doubles every 10 years after the age of 55. It ranks third among the causes of death in the population over 65 years of age, after heart diseases and cancer. Since stroke occurs as a sudden event, this event that happens to a person unexpectedly is considered as a crisis situation [3–5]. MRI images contain important information for the classification of stroke severity. On the other hand, it is very difficult to interpret the scan results as small changes in these images are spots that indicate the severity of the stroke. This complexity corresponds to a significant amount of time spent manually analyzing images. The mental efforts required to determine the severity of a stroke create exhaustion, and fatigue can contribute to human error, which affects diagnostic quality. On the outside of fatigue, all human-based classification approaches have interobserver and intraobserver variability. Clinical analyst education is a technique that can help to mitigate some of these errors. However, educating someone to be an expert takes time and money, making regular stroke risk prediction activities uneconomical. Researchers have developed methods for automatically detecting the severity of stroke disease to overcome these issues [6]. Normal and stroke brain computed tomography (CT) images are given in Fig. 7.1.

Besides, scientists and engineers have frequently used machine learning (ML) and artificial intelligence methods for the detection and classification of stroke. Al-Qazzaz et al. [7] suggested an autonomous machine interface structure to determine rehabilitation changes and present a new method based on BCI. The EEG samples from poststroke subjects with upper extremity hemiparesis were examined. Additionally, Random Forest (RF), Support vector machine (SVM), and k-NN classifiers were used to classify EEG signals. Sung et al. [8] proposed an ML program that could detect subjects with suspected stroke during emergency department triage. The application can be integrated into an electronic triage system and used to initiate code strokes. To develop stroke classification models, researchers investigated SVM, RF, k-NN, C4.5, CART, and logistic regression (LR). In the experimental study, the accuracy values of C4.5, CART, RF, k-NN, SVM, and LR classifiers were obtained 81.2%, 81.1%, 82.6%, 80.6%, 81.3%, and 82.0%, respectively. Ref. [9] presents a new deep learning-based technique for segmenting stroke lesions in CT perfusion maps. The suggested approach was tested using the ISLES 2018 dataset. The positive predictive value and sensitivity (SEN) value of the proposed method were obtained as 68% and 67%, respectively. In Ref. [10] a method generated from the

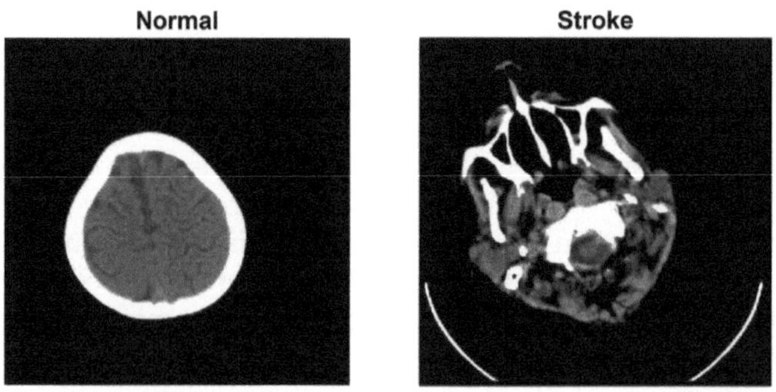

FIGURE 7.1 Normal and stroke brain CT image samples. *CT*, Computed tomography.

primary frequencies of CT samples of the brain is used to categorize hemorrhagic and ischemic strokes. The major benefit of this method is that it only makes use of the image as a parameter. The results mentioned a high-value rate with SPE 99.1%, SEN 97%, and ACC 98% in stroke classification.

In this chapter, deep learning models are employed for stroke classification using brain CT images. Moreover, the Brain Stroke CT Image Dataset was used for stroke classification. The chapter is arranged as follows: studies in brain stroke detection are detailed in Part 2. The deep learning techniques used in the chapter are described in Part 3. The experimental results are reported in Part 4. In Parts 5, discussion and concluding remarks are given, respectively.

7.2 Literature survey in brain stroke detection

In recent years, many studies have been carried out on stroke detection in the field of computer science. Studies on this subject are generally on the classification of CT and MR images through deep learning and ML algorithms. Bento et al. [11] proposed a model to identify subjects with carotid artery atherosclerosis from a cohort of magnetic resonance brain samples based on computer aid. They used the SVM classifier after identifying a set of handcrafted and convolutional discriminant features to apply the automatic identification task. According to their results, they stated that it has an accuracy rate of 97.5%, a sensitivity of 96.4%, and a specificity of 97.9%. Filho et al. [12] developed a method for extracting features based on radiological density models of the brain, called Brain Tissue Density Analysis. They applied this method to CT images to classify and identify the occurrence of stroke diseases. They applied multi-layer perceptron, optimum path forest, SVM, k-NN, and Bayesian classifiers to classify CT images. They reported that the mean accuracy values for all classifiers were above 95%. Vargas et al. [13] proposed a model to predict the presence and laterality of perfusion deficiency in patients with acute ischemic stroke. In the proposed model, a long-term recurrent convolutional network is built, consisting of a convolutional neural network (CNN) stacked on top of the long short-term memory layer. They reported that the best accuracy score in the validation data in experimental studies was 85.8%. Gautam and Raman [14] proposed a new classification model that uses image fusion and deep learning to define cerebral palsy. They preprocessed the CT images used as input data with quadtree-based image fusion method and used fusion technique to increase the contrast of the stroke district. Next, they proposed a new CNN structure for the classification of cerebral palsy from CT images into three and two (ischemic and hemorrhagic) categories. The classification accuracy of their proposed method is between 98.33% and 98.77%. Kanchana and Menaka [15] proposed a new histogram-based model for the detection of ischemic stroke, using optimal feature group selection to classify abnormal and normal districts on CT images and segment the ischemic stroke lesion. They used SVM, LR, RF, and neural network (NN) as classification methods. In experimental studies, they reported that they reached classification accuracy ranging from 88.77% to 99.79% in the detection of ischemic stroke lesion. Raghavendra et al. [16] proposed a probabilistic NN model using nonlinear features for early detection of intracerebral hemorrhage on CT images. The model they developed reported the best accuracy rate of 97.37% in distinguishing between normal and hemorrhagic subjects. Herzog et al. [17] proposed a CNN model that incorporates Bayesian uncertainty into the analysis procedure to diagnose ischemic stroke patients based on MR images. In a cohort of 511 patients, their proposed model was improved by 2% over a non-Bayesian counterpart, reporting 95.33% accuracy at the image level. Badriyah et al. [18] employed

RF, Naive Bayes, k-NN, LR, Decision Tree, SVM, multilayer perceptron neural network, and deep learning, to classify stroke into its two subtypes (ischemic and hemorrhage). According to their experimental results, they reported that the RF algorithm gave the best classification score with 95.97% accuracy.

7.3 Deep learning methods

In this chapter, a model based on the CNN has been proposed to classify it as stroke and normal, which consists of two classes in total. The proposed CNN structure is given in Fig. 7.2.

7.3.1 AlexNet

Even though it is stated that Yann LeCun employed deep learning for the first time in an article published in 1998, it was first heard worldwide in 2012. The AlexNet model, designed with a deep learning architecture, won the ImageNet competition held that year. The study was announced with the paper named "ImageNet Classification with Deep Convolutional Networks." The computerized object recognition error rate was lowered from 26.2% to 15.4% using this deep learning model. The architecture is consisting of five convolution layers, a pooling layer, and three fully connected layers (FCL) given in Fig. 7.3. The first convolutional layer filters the input image

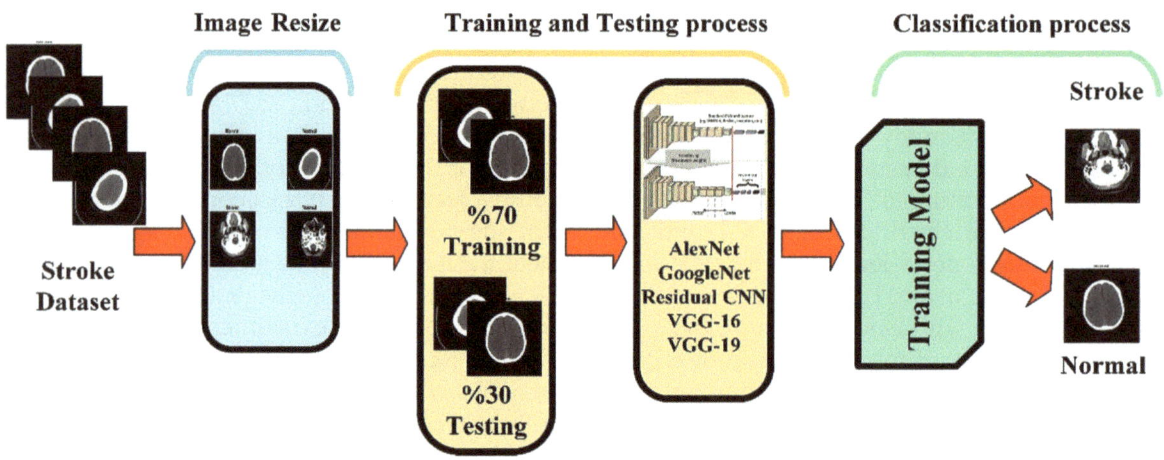

FIGURE 7.2 Block diagram of the brain stroke detection architecture.

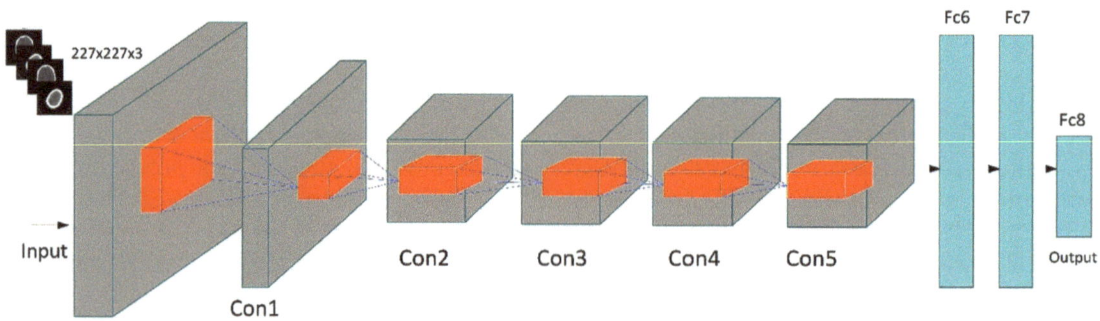

FIGURE 7.3 AlexNet architecture.

(227 × 227 × 3) with 96 cores of 11 × 11 × 3 size. The second convolutional layer filters the output of the first convolutional layer, which is locally normalized and pooled with 256 cores of size 5 × 5 × 96. In the third convolutional layer, there is no local response normalization and no pooling layer. There are 384 cores with a size of 3 × 3 × 256. In the fifth convolutional layer, the output will be pooled. There are 256 cores with the size of 3 × 3 × 384. Each of the FCL contains

TABLE 7.1 Layers of AlexNet architecture.

Layer	#filters/neurons	Filter size	Stride	Padding	Size of feature map	Activation function
Input	–	–	–	–	227 × 227 × 3	–
Conv 1	96	11 × 11	4	–	55 × 55 × 96	ReLU
Max Pool 1	–	3 × 3	2	–	27 × 7 × 96	–
Conv 2	256	5 × 5	1	2	27 × 27 × 256	ReLU
Max Pool 2	–	3 × 3	2	–	13 × 13 × 256	–
Conv 3	384	3 × 3	1	1	13 × 13 × 384	ReLU
Conv 4	384	3 × 3	1	1	13 × 13 × 384	ReLU
Conv 5	256	3 × 3	1	1	13 × 13 × 256	ReLU
Max Pool 3	–	3 × 3	2	–	6 × 6 × 256	–
Dropout 1	Rate = 0.5	–	–	–	6 × 6 × 256	–

```
%% Load AlexNet
%Pretrained model "AlexNet "
net = alexnet();
%% AlexNet training options
options = trainingOptions('sgdm', ...
    'MiniBatchSize',32, ...
    'MaxEpochs',100, ...
    'Momentum',0.8, ...
    'InitialLearnRate',1e-4, ...
    'L2Regularization',1e-6,...
    'Shuffle','every-epoch', ...
    'Verbose',true);
[net,info] = trainNetwork(imdsTrain,layers,options);
%% Alexnet Classification
[YPred,scores] = classify(net,imdsTest);% Testing for accuracy
testLabels = imdsTest. Labels;
trainLabels = imdsTrain. Labels;
Accuracy_AlexNet = mean(YPred == testLabels)*100;% Alexnet Classification result
```

4096 neurons. Additionally, AlexNet is designed to classify 1000 objects [19,20].

The filters are 11 × 11 in size and the number of step shifts is 4 as shown in Table 7.1

7.3.2 GoogleNet

Because of the early modules in its construction, GoogleNet is a CNN type with an intricate design. It contains 22 layers with a 5.7% error rate. Besides, this architecture, a deep CNN, obtained a high classification performance in competition. It is frequently considered as one of the primary CNN structures to refrain from stacking convolutional and accumulated layers sequentially. It also has a significant situation for stacking all layers and incorporating assorted filters will involve time-consuming computations [21,22]. Block diagram of the GoogleNet is shown in Fig. 7.4.

```
%% Load GoogleNet
%Pretrained model "GoogleNet "
net = googlenet();
%% GoogleNet
options = trainingOptions('sgdm', ...
    'MiniBatchSize',32, ...
    'MaxEpochs',100, ...
    'Momentum',0.8, ...
    'InitialLearnRate',1e-4, ...
    'L2Regularization',1e-6,...
    'Shuffle','every-epoch', ...
    'Verbose',true);
[net,info] = trainNetwork(imdsTrain,layers,options);
%% GoogleNet Classification
[YPred,scores] = classify(net,imdsTest);% Testing for accuracy
testLabels = imdsTest. Labels;
trainLabels = imdsTrain. Labels;
Accuracy_GoogleNet = mean(YPred == testLabels)*100;% GoogleNet Classification result
```

7.3.3 Residual convolutional neural network

Residual Network, known as ResNet in short in the literature, is a type of NN introduced by He et al. [23], in 2015 and titled "Deep Residual Learning for Image Recognition." To solve a mostly complex problem, some additional layers have been added resulting in improved accuracy and performance in deep neural networks. The intuition behind adding more layers is that those layers learn more and more complex features. Thus, in the condition of recognizing samples, the first layer can find out to determine edges, the second layer can find out to determine textures, and likewise, the third layer can find out to determine objects. However, He et al. experimentally showed that there is the highest threshold for depth with the conventional deep learning structure. Among the various methods clarifying why deeper networks do not outperform their shallow counterparts, it is occasionally preferable to seek empirical results for clarification and work backward from there. Residual block as shown in Fig. 7.5 is a new NN layer where the challenge of training very deep networks was decreased.

The most important thing to understand from the figure shown above is the identity mapping "skip connection." There are no parameters in this identity mapping, and it's solely used to enhance the last from the former layer to the following layer. But from time to time x and $F(x)$ will not have the equal dimension. Recall that a convolution process generally decreases the spatial resolution of a sample, for example, 3×3 convolution on a 32×32 image consequences in a 30×30 image. The identity mapping is multiplied by a linear projection W to extend the shortcut channels to couple the residual. This consents the input x and F(x) to be gathered as input to the next layer as shown in Eq. (7.1).

```
%% Load Residual Leaky CNN
%Pretrained model "Residual Leaky CNN"
netWidth = 8;
layers = [
    imageInputLayer ([224 224 3],'Name','input')
    convolution2dLayer(3,netWidth,'Padding','same','Name','convInp')
    batchNormalizationLayer('Name','BNInp')
    leakyReluLayer('Name','reluInp')

%% Residual Leaky CNN training options
options = trainingOptions ('sgdm', ...
    'MiniBatchSize',32, ...
    'MaxEpochs',100, ...
    'Momentum',0.8, ...
    'InitialLearnRate',1e-4, ...
    'L2Regularization',1e-6,...
    'Shuffle','every-epoch', ...
    'Verbose', true);
[net_Residual, info] = trainNetwork (imdsTrain, lgraph, options);

%% Residual Leaky CNN classification
[YPred, scores] = classify (net_Residual, imdsTest);
testLabels = imdsTest.Labels;
Accuracy = mean(YPred == testLabels)*100;% Residual Leaky CNN classification result
```

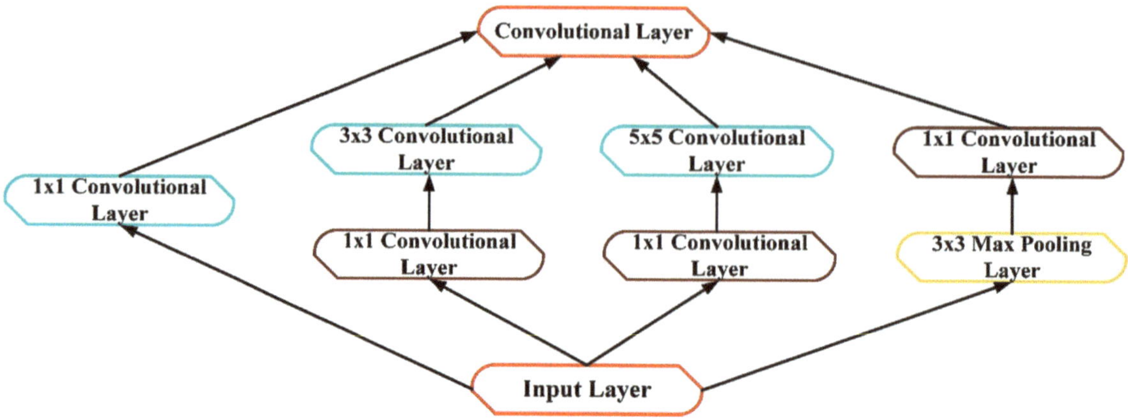

FIGURE 7.4 GoogleNet architecture.

$$y = F(x, \{W_i\}) + W_s x \quad (7.1)$$

where $F(x)$ and the equation used when x has a different size such as 32×32 and 30×30. Here, the term W_s can be applied with 1×1 convolutions.

7.3.4 VGG-16

VGG-16 is a CNN architecture with 16 layers that uses a size 3×3 filter for all convolution layers, making it the smallest size filter. The architecture is fed RGB images at a resolution of 224×224 size. A set of convolutional layers are used to extract the image's features. The convolution stride is 1. After each convolution operation, the spatial padding of the

```
%% Load VGG-16
%Pretrained model "VGG-16 "
net = vgg16();
%% VGG-16 training options
options = trainingOptions('sgdm', ...
    'MiniBatchSize',32, ...
    'MaxEpochs',100, ...
    'Momentum',0.8, ...
    'InitialLearnRate',1e-4, ...
    'L2Regularization',1e-6,...
    'Shuffle','every-epoch', ...
    'Verbose',true);
[net,info] = trainNetwork(imdsTrain,layers,options);

%% VGG-16 Classification
[YPred,scores] = classify(net,imdsTest);% Testing for accuracy
testLabels = imdsTest. Labels;
trainLabels = imdsTrain. Labels;
Accuracy_VGG-16 = mean(YPred == testLabels)*100;% VGG-16 Classification result
```

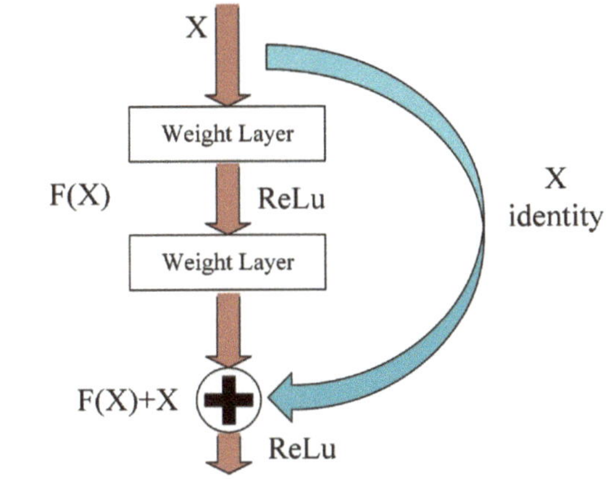

FIGURE 7.5 The residual learning: a building block.

convolutional layer input must ensure its spatial size. The convolution output is used for nonlinear activation. Five pooling layers are responsible for spatial pooling. A 2×2 size filter and stride 2 are used for max-pooling. Three FCL are constructed after a succession of convolutional and max-pooling layers. The softmax layer is the final layer [24]. The architecture of VGG-16 is shown in Fig. 7.6.

7.3.5 VGG-19

VGG-19 is a 19-layer CNN that has been pretrained. The model has trained on over a million data which contains roughly 10,000

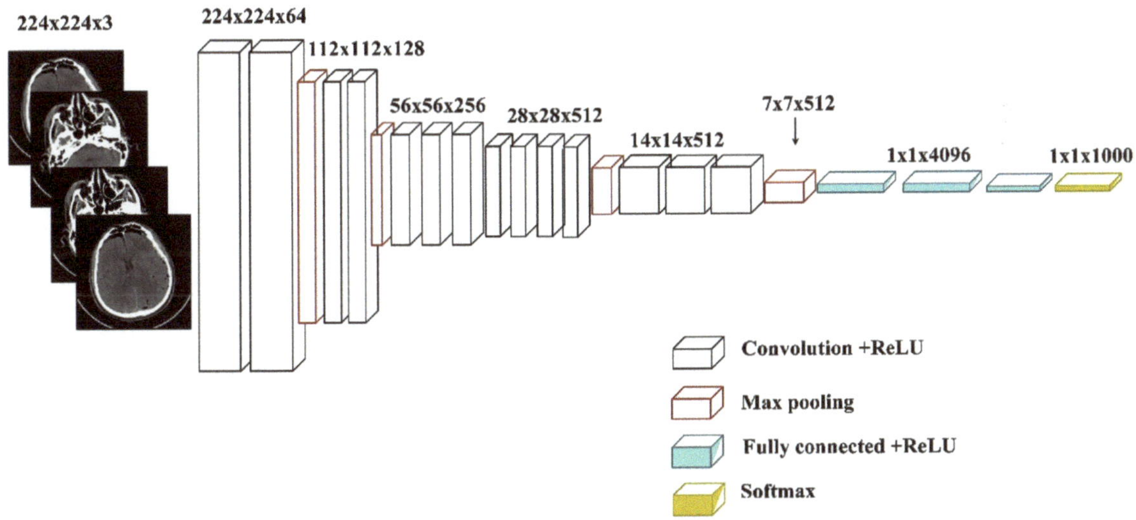

FIGURE 7.6 VGG-16's structure.

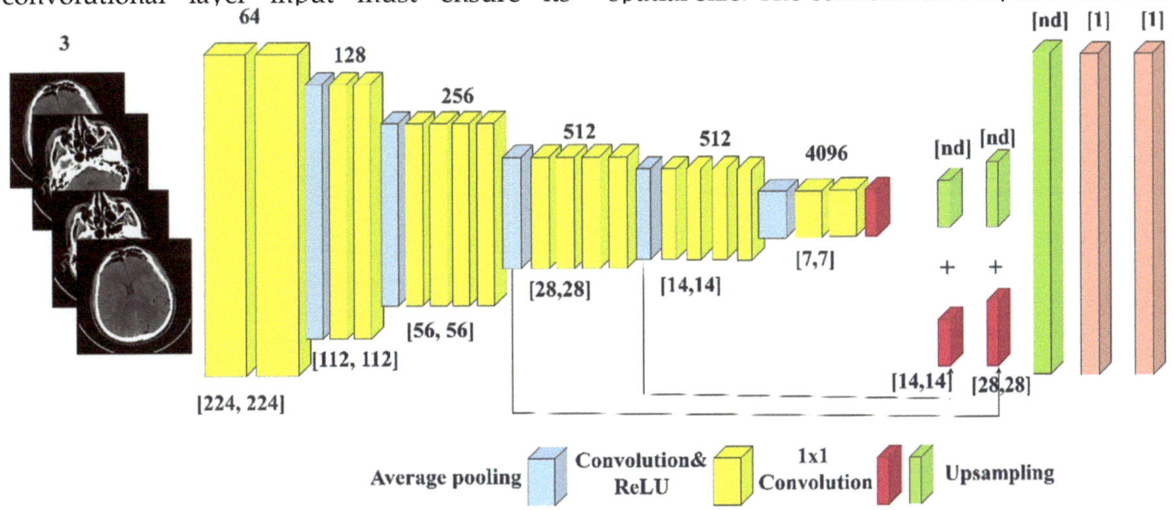

FIGURE 7.7 VGG-19's structure.

```
%% Load VGG-19
%Pretrained model "VGG-19 "
net = vgg19();
%% VGG-19 training options
options = trainingOptions('sgdm', ...
    'MiniBatchSize',32, ...
    'MaxEpochs',100, ...
    'Momentum',0.8, ...
    'InitialLearnRate',1e-4, ...
    'L2Regularization',1e-6,...
    'Shuffle','every-epoch', ...
    'Verbose',true);
[net,info] = trainNetwork(imdsTrain,layers,options);
%% VGG-19 Classification
[YPred,scores] = classify(net,imdsTest);% Testing for accuracy
testLabels = imdsTest. Labels;
trainLabels = imdsTrain. Labels;
Accuracy_VGG-19 = mean(YPred == testLabels)*100;% VGG-19 Classification result
```

object types. There are 16 convolution layers, 3 FCL, 5 max pool layers, and a softmax layer in this VGG-19 pretrained model. The network is utilized for feature extraction from the input layer to the last max-pooling layer, while the rest of the network is used for classification [25]. The architecture of VGG-19 is given in Fig. 7.7.

7.4 Experimental results

The experiments were executed on 2.60 GHz CPU, 16GB RAM, and 2 GB RAM NVIDIA GTX960M graphic card hardware in the computer. The simulation environment was MATLAB (2019a).

To evaluate the proposed model, we have used the confusion matrix and commonly used performance metrics, which are obtained from this matrix such as accuracy (ACC), sensitivity (SEN), specificity (SPE), F-score, and receiver operating characteristic (ROC). It comprises four components: true positive (TP), true negative (TN), false positive (FP), and false negative. Confusion matrix in Fig. 7.8 and Eq. (7.2–7.5) for these components are given below.

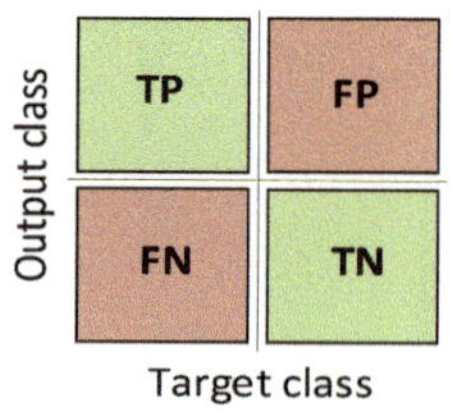

FIGURE 7.8 Confusion matrix representation.

$$ACC = \frac{TP + TN}{TP + TN + FP + FN} \quad (7.2)$$

$$SEN = \frac{TP}{TP + FN} \quad (7.3)$$

$$SPE = \frac{TN}{TN + FP} \quad (7.4)$$

$$F - score = \frac{2 \times TP}{2 \times TP + FP + FN} \quad (7.5)$$

Hereby, TP and TN indicate the sample of true estimated positive and negative samples, whereas FP and FN show the number of false predicted positive and negative samples. Additionally, ROC has been considered to assess the deep model efficiency. The training of the deep models was achieved in 100 epochs, the mini-batch size was 32 as can be seen from the given MATLAB code of each model in Section 7.3.

7.4.1 Dataset

The Brain Stroke CT Image Dataset [26] contains a total of 2501 CT images of 130 healthy (normal) and stroke-diagnosed subjects. All images in the dataset are 650 × 650 pixels and are in JPEG format. A total of 1551 of the images in the dataset belong to healthy people, and 950 of them belong to patients diagnosed with stroke. The figures of the brain stroke are shown in Fig. 7.9, the top of the image depicts images of cases which do not have any problem, that is, healthy, and the figures of strokes are depicted in the below row.

The efficiency of CNN structures was investigated in an experimental investigation for the brain stroke images. For the training and testing of pretrained models, the dataset is divided into 70% training and 30% testing. The confusion matrix and ROC curves of the pretrained CNN models are given in Fig. 7.10 and Fig. 7.11.

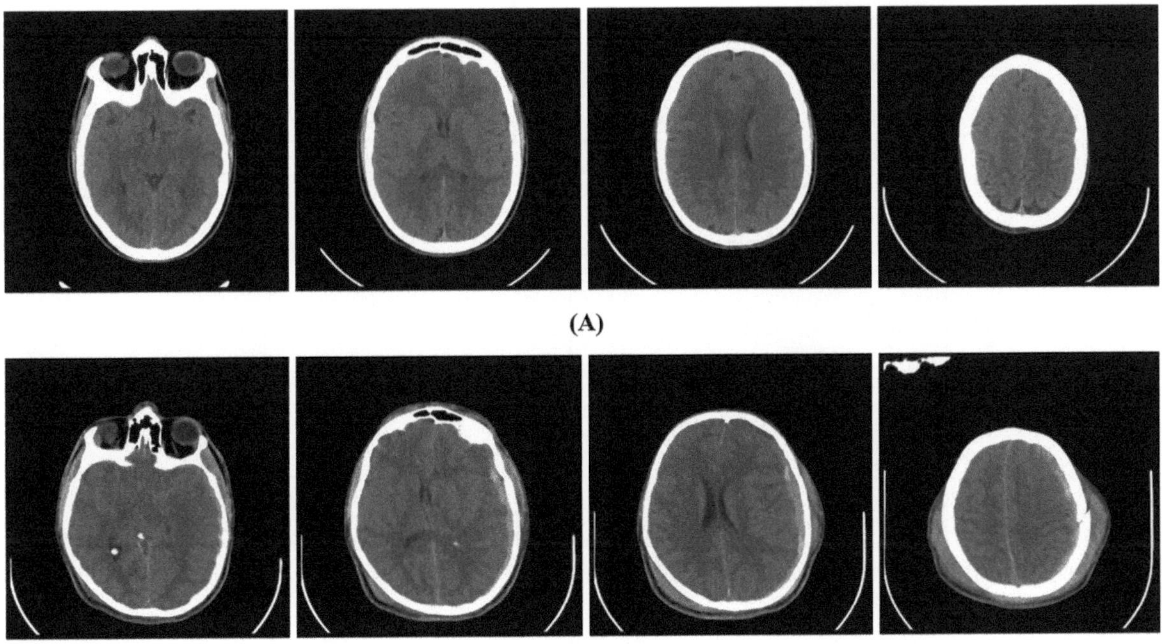

(A)

(B)

FIGURE 7.9 CT images of (A) normal and (B) stroke. *CT*, Computed tomography.

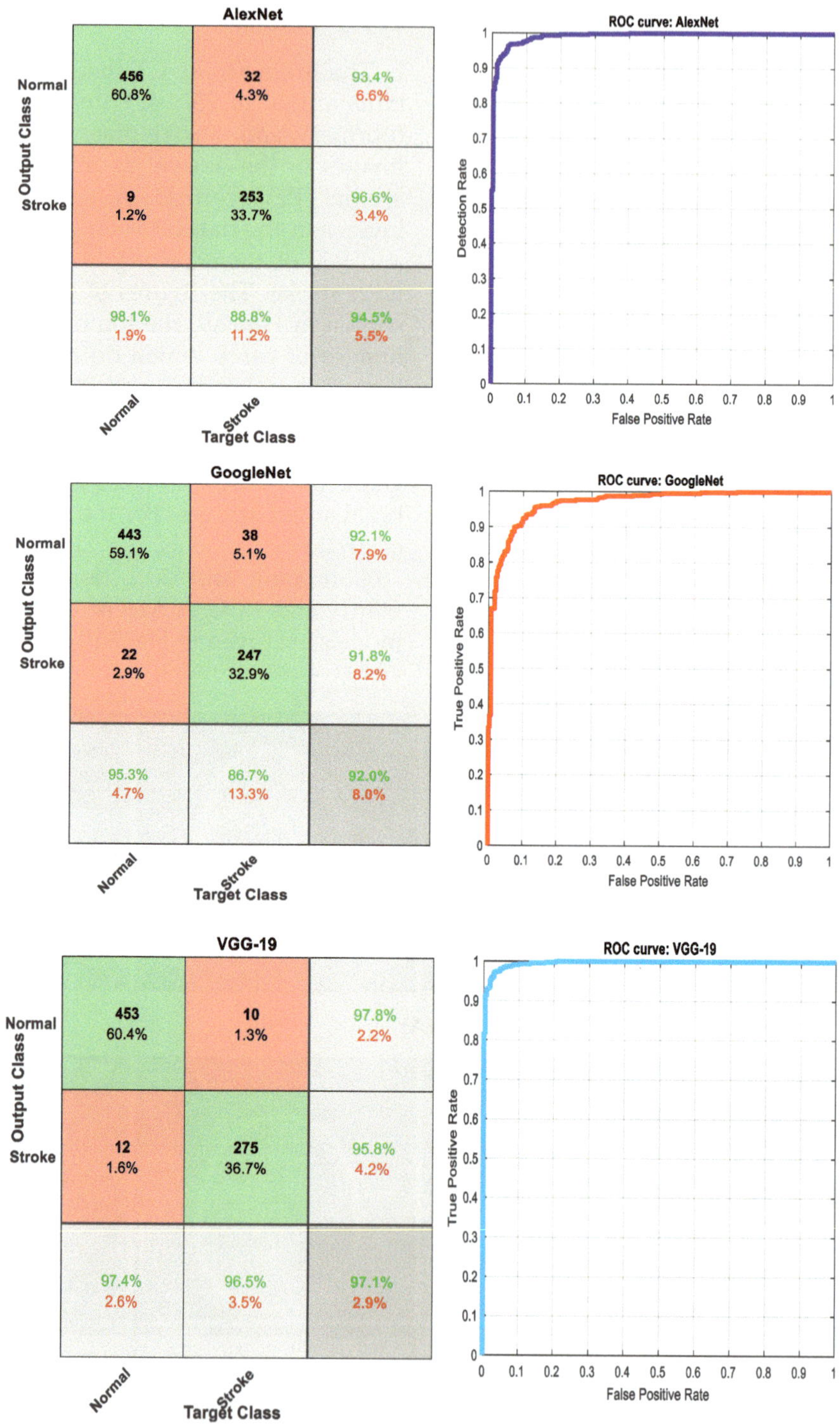

FIGURE 7.10 Confusion matrix and ROC of AlexNet, GoogleNet, and VGG-19. *ROC*, Receiver operating characteristic.

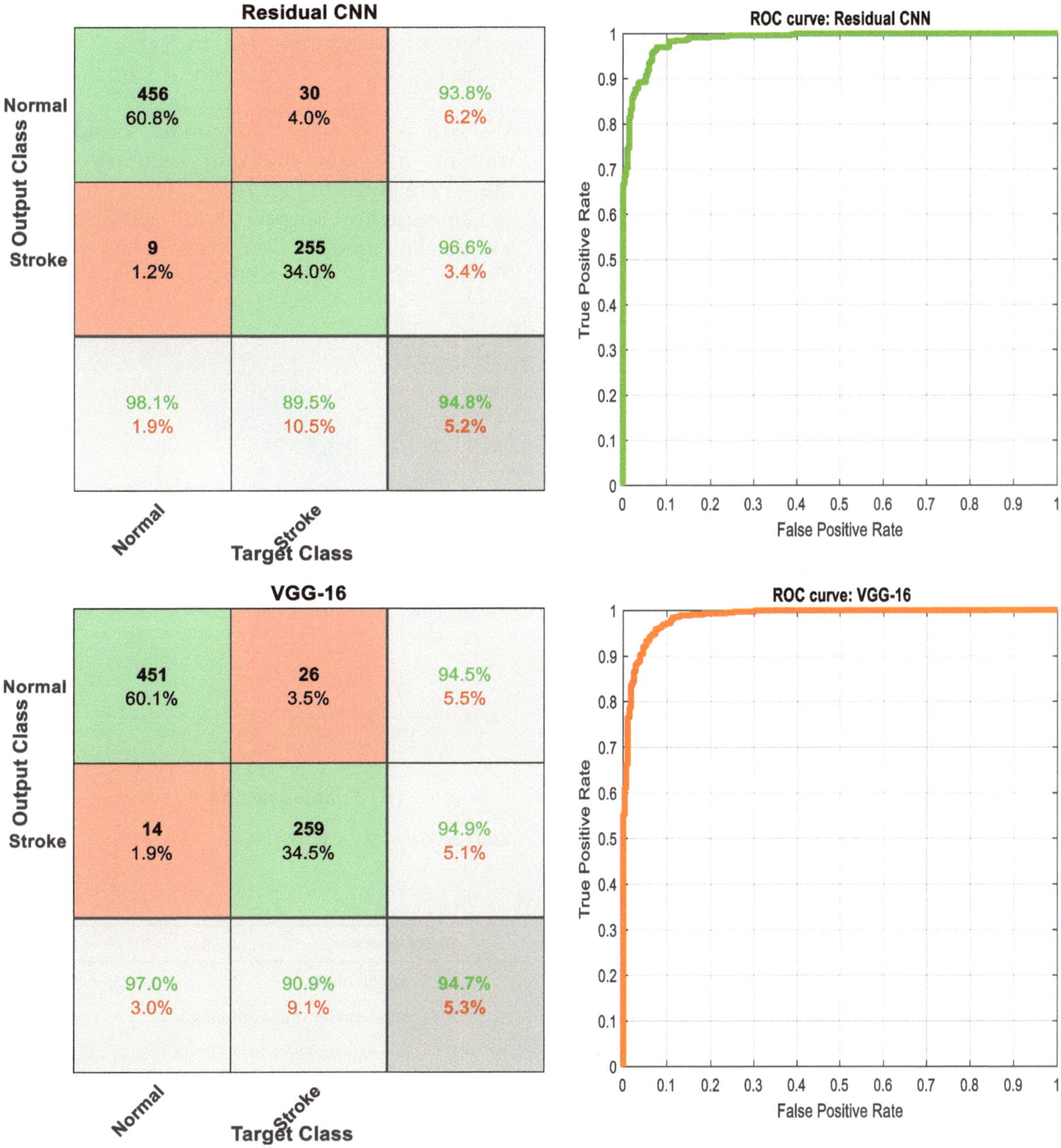

FIGURE 7.11 Confusion matrix and ROC of Residual CNN and VGG-16. *ROC*, Receiver operating characteristic.

TABLE 7.2 Stroke classification performance evaluation results.

CNN models	ACC	SEN	SPE	F-score
AlexNet	94.53%	98.06%	88.77%	93.18%
GoogleNet	92.00%	95.26%	86.66%	90.76%
Residual CNN	94.80%	98.06%	89.47%	93.57%
VGG-16	94.66%	96.98%	90.87%	93.83%
VGG-19	*97.06%*	*97.41%*	*96.49%*	*96.95%*

All scores of the classifiers pretrained CNN architectures are reported in Table 7.2. With respect to Table 7.2, the highest stroke classification performance was reached 97.06% with VGG-19 pretrained CNN model. Besides, the training loss and training accuracy of the model are given in Fig. 7.12.

Consequently, the maximum accuracy value for stroke classification was achieved ACC 97.06%, SEN 97.41%, SPE 96.49%, and F-Score

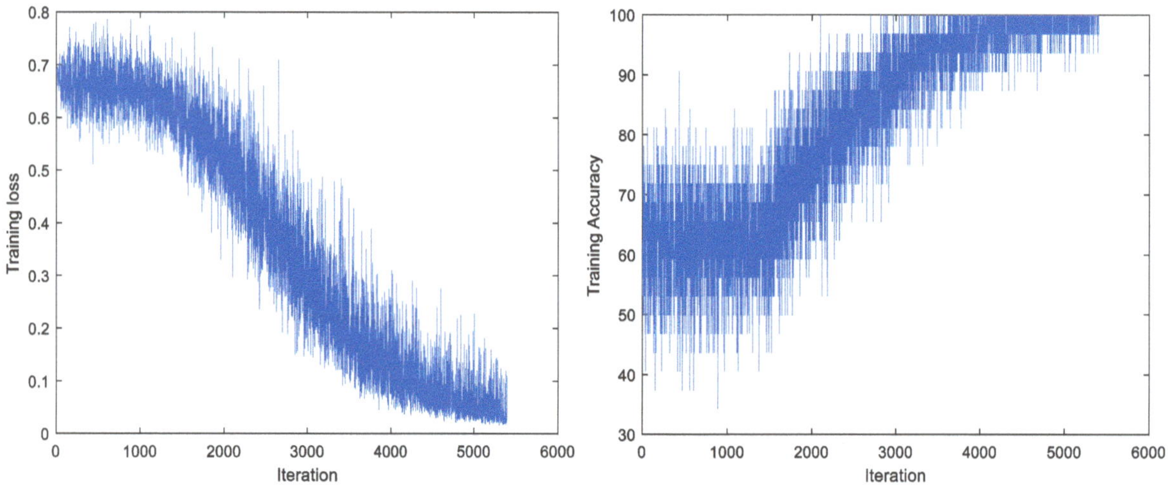

(A) Training loss (B) Training accuracy

FIGURE 7.12 (A) Training loss and (B) training accuracy of CNN model. *CNN*, Convolutional neural network.

TABLE 7.3 Comparison of the proposed method with previous studies.

References	Methods	Image datasets	Accuracy
Chin et al. (2017) [27]	Deep learning (CNN)	CT image dataset	90%
Karthik et al. [28]	Fully convolutional network (FCN)	MRI image dataset with 4,284 samples	70%
Liu et al. [29]	Support vector machine (SVM)	CT-scan image dataset with 1,157 samples	83.3%
Gaidhani et al. [30]	Deep learning models (LeNet and SegNet)	MRI scan with 400 samples	96%–97%
Badriyah et al. [31]	Random Forest	CT scan images from 102 patients	95.97%
This study	*AlexNet* *GoogleNet* *Residual CNN* *VGG-16* *VGG-19*	*CT image dataset with 2501 samples*	*94.53%* *92.00%* *94.80%* *94.66%* *97.06%*

96.95% with VGG-19, respectively. In the experimental study, it was observed that the training of the model was completed in 87 minutes and 42 seconds. In addition, at the end of 4400 iterations, it was observed that the training accuracy became balanced.

In Table 7.3, the conventional ML models, CNN models, and the proposed method are compared according to the performance criteria, which are explained in Section 7.4. Gaidhani et al. [30] suggested a technique to classify brain stroke MRI samples as healthy and unhealthy. The LeNet CNN was used for stroke classification. The classification score in the experimental study was obtained in the range of 96%–97%. Liu et al. [29], used normally accessible data to estimate hematoma expansion in spontaneous intracerebral hemorrhage (ICH). For this purpose, 1157 subjects with ICH therein were examined. Consequently, an overall accuracy of 83% in the prediction of hematoma expansion was obtained. Table 7.3 shows that the proposed model's classification accuracy results in satisfactory results.

7.5 Conclusion

Stroke is a complex condition with around 150 different causes. Early diagnosis of stroke is vitally important. The long-term effects of stroke can be minimized with the help of early diagnosis. Deep learning models can no way be used as a decision support system for clinicians. But with its image analysis power, it can make a big impact. Computer-aided, especially deep learning-based medical image analysis methods have increased in recent years. In this chapter, five deep CNN approaches are considered in stroke classification.

With this chapter, a pretrained CNN models that can distinguish between stroke and normal on brain stroke CT images have emerged. The dataset used consists of two classes. The classification accuracy of VGG-19 CNN model has been superior compared to the other transfer learning models. The proposed model was tested by operating the Brain Stroke CT Image Database. As a result of the best classification, the values of accuracy 97.06%, sensitivity 97.41%, specificity 96.49%, and 96.95% with VGG-19 CNN model were obtained.

References

[1] M. Nishio, S. Koyasu, S. Noguchi, T. Kiguchi, K. Nakatsu, T. Akasaka, et al., Automatic detection of acute ischemic stroke using non-contrast computed tomography and two-stage deep learning model, Comput. Methods Prog. Biomed. 196 (2020) 105711.

[2] A. Gautam, B. Raman, Towards effective classification of brain hemorrhagic and ischemic stroke using CNN, Biomed. Signal. Process. Control. 63 (2021) 102178. no. August 2020.

[3] E. Altunışık, A. Arık, Decreased stroke applications during pandemic: collateral effects of covid-19, Turk. Noroloji Derg. 27 (2) (2021) 171–175.

[4] V. Abramova, A. Clèrigues, A. Quiles, D.G. Figueredo, Y. Silva, S. Pedraza, et al., Hemorrhagic stroke lesion segmentation using a 3D U-Net with squeeze-and-excitation blocks, Comput. Med. Imaging Graph. 90 (2021).

[5] F. Er, Ş. Yıldırım, İnme sonrası hasta ve Bakım verenlerin aile ilişkilerinin Değerlendirilmesi, Celal Bayar Üniversitesi Sağlık Bilim. Enstitüsü Derg. 6 (3) (2019) 182–189.

[6] U. Rajendra Acharya, K.M. Meiburger, O. Faust, J. En Wei Koh, S. Lih Oh, E.J. Ciaccio, et al., Automatic detection of ischemic stroke using higher order spectra features in brain MRI images, Cogn. Syst. Res. 58 (2019) 134–142.

[7] N.K. Al-Qazzaz, Z.A.A. Alyasseri, K.H. Abdulkareem, N.S. Ali, M.N. Al-Mhiqani, C. Guger, EEG feature fusion for motor imagery: a new robust framework towards stroke patients rehabilitation, Comput. Biol. Med. 137 (August) (2021) 104799.

[8] S.F. Sung, L.C. Hung, Y.H. Hu, Developing a stroke alert trigger for clinical decision support at emergency triage using machine learning, Int. J. Med. Inform. 152 (February) (2021) 104505.

[9] M. Soltanpour, R. Greiner, P. Boulanger, B. Buck, Improvement of automatic ischemic stroke lesion segmentation in CT perfusion maps using a learned deep neural network, Comput. Biol. Med. 137 (August) (2021) 104849.

[10] S.A. Peixoto, P.P. Rebouças Filho, Neurologist-level classification of stroke using a Structural Co-Occurrence Matrix based on the frequency domain, Comput. Electr. Eng. 71 (August) (2018) 398–407.

[11] M. Bento, R. Souza, M. Salluzzi, L. Rittner, Y. Zhang, R. Frayne, Automatic identification of atherosclerosis subjects in a heterogeneous MR brain imaging data set, Magn. Reson. Imaging 62 (June) (2019) 18–27.

[12] P.P. Rebouças Filho, R.M. Sarmento, G.B. Holanda, D. de Alencar Lima, New approach to detect and classify stroke in skull CT images via analysis of brain tissue densities, Comput. Methods Prog. Biomed. 148 (2017) 27–43.

[13] J. Vargas, A. Spiotta, A.R. Chatterjee, Initial experiences with artificial neural networks in the detection of computed tomography perfusion deficits, World Neurosurg. 124 (2019) e10–e16.

[14] A. Gautam, B. Raman, Towards effective classification of brain hemorrhagic and ischemic stroke using CNN, Biomed. Signal. Process. Control. 63 (April 2020) (2021) 102178.

[15] R. Kanchana, R. Menaka, Ischemic stroke lesion detection, characterization and classification in CT images with optimal features selection, Biomed. Eng. Lett. 10 (3) (2020) 333–344.

[16] U. Raghavendra, T.-H. Pham, A. Gudigar, V. Vidhya, B.N. Rao, S. Sabut, et al., Novel and accurate non-linear index for the automated detection of haemorrhagic brain stroke using CT images, Complex. Intell. Syst. 7 (2) (2021) 929–940.

[17] L. Herzog, E. Murina, O. Dürr, S. Wegener, B. Sick, Integrating uncertainty in deep neural networks for MRI based stroke analysis, Med. Image Anal. 65 (2020) 1–21.

[18] T. Badriyah, N. Sakinah, I. Syarif, D.R. Syarif, Machine Learning Algorithm for Stroke Disease Classification, 2020 International Conference on Electrical, Communication, and Computer Engineering (ICECCE) 1 (5) (2020). Available from: doi:10.1109/ICECCE49384.2020.9179307. In this issue.

[19] J. Chen, Z. Wan, J. Zhang, W. Li, Y. Chen, Y. Li, et al., Medical image segmentation and reconstruction of prostate tumor based on 3D AlexNet, Comput. Methods Prog. Biomed. 200 (2021) 105878.

[20] Ö. İnik, E. Ülker, Deep learning and deep learning models used in image analysis, Gaziosmanpasa J. Sci. Res. 6 (3) (2017) 85–104.

[21] S. Deepak, P.M. Ameer, Retrieval of brain MRI with tumor using contrastive loss based similarity on GoogLeNet encodings, Comput. Biol. Med. 125 (June) (2020) 103993.

[22] A. Diker, Sıtma Hastalığının Sınıflandırılmasında Evrişimsel Sinir Ağlarının Performanslarının Karşılaştırılması, BEÜ Fen. Bilim. Derg. 9 (4) (2020) 1825–1835.

[23] K. He, X. Zhang, S. Ren, J. Sun, Deep residual learning for image recognition, in: Proc. IEEE Comput. Soc. Conf. Comput. Vis. Pattern Recognit., vol. 2016–Dec, pp. 770–778, 2016.

[24] P. Saha, M.S. Sadi, O.F.M.R.R. Aranya, S. Jahan, F.-A. Islam, COV-VGX: An automated COVID-19 detection system using X-ray images and transfer learning, Inform. Med. Unlocked 26 (2021) 100741.

[25] S. Agarwal, A. Rattani, C.R. Chowdary, A comparative study on handcrafted features v/s deep features for open-set fingerprint liveness detection, Pattern Recognit. Lett. 147 (2021) 34–40.

[26] A. Rahman, Brain Stroke CT Image Dataset, Kaggle, <https://www.kaggle.com/afridirahman/brain-stroke-ct-image-dataset>, 2021. (accessed 10.11.21)

[27] C.L. Chin, B.J. Lin, G.R. Wu, T.C. Weng, C.S. Yang, R. C. Su, et al., An automated early ischemic stroke detection system using CNN deep learning algorithm, in: Proc. - 2017 IEEE 8th Int. Conf. Aware. Sci. Technol. iCAST 2017, vol. 2018–January, no. iCAST, pp. 368–372, 2017.

[28] R. Karthik, U. Gupta, A. Jha, R. Rajalakshmi, R. Menaka, A deep supervised approach for ischemic lesion segmentation from multimodal MRI using Fully Convolutional Network, Appl. Soft Comput. J. 84 (2019) 105685.

[29] J. Liu, H. Xu, Q. Chen, T. Zhang, W. Sheng, Q. Huang, et al., Prediction of hematoma expansion in spontaneous intracerebral hemorrhage using support vector machine, EBioMedicine 43 (2019) 454–459.

[30] B.R. Gaidhani, R. Rajamenakshi, S. Sonavane, Brain stroke detection using convolutional neural network and deep learning models, in: 2019 2nd Int. Conf. Intell. Commun. Comput. Tech. ICCT 2019, pp. 242–249, 2019.

[31] T. Badriyah, N. Sakinah, I. Syriaf, Machine Learning Algorithm for Classification, J. Phys. Conf. Ser. 1994 (1) (2021) 12–13.

CHAPTER

8

A deep learning approach for COVID-19 detection from computed tomography scans

Ashutosh Varshney[1] *and Abdulhamit Subasi*[2,3]

[1]Department of Computer Science and Engineering, Indian Institute of Technology, Kharagpur, West Bengal, India [2]Institute of Biomedicine, Faculty of Medicine, University of Turku, Turku, Finland [3]Department of Computer Science, College of Engineering, Effat University, Jeddah, Saudi Arabia

OUTLINE

Applications of Artificial Intelligence in Medical Imaging
DOI: https://doi.org/10.1016/B978-0-443-18450-5.00011-6

8.1 Introduction

Coronavirus disease 2019 (COVID-19) has affected more than 17 million people all around the world and caused 681 K deaths worldwide, as of August 2 in 2020 [1]. The estimated viral reproduction number tells that an infected individual can transmit this deadly disease to around 2.5 noninfected individuals [1] with low immunity, indicating a high risk of massive spread of the disease. Therefore it is crucial to have fast testing of suspected individuals as early as possible for quarantine and treatment purposes.

The major problem in disease control is the lack of sufficient test kits available for testing. The current tests are mostly based on reverse transcription-polymerase chain reaction (RT-PCR). The RT-PCR tests are not very accurate and might sometimes give false-positive results. It is reported that many "suspected" cases with typical clinical characteristics of COVID-19 and identical specific computed tomography (CT) images were not diagnosed [2]. The test takes around 6 hours which is very slow compared to the disease spreading rate. Thus the shortage of RT-PCR kits and their inaccurate results motivates us to study an alternative testing procedure, which can be made widely available and is faster, cheaper, and more feasible than RT-PCR, in particular, CT scans.

The CT images of various viral pneumonia and other lung diseases are more or less similar to that of COVID-19, therefore it becomes difficult for radiologists to diagnose the disease as well as to distinguish it from other viral pneumonia [3]. There are several artificial intelligence techniques developed to extract shape and spatiotemporal features from images and use it for disease prediction. There has been recent development in medical imaging techniques using deep learning, especially convolutional neural network (CNN).

Several features are used for identifying viral pathogens on the basis of imaging patterns, which are associated with their specific pathogenesis. The main identifying features of COVID-19 are the white glassy patches called the ground-glass opacity [3]. Our intuition therefore was to make use of CNNs to identify these unique and COVID-specific features in the CT scans which might not be distinguishable through the naked eye. Hence, the purpose of our research was to study the diagnostic performance of various deep learning models using CT images to screen for COVID-19.

8.2 Literature review

The recent advances in medical imaging and disease prediction have witnessed a great rise in machine learning methods. But these methods, to work accurately, require features to build upon which are themselves not very easily extractable. Therefore there has been high amount of progress in deep learning models because they can extract features or can make use of some pretraining beforehand [4] and the whole process can be molded into one single step.

There have been several previous attempts of using deep learning methods for detecting COVID-19 from chest CT and X-Ray scans. Shin et al. [5] used deep CNN to classify the interstitial lung disease in CT images. Pezeshk et al. [6] used 3D CNN to detect the pulmonary nodules in chest CT. Li et al. [7] developed a 3D deep learning architecture COVNet for COVID-19 detection. Panwar et al. [8] used a Grad-CAM based color visualization along with a transfer learning approach for COVID-19 detection. Meng et al. [9] showed that CT scans of patients with COVID-19 have definite characteristics. Das et al. [10] also tried to use a transfer learning approach using Xception architecture. Similarly, Lalmuanawma et al. [11] also showed that machine learning and AI-based methods can prove to be helpful in automatic detection of COVID-19. Ardakani et al. [12] explored the possibility of CNNs being used, whereas Panwar et al. [13] proposed nCOVnet which makes use of X-Ray scans for the classification.

Fan et al. [14] proposed a semisupervised COVID-19 Lung Infection Segmentation Deep Network (Inf-Net). Minaee et al. [15] also worked in the direction of X-Rays for the detection and prepared a rich dataset of X-Ray scans. Brunese et al. [16] also show that imaging techniques such as X-Rays can prove to be beneficial for the disease detection. Nour et al. [17] used simple shallow CNNs instead of pretrained networks for COVID-19 diagnosis. Tuncer et al. [18] tried residual exemplar LBP and iterative ReliefF, while Hassantabar et al. [19] used CNNs on lung X-Rays for the detection of infected tissue.

Similar experiments were carried out by Mahmud et al. [20] where they used an architecture with depthwise dilated convolutions. Alakus et al. [21] also provide a comparative study of various deep learning approaches. Curvelet transform was also one of the techniques which was used by Altan and Karasu [22]. Similar experiments were carried out by Sufian et al. [23] using transfer learning. Ozturk et al. [24] proposed heatmaps, which can help radiologists to locate the affected regions in chest X-Rays. Khan et al. [25] used Xception architecture, whereas Shaban et al. [26] used K-Nearest Neighbor (K-NN) classifier for the same task.

8.3 Subjects and data acquisition

The dataset used for the experiments comprised 1252 CT scan images of COVID-19-infected patients and 1229 CT scan images of non-COVID patients. It was divided into 60% training, 20% validation, and 20% test sets and the images were normalized. The image size used was $200 \times 200 \times 3$. These data had been collected from real patients in hospitals from Sao Paulo, Brazil. Images from the dataset are given in Fig. 8.1, in which the left one is the CT scan of a COVID-19 patient, while the right one is the CT scan of a non-COVID patient.

8.4 Proposed architecture and transfer learning

We experimented with eight different CNNs pretrained on ImageNet as our base models. The top layer of all the pretrained models was removed. For transfer learning, the "number of features" of the final fully connected layer was kept 2. The models were trained using categorical cross entropy loss and Adam optimizer with a learning rate of 0.002 and batch size of 64 for 50 epochs. The dataset was shuffled at every epoch.

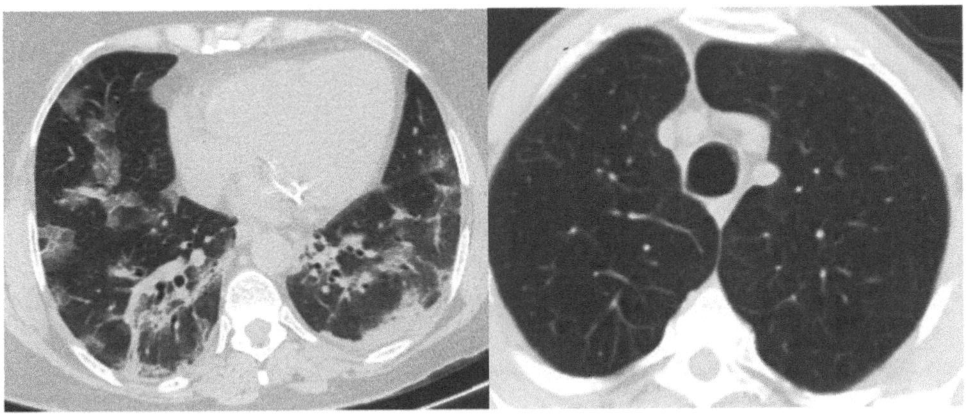

FIGURE 8.1 Images from dataset.

The same training and validation datasets were selected for all networks to facilitate the performance comparison of networks. Fig. 8.2 depicts our proposed architecture. The input image is passed through a convolutional layer and then through a pretrained model. The output is then passed through a series of pooling, batch normalization, and dense layers to get the final output.

```
def select_pretrained_model(name):
    # function to select one of the pre-trained ImageNet models
    pretrainedbase = None
    if name == "DenseNet121":
      pretrainedbase = DenseNet121(weights='imagenet', include_top=False)
    if name == "VGG16":
      pretrainedbase = VGG16(weights='imagenet', include_top=False)
    if name == "VGG19":
      pretrainedbase = VGG19(weights='imagenet', include_top=False)
    if name == "ResNet50":
      pretrainedbase = ResNet50(weights='imagenet', include_top=False)
    if name == "InceptionV3":
      pretrainedbase = InceptionV3(weights='imagenet', include_top=False)
    if name == "InceptionResNetV2":
      pretrainedbase = InceptionResNetV2(weights='imagenet', include_top=False)
    if name == "Xception":
      pretrainedbase = Xception(weights='imagenet', include_top=False)
    if name == "MobileNet":
      pretrainedbase = MobileNet(weights='imagenet', include_top=False)
    return pretrainedbase

def build_model(pretrainedbase, num_features):
    # function to build the complete model
    input = Input(shape=(200,200,3))
    x = Conv2D(3, (3, 3), padding='same')(input)
    x = pretrainedbase(x)
    x = GlobalAveragePooling2D()(x)
    x = BatchNormalization()(x)
    x = Dropout(0.5)(x)
    x = Dense(256, activation='relu')(x)
    x = BatchNormalization()(x)
    x = Dropout(0.5)(x)
    output = Dense(num_features,activation = 'softmax', name='root')(x)
    model = Model(input,output)
    return model

def transfer_learning(modelname):
    # function to train model using transfer learning
    base = select_pretrained_model(modelname)
    model = build_model(base, 2)
    optimizer = Adam(lr=0.002, beta_1=0.9, beta_2=0.999, epsilon=0.1, decay=0.0)
    model.compile(loss='categorical_crossentropy', optimizer=optimizer,
                                              metrics=['accuracy']) model.summary()
    checkpoint = ModelCheckpoint('model.h5', verbose=1, save_best_only=True)
    history = model.fit(X_train, Y_train, validation_split=0.25, epochs=50,
                                   batch_size=64, verbose=2,callbacks=[checkpoint])
    return history
```

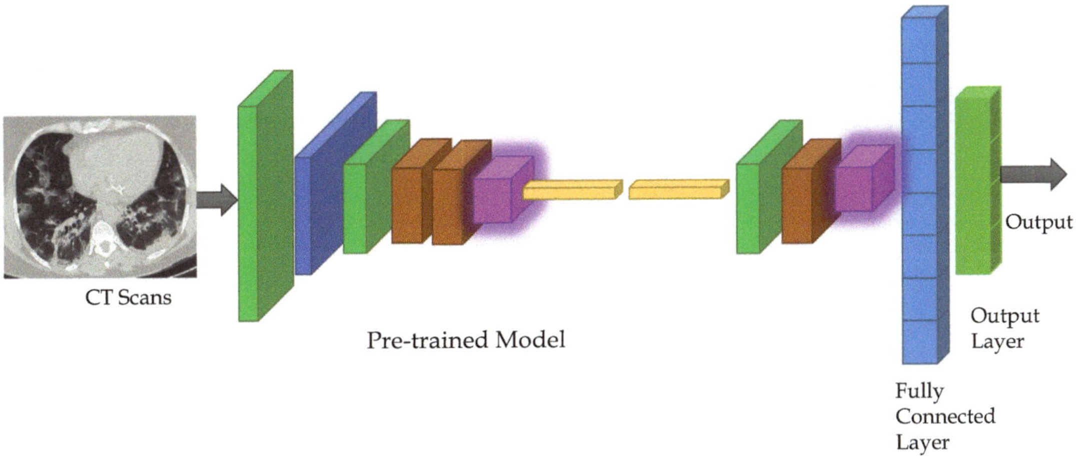

FIGURE 8.2 Proposed architecture.

8.4.1 ResNet

Residual blocks help in tackling the vanishing gradient problem without compromising on the number of hidden layers of the architecture being used. Residual networks preserve what they have learned and keep on adding to the preserved weights by using an identity mapping that is added to the activations after one step. Thus the network keeps on adding whenever it has learnt something new to its previous experience.

The residual block first creates a copy of the weights *x* and preserves it to be used in the skip connection. Then an activation is applied to the weights *x* to get *F(x)*. After the activation, the preserved weights *x* through the skip connection is added to *F(x)* which completes the basic working of a residual block [27]. ResNet is made up of several such residual blocks and are very deep and have proven to be very successful for image recognition and other such tasks.

8.4.2 DenseNet

In a traditional deep learning architecture, a series of operations, which may include activations such as ReLU, softmax, or operations such as convolutions, pooling, or batch normalization are applied at each layer. The equation for this would be:

$$x_l = H_l(x_{l-1})$$

ResNets extended this behavior including the skip connection, reformulating this equation into:

$$x_l = H_l(x_{l-1}) + x_{l-1}$$

DenseNets differ from ResNets mainly at this step. Instead of summing up the history of the weights learned, it is concatenated together. Consequently, the equation transforms again into:

$$x_l = H_l([x_0, x_1, \ldots, x_{l-1}])$$

```
history = transfer_learning("ResNet50")  ####change the model name here
model = load_model('./model.h5')
final_loss, final_accuracy = model.evaluate(X_test, Y_test)
print('Final Loss: {}, Final Accuracy: {}'.format(final_loss, final_accuracy))
plot_model_accuracy()
plot_model_loss()
```

DenseNets comprise dense blocks which are based upon this idea of collective knowledge. Since the feature maps are getting concatenated, the channel dimension keeps on increasing at each layer. If we make H_l to produce k feature maps every time, then for the lth layer:

$$k_l = k_0 + k(l - 1)$$

This hyperparameter k, called the growth rate, is a measure of the amount of information that is being added to the network in the form of feature maps at each layer [28].

```
history = transfer_learning("DenseNet121")  ####change the model name here
model = load_model('./model.h5')
final_loss, final_accuracy = model.evaluate(X_test, Y_test)
print('Final Loss: {}, Final Accuracy: {}'.format(final_loss, final_accuracy))
plot_model_accuracy()
plot_model_loss()
```

8.4.3 MobileNet

MobileNets are based on a streamlined architecture that uses depth-wise separable convolutions to build lightweight deep neural networks. They are known to be memory efficient and lightweight because of a smaller number of parameters to be trained upon. They make use of dense blocks efficiently by having large number of output feature maps with very few convolution operations. The growth rate k is also kept low which further optimizes the computation and makes it much more memory efficient [29].

```
history = transfer_learning("MobileNet")  ####change the model name here
model = load_model('./model.h5')
final_loss, final_accuracy = model.evaluate(X_test, Y_test)
print('Final Loss: {}, Final Accuracy: {}'.format(final_loss, final_accuracy))
plot_model_accuracy()
plot_model_loss()
```

8.4.4 Xception

Xception just like MobileNet makes use of depthwise separable convolutions to make lightweight deep learning networks. It was developed by Google researchers. Inception module can be summarized as a depthwise convolution followed by a pointwise convolution. Thus the depthwise separable convolution used here is just an Inception module with large number of towers. The data first goes through the entry flow, then through the middle flow, which is repeated eight times and finally through the exit flow [30].

```
history = transfer_learning("Xception")  ####change the model name here
model = load_model('./model.h5')
final_loss, final_accuracy = model.evaluate(X_test, Y_test)
print('Final Loss: {}, Final Accuracy: {}'.format(final_loss, final_accuracy))
plot_model_accuracy()
plot_model_loss()
```

8.4.5 Visual geometry group (VGG)

The visual geometry group (VGG) architecture is very simple to understand. The input image is of size 224×224 and is then passed through a series of five convolution blocks consisting of 3×3 kernels with stride 1 and ReLU activations. The convolution blocks are followed by max-pooling. Finally, it is passed through three fully connected layers with last one having a dimension of 1000 corresponding to 1000 different image classes [31]. VGG architectures have also proved to be very successful in ImageNet challenge.

```
history = transfer_learning("VGG19")  ####change the model name here
model = load_model('./model.h5')
final_loss, final_accuracy = model.evaluate(X_test, Y_test)
print('Final Loss: {}, Final Accuracy: {}'.format(final_loss, final_accuracy))
plot_model_accuracy()
plot_model_loss()
```

8.4.6 Inception/GoogLeNet

In most of the standard network architectures, the intuition is not clear why and when to perform the max-pooling operation, when to use the convolutional operation. For example, in AlextNet, we have the convolutional operation and max-pooling operation following each other, whereas in VGGNet, we have three convolutional operations in a row and then one max-pooling layer. Thus the idea behind GoogLeNet is to use all the operations at the same time. It computes multiple kernels of different size over the same input map in parallel, concatenating their results into a single output. The intuition is that convolution filters of different sizes will handle objects at multiple scale better. This is called an Inception module [32].

```
history = transfer_learning("InceptionV3")  ####change the model name here
model = load_model('./model.h5')
final_loss, final_accuracy = model.evaluate(X_test, Y_test)
print('Final Loss: {}, Final Accuracy: {}'.format(final_loss, final_accuracy))
plot_model_accuracy()
plot_model_loss()
```

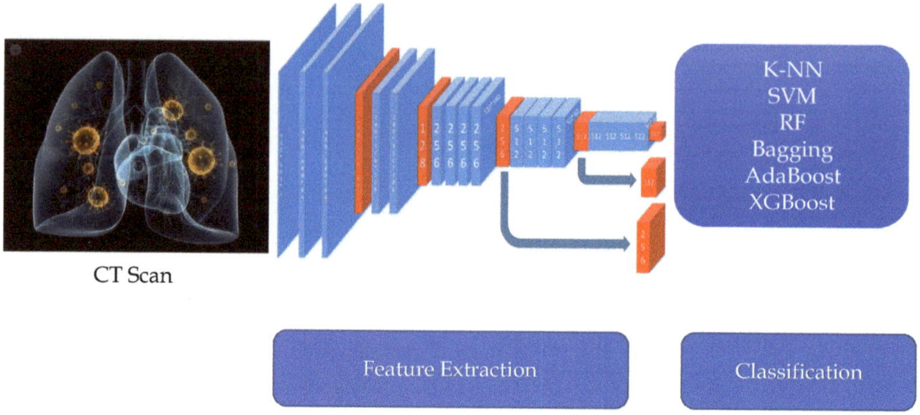

FIGURE 8.3 COVID-19 detection using deep feature extraction and conventional machine learning.

8.5 COVID-19 detection with deep feature extraction

We also tried out with six different machine learning classifiers on the extracted features from our proposed architecture. For K-NN classifier, the optimal value of K was found to be 3 after fine-tuning. Similarly, the number of estimators used in Random Forest, AdaBoost, XGBoost, and Bagging was found to be 1000. The learning rate for XGBoost was kept 1. The *num_features* of the final fully connected layer were kept 256 (Fig. 8.3).

8.5.1 K-nearest neighbors

The K-NN algorithm is very simple. K objects that are nearest to the test object are found using some distance metric from the training set. Then, the class of the test object is decided by a majority vote count of the K neighbors found. The choice of the distance metric is another important consideration. The metric is selected such that the smaller the distance between two objects, the more is the likelihood of them having the same class. Thus, for example, if K-NN is being applied to

```
def featureextraction(modelname):
    base = select_pretrained_model(modelname)
    model = build_model(base,500)
    X_T = np.array([model.predict(i.reshape(1,200,200,3)).reshape(500) for i in X_Train])
    return X_T
```

classify documents, then it may be better to use the cosine similarity rather than Euclidean distance [33].

```
#Import k-NN Model
from sklearn.neighbors import KNeighborsClassifier
#Create the Model
clf = KNeighborsClassifier(n_neighbors=3)
#Train the model with Training Dataset
clf.fit(X_train,Y_train)
#Test the model with Testset
Y_pred = clf.predict(X_test)
#Print performance
print(clf.score(X_train,Y_train))
print(clf.score(X_val,Y_val))
print(clf.score(X_test,Y_test))
print_confusion_matrix(Y_test,Y_pred)
print_scores(Y_test,Y_pred)
```

8.5.2 Support vector machine

Support vector machine (SVM) is one of the most robust and accurate methods among all well-known machine learning algorithms. The main aim of SVM is to find a function that can differentiate the two different classes accurately in a binary classification problem. For a linearly separable dataset, a linear classification function can be visualized as a separating hyperplane that passes through the middle of the two classes, separating the two. Since there can be many such hyperplanes, SVM helps in deciding the hyperplane which maximizes the separation between the two classes. This separation is also called the margin sometimes. The margin can be defined as the distance between the two objects of different classes as classified by the classifier that are nearest to each other. This definition of margin enforces the classifier to select the best boundary among all the possible candidates. This helps in making the classifier much more generalized to unseen data as the margin is set to maximum possible [33].

```
#Import SVM Model
from sklearn import svm
#Create the Model
clf = make_pipeline(StandardScaler(), SVC())
#Train the model with Training set
clf.fit(X_train, Y_train)
#Test the model with Test set
Y_pred = clf.predict(X_test)
#Print performance
print(clf.score(X_train,Y_train))
print(clf.score(X_val,Y_val))
print(clf.score(X_test,Y_test))
print_confusion_matrix(Y_test,Y_pred)
print_scores(Y_test,Y_pred)
```

8.5.3 Random Forests

Random Forests as the name suggest is an ensemble machine learning algorithm where several decision trees are combined together to form a "forest." The main idea behind the algorithm is that these trees are uncorrelated from each other and thus overall, they reduce the error percentage. Decision trees work by splitting the data at each step and thus selecting the most favorable splits that lead to correct classification. Also, the dataset for each tree is forced to be selected from a random subset of features, thus enforcing more randomness and generalization and thereby reducing the collective error. Thus this idea of combining predictions from several decision trees increases the overall accuracy on the dataset [34,35].

```
#Import Random Forest Ensemble Model
from sklearn.ensemble import RandomForestClassifier
#Create the Model
clf = RandomForestClassifier(n_estimators=100)
#Train the model with Training set
clf.fit(X_train,Y_train)
#Test the model with Test set
Y_pred = clf.predict(X_test)
#Print performance
print(clf.score(X_train,Y_train))
print(clf.score(X_val,Y_val))
print(clf.score(X_test,Y_test))
print_confusion_matrix(Y_test,Y_pred)
print_scores(Y_test,Y_pred)
```

8.5.4 Bagging

The decisions of different models can be combined into a single prediction to get better results which is done in various ensemble methods. The most obvious way is to take a majority vote count in case of classification or just taking the average of all model predictions if a numeric output is desired. In Bagging the models have equal weight; while in boosting weighting is used to achieve more impact to the more effective ones depending on how successful their predictions were in the past. In Bagging, numerous training datasets of the same size are selected at random from the problem domain to build a decision tree for each dataset. These trees are practically identical to make the same prediction for each new test instance. But this hypothesis is generally not true, especially if the training datasets are fairly small. This inevitably indicates that there are test instances for the prediction of some of the models which were not correct. Bagging tries to eliminate the instability of machine learning methods by sampling the instances randomly with replacement from the original dataset to create a new one of the same size. This sampling procedure unavoidably deletes some of the examples and duplicates the rest. So, Bagging uses each one of these resultant datasets in the learning algorithm, and the outputs generated from them vote for the class to be predicted [36].

```
#Import Bagging Ensemble Model
from sklearn.ensemble import BaggingClassifier
from sklearn import tree
#Create a Bagging Ensemble Model
clf = BaggingClassifier(n_estimators=1000)
#Train the model using the training set
clf.fit(X_train,Y_train)
#Predict the response for test set
Y_pred = clf.predict(X_test)
#Print performance
print(clf.score(X_train,Y_train))
print(clf.score(X_val,Y_val))
print(clf.score(X_test,Y_test))
print_confusion_matrix(Y_test,Y_pred)
print_scores(Y_test,Y_pred)
```

8.5.5 AdaBoost

AdaBoost, which is the short form of Adaptive Boosting, is a very simple yet elegant ensemble learning technique in which weights are assigned to the data points according to the prediction of the classifier. Initially, all the data points are assigned equal weights and the classifier is used to make the prediction. Now, the predictions are compared with the ground truth labels and the weights for the correct predictions are decreased while those for the incorrect ones are increased. In the next iteration, the classifier is built upon this newly weighted dataset. This time, the difference is that since the earlier misclassified data points have been given more weights, the classifier is a bit "harder" to them as compared to the data points with low weight. After a series of iterations of this algorithm, some hard instances become harder, while some of them become correct which varies from model to model but overall, the ensemble method has its accuracy increased. Since the algorithm adapts to the dataset by assigning and modifying weights according to the current predictions, it is called Adaptive Boosting [36].

```
#Import Adaboost ensemble model
from sklearn.ensemble import AdaBoostClassifier
#Create an Adaboost Ensemble Model
clf = AdaBoostClassifier(n_estimators=1000)
#Train the model using the training set
clf.fit(X_train,Y_train)
#Predict the response for test set
Y_pred = clf.predict(X_test)
#Print performance
print(clf.score(X_train,Y_train))
print(clf.score(X_val,Y_val))
print(clf.score(X_test,Y_test))
print_confusion_matrix(Y_test,Y_pred)
print_scores(Y_test,Y_pred)
```

8.5.6 XGBoost

XGBoost, also known as gradient boosting, is a much recent ensemble boosting algorithm just like AdaBoost. The main difference lies in the way it boosts the predictions. Instead of assigning weights according to the predicted values, the algorithm moves forward on the results of a loss function. The loss function calculates the loss value using the current predictions and the ground truth labels. The algorithm tries to minimize this loss function by computing the gradients with respect to it and taking one step in its suggested direction. The steps are taken according to a learning rate, which is a hyperparameter for the algorithm. After repeated iterations, the algorithm has its accuracy increased with the loss being settled to its minimal value [37].

```
% pip install xgboost
#Import XGBoost ensemble model
from xgboost import XGBClassifier
# Create XGB model
clf = GradientBoostingClassifier(n_estimators=1000,learning_rate=1)
#Train the model using the training set
clf.fit(X_train,Y_train)
# make predictions for test data
Y_pred = clf.predict(X_test)
#Print performance
print(clf.score(X_train,Y_train))
print(clf.score(X_val,Y_val))
print(clf.score(X_test,Y_test))
print_confusion_matrix(Y_test,Y_pred)
print_scores(Y_test,Y_pred)
```

8.6 Results and discussions

8.6.1 Performance evaluation measures

Performance on the training set is definitely not a good indicator of performance on an independent test set. The question of predicting performance based on limited data is an interesting, and still controversial, one. The classifier's performance is measured in terms of the number of samples for which the prediction was correct. This gives rise to accuracy of the classifier, the higher the accuracy, the more correct it is. But to assess the actual performance of the classifier, the accuracy on the training set is not a good measure. Therefore the need of a test set comes into picture, which is also a representative of the same dataset but which the classifier has never seen before. More formally, the training and test sets are totally disjoint. The classifier is trained on the training set to find a good classifier for the problem and then its performance is evaluated by measuring its accuracy on the test set. Sometimes, the dataset is divided into three disjoint sets called train, test, and validation. Here, the validation set is used to fine-tune the hyperparameters involved in the training of the classifier, while the purpose of the rest two remains the same. The training set is kept the largest to facilitate good training, while the test set is kept small [36].

8.6.1.1 F1 measure and confusion matrix

The true positives (TP) and true negatives (TN) refers to the number of samples for which the classification was correct. A false positive (FP) is when the outcome is incorrectly predicted as yes (or positive) when it is actually no (negative). A false negative (FN) is when the outcome is incorrectly predicted as negative when it is actually positive [36]. Information retrieval researchers define parameters called recall and precision:

$$Recall = \frac{TP}{TP + FN}$$

$$Precision = \frac{TP}{TP + FP}$$

Then F1 measure can be formulated as:

$$F1 = \frac{2 * Precision * Recall}{Precision + Recall}$$

Another important measure is the confusion matrix which can be formulated as:

$$Confusion\ Matrix = \begin{bmatrix} TP & FP \\ FN & TN \end{bmatrix}$$

8.6.1.2 *Receiver operating characteristic (ROC) analysis*

Receiver operating characteristic (ROC) curves depict the performance of a classifier without taking into account the actual error rate or cost. The curve is plotted by plotting "true positive rate" on the y-axis and "true negative rate" on the x-axis. Formally:

$$TP\ Rate = 100 \times TP/(TP \times FN)$$

$$FP\ Rate = 100 \times FP/(FP \times TN)$$

The area under the ROC curve (AUC) is a measure of the probability that the classifier ranks a positively predicted instance in front of a negatively predicted instance. Therefore this area is sometimes used as a performance metric in various classification tasks. Several methods are commonly employed for computing the area under the ROC curve [36].

8.6.1.3 *Kappa statistic*

Cohen's kappa statistic is a measure which takes into account the samples that were classified just by chance by the classifier using the concepts of expectation [36]. The kappa statistic (κ) always considers the predictions to be by chance and probabilistic. The probabilities of getting a correct prediction by chance and confidence are calculated using the concept of expectation of a random variable. The kappa statistic is the most frequently used statistic for the evaluation of categorical data when there is no independent means of assessing the probability of chance agreement between two or more observers. A kappa value of 0 designates agreement equivalent to chance, whereas a kappa value of 1 designates perfect agreement [38,39]. Cohen [40] defined the kappa statistic as an agreement index and defined as the following:

$$K = \frac{P_0 - P_e}{1 - P_e}$$

Where P_0 is observed agreement and P_e measures the agreement expected by chance [41].

8.6.2 Experimental results

8.6.2.1 *Transfer learning*

Table 8.1 shows the experiment results for transfer learning performed on various pretrained CNNs using our proposed architecture. We were able to achieve a high-test accuracy of

TABLE 8.1 Transfer learning experiment results.

Model	Accuracy	F1 Score	Kappa	AUC
DenseNet121	97.58%	0.973	0.947	0.974
VGG16	95.97%	0.971	0.943	0.971
ResNet50	95.37%	0.953	0.921	0.953
InceptionResNetV2	96.17%	0.945	0.891	0.946
InceptionV3	95.97%	0.972	0.944	0.972
MobileNet	96.17%	0.947	0.895	0.947
Xception	95.77%	0.941	0.883	0.941
VGG19	98.30%	0.982	0.964	0.982

98.30% with VGG19. F1 score, AUC, and Cohen's kappa were also found to be the highest with VGG19 with values of 0.982, 0.982, and 0.964, respectively. It should also be noted that all the models were able to yield test accuracy of more than 95% with F1 score greater than 0.94, AUC area greater than 0.94, and kappa value greater than 0.88.

```
def print_scores(Y_test,Y_pred):
    print("Confusion Matrix: ", confusion_matrix(Y_test,Y_pred))
    print("F1 score: ", f1_score(Y_test,Y_pred))
    print("Kappa: ", cohen_kappa_score(Y_test,Y_pred))
    print("ROC area: ", roc_auc_score(Y_test,Y_pred))
```

8.6.2.2 *Feature extraction with pretrained models*

K-NN, SVM, Random Forest, AdaBoost, XGBoost, and Bagging classifiers were applied to features extracted from each pretrained model using the proposed architecture as the second part of our experiments. The dimension of the extracted features was kept 256 after fine-tuning.

8.6.2.2.1 K-nearest neighbors

K-NN algorithm is one of the simplest classification algorithms. Even with such simplicity, it gave highly competitive results. The highest test accuracy achieved with the K-NN classifier was 91.75% with VGG16. The F1 score, AUC, and kappa for VGG16 were 0.916, 0.917, and 0.835 which are also pretty high compared to other architectures. Except for InceptionV3 and InceptionResNetV2, all other models were able to achieve a test accuracy of more than 80%, F1 score greater than 0.82, AUC greater than 0.82, and kappa value greater than 0.65. The performance of MobileNet, VGG19, and DenseNet121 were good and comparable to our best performing candidate VGG16 (Table 8.2).

8.6.2.2.2 Support vector machine

We observed that the results for the SVM classifier were slightly weaker as compared to K-NN classifier for VGG16 as the accuracy decreased from 91.75% to 88.93%, whereas it was slightly better for VGG19 and ResNet50 as it showed an increase. SVM classifier proved to be the best for VGG19 and ResNet50 as the test accuracy of 87.92% and 85.91% was the highest for them among all other methods. VGG16 achieved the highest test accuracy of 88.93% in case of SVM. The F1 score of 0.888, AUC of 0.889, and kappa value of 0.778 were also highest for VGG16. It is also worthy to mention that the performance of MobileNet and DenseNet121 was almost as good as VGG19 (Table 8.3).

TABLE 8.2 Feature extraction along with K-NN classifier.

	Accuracy	F1 score	Kappa	AUC
MobileNet	87.93%	0.873	0.758	0.879
Xception	82.89%	0.828	0.658	0.829
InceptionV3	78.67%	0.791	0.574	0.787
InceptionResNetV2	78.87%	0.785	0.577	0.788
VGG19	87.73%	0.875	0.745	0.877
VGG16	91.75%	0.916	0.835	0.917
DenseNet121	86.31%	0.86	0.762	0.863
ResNet50	83.90%	0.832	0.678	0.838

TABLE 8.3 Deep feature extraction along with SVM classifier.

	Accuracy	F1 score	Kappa	AUC
MobileNet	86.92%	0.869	0.738	0.869
Xception	79.07%	0.786	0.581	0.79
InceptionV3	75.85%	0.763	0.517	0.758
InceptionResNetV2	78.87%	0.789	0.578	0.788
VGG19	87.92%	0.882	0.758	0.879
VGG16	88.93%	0.888	0.778	0.889
DenseNet121	85.31%	0.853	0.706	0.853
ResNet50	85.91%	0.864	0.718	0.859

TABLE 8.4 Deep feature extraction along with Random Forest classifier.

	Accuracy	F1 score	Kappa	AUC
MobileNet	81.69%	0.8169	0.634	0.817
Xception	77.86%	0.786	0.558	0.779
InceptionV3	67.80%	0.676	0.356	0.678
InceptionResNetV2	78.47%	0.784	0.569	0.784
VGG19	81.08%	0.816	0.622	0.811
VGG16	84.31%	0.846	0.686	0.843
ResNet50	85.11%	0.85	0.702	0.851
DenseNet121	81.08%	0.811	0.622	0.811

8.6.2.2.3 Random Forest

In the case of Random Forest, ResNet50 performed the best with an accuracy of 85.11%, beating VGG16 by about a difference of 0.8% which is not a large margin. Similarly, the F1 score, AUC, and kappa values were also greater in case of ResNet50 compared to VGG16 by a very small margin. InceptionV3 and InceptionResNetV2 showed extremely poor compared to others, following the trend of the results of the K-NN and SVM classifier. Similar to SVM, the performance of MobileNet, DenseNet121, and VGG19 was also very good with the test accuracy being greater than 80% for all the three (Table 8.4).

TABLE 8.5 Deep feature extraction along with AdaBoost.

	Accuracy	F1 score	Kappa	AUC
MobileNet	77.66%	0.782	0.553	0.777
Xception	74.04%	0.746	0.481	0.741
InceptionV3	64.19%	0.645	0.284	0.642
InceptionResNetV2	73.84%	0.734	0.476	0.738
VGG19	80.28%	0.797	0.605	0.803
VGG16	82.49%	0.825	0.649	0.825
DenseNet121	78.67%	0.784	0.574	0.786
ResNet50	78.87%	0.791	0.577	0.788

8.6.2.2.4 AdaBoost and XGBoost

AdaBoost and XGBoost are the two famous boosting algorithms that are used widely. AdaBoost is sequentially growing decision trees as weak learners and punishing incorrectly predicted samples by assigning a more significant weight to them after each round of prediction. This way, the algorithm is learning from previous mistakes. On the other hand, in XGBoost, also known as gradient boosting, the boosting is done along with the gradient changes. The compared results of AdaBoost and XGBoost are shown in Table 8.5 and 8.6. It was observed that the performance of XGBoost was better than AdaBoost with a margin of 2%−5% for every model except DenseNet121, which clearly shows that XGBoost is better at classifying the CT scan features obtained from our proposed model. For both the classifiers, VGG16 performed the best with a test accuracy of 82.49% and 85.71%, respectively. The F1 score, ROC AUC, and kappa values also show the same pattern. Following the trends, for both classifiers we again observed that ResNet50, VGG19, and MobileNet gave

TABLE 8.6 Deep feature extraction along with XGBoost.

	Accuracy	F1 score	Kappa	AUC
MobilNet	81.08%	0.816	0.622	0.811
Xception	77.06%	0.771	0.541	0.771
InceptionV3	68.41%	0.688	0.368	0.684
InceptionResNetV2	74.64%	0.746	0.493	0.746
VGG19	82.49%	0.821	0.649	0.825
VGG16	85.71%	0.857	0.714	0.857
DenseNet121	77.46%	0.776	0.549	0.775
ResNet50	83.90%	0.839	0.678	0.839

TABLE 8.7 Deep feature extraction along with Bagging.

	Accuracy	F1 score	Kappa	AUC
MobileNet	80.88%	0.809	0.6177	0.809
Xception	77.86%	0.782	0.557	0.778
InceptionV3	68.61%	0.691	0.372	0.686
InceptionResNetV2	78.47%	0.787	0.569	0.785
VGG19	82.89%	0.828	0.658	0.829
VGG16	85.51%	0.858	0.71	0.855
DenseNet121	78.87%	0.793	0.577	0.788
ResNet50	80.48%	0.804	0.609	0.805

comparable performance. VGG19 was the second-best performer after VGG16. It should be noted that both these methods could not beat K-NN, SVM, and Random Forest as the obtained scores were lesser compared to them.

8.6.2.2.5 Bagging

Table 8.7 shows the results of the Bagging classifier, which proved to be on average with the XGBoost algorithm. The results for all the models were almost similar with no substantial changes observed in the test accuracies. The best test accuracy of 85.51% obtained is for VGG16 compared to a value of 85.71% for XGBoost. The F1 score, ROC AUC, and kappa are almost same for the cases. The results show that both Boosting and Bagging algorithms are equally efficient in classifying the extracted features.

For almost all the machine learning classifiers, VGG16 proved to give the highest test accuracy among all the models. The highest value obtained was 91.75% with the K-NN classifier. The F1 score, AUC, and Cohen's kappa were also the highest in this case. Among the machine learning classifiers, K-NN seemed to work the best followed by SVM, XGBoost, Bagging, Random Forest, and AdaBoost, respectively. It should also be noted that VGG19 and ResNet50 gave great results with the maximum value of 87.92% and 85.91% with SVM, respectively. They both lagged behind VGG16 by only a margin of 2%–4% for all the classifiers used. InceptionV3 performed the worst, whereas the performance of MobileNet and DenseNet121 was comparable to VGG16 and VGG19.

8.6.3 Discussion

Our proposed methods and architecture beat various other deep learning models. Our transfer learning approach yielded an accuracy of 98.30% compared to an accuracy of 95.12% by DeTrac [42] and 96% by Bai et al. [43]. Our model also performed better on the AUC, with a value of 0.982 compared to 0.96 by Li et al. [7]. Similarly, our machine learning-based approach yielded an accuracy of 91.75%, which is by far better than the deep learning approaches, such as accuracy of 90.1% by Zheng et al. [44]. AUC obtained by using SVM on our feature extractor is 0.889, which is better than 0.862 reported by Mei et al. [45], which shows that our proposed architecture is better at extracting features. These results support other unique feature extraction-based methods [46,47] and novel techniques [48,49] in this area.

8.7 Conclusion

A deep learning approach for COVID-19 detection from CT images has been applied to distinguish infection of COVID-19 from other common pneumonia diseases. A variety of deep and machine learning methods have been used to extract and identify images of COVID-19 to ensure a clinical diagnosis ahead of the pathogenic test for the disease control. The proposed architecture showed that the ImageNet pretrained models yielded 91.75% test accuracy using classical machine learning algorithms with a test accuracy of 98.30% using transfer learning. The present proposal increased the existing accuracy of the AUC area from 0.96 to a current value of 0.982. Similarly, the present machine learning-based approach increased the result to 91.75% rather than 90.1%. This study showed that the present architecture model is very promising in characterizing and diagnosing COVID-19 infections.

References

[1] WHO, Coronavirus disease (COVID-19) situation reports. <https://www.who.int/emergencies/diseases/novel-coronavirus-2019/situation-reports>, 2020 (accessed 08.08.20).

[2] A. Tahamtan, A. Ardebili, Real-time RT-PCR in COVID-19 detection: issues affecting the results, Expert Rev. Mol. Diagn. (2020) 1–2. Available from: https://doi.org/10.1080/14737159.2020.1757437.

[3] S. Wang, et al., A deep learning algorithm using CT images to screen for Corona Virus Disease (COVID-19), Infectious Diseases (except HIV/AIDS) (2020). Available from: https://doi.org/10.1101/2020.02.14.20023028. preprint.

[4] Y. Yu, H. Lin, J. Meng, X. Wei, H. Guo, Z. Zhao, Deep transfer learning for modality classification of medical images, Information 8 (3) (2017). Available from: https://doi.org/10.3390/info8030091. Art. no. 3.

[5] H.-C. Shin, et al., Deep convolutional neural networks for computer-aided detection: CNN architectures, dataset characteristics and transfer learning, IEEE Trans. Med. Imaging 35 (5) (2016) 1285–1298. Available from: https://doi.org/10.1109/TMI.2016.2528162.

[6] A. Pezeshk, S. Hamidian, N. Petrick, B. Sahiner, 3-D convolutional neural networks for automatic detection of pulmonary nodules in chest CT, IEEE J. Biomed. Health Inform. 23 (5) (2019) 2080–2090. Available from: https://doi.org/10.1109/JBHI.2018.2879449.

[7] L. Li, et al., Using artificial intelligence to detect COVID-19 and community-acquired pneumonia based on pulmonary CT: evaluation of the diagnostic accuracy, Radiology 296 (2) (2020) E65–E71. Available from: https://doi.org/10.1148/radiol.2020200905.

[8] H. Panwar, P.K. Gupta, M.K. Siddiqui, R. Morales-Menendez, P. Bhardwaj, V. Singh, A deep learning and Grad-CAM based color visualization approach for fast detection of COVID-19 cases using chest X-ray and CT-scan images, Chaos Solitons Fractals (2020) 110190. Available from: https://doi.org/10.1016/j.chaos.2020.110190.

[9] H. Meng, et al., CT imaging and clinical course of asymptomatic cases with COVID-19 pneumonia at admission in Wuhan, China, J. Infect. 81 (1) (2020) e33–e39. Available from: https://doi.org/10.1016/j.jinf.2020.04.004.

[10] N. Narayan Das, N. Kumar, M. Kaur, V. Kumar, D. Singh, Automated deep transfer learning-based approach for detection of COVID-19 infection in chest X-rays, IRBM (2020). Available from: https://doi.org/10.1016/j.irbm.2020.07.001.

[11] S. Lalmuanawma, J. Hussain, L. Chhakchhuak, Applications of machine learning and artificial intelligence for Covid-19 (SARS-CoV-2) pandemic: a review, Chaos Solitons Fractals 139 (2020) 110059. Available from: https://doi.org/10.1016/j.chaos.2020.110059.

[12] A.A. Ardakani, A.R. Kanafi, U.R. Acharya, N. Khadem, A. Mohammadi, Application of deep learning technique to manage COVID-19 in routine clinical practice using CT images: Results of 10 convolutional neural networks, Comput. Biol. Med. 121 (2020) 103795. Available from: https://doi.org/10.1016/j.compbiomed.2020.103795.

[13] H. Panwar, P.K. Gupta, M.K. Siddiqui, R. Morales-Menendez, V. Singh, Application of deep learning for fast detection of COVID-19 in X-Rays using nCOVnet, Chaos Solitons Fractals 138 (2020) 109944. Available from: https://doi.org/10.1016/j.chaos.2020.109944.

[14] D.-P. Fan, et al., Inf-Net: Automatic COVID-19 lung infection segmentation from CT images, ArXiv200414133 Cs Eess, May 2020, <http://arxiv.org/abs/2004.14133> (accessed 09.08.20).

[15] S. Minaee, R. Kafieh, M. Sonka, S. Yazdani, G. Jamalipour Soufi, Deep-COVID: Predicting COVID-19 from chest X-ray images using deep transfer learning, Med. Image Anal. 65 (2020) 101794. Available from: https://doi.org/10.1016/j.media.2020.101794.

[16] L. Brunese, F. Mercaldo, A. Reginelli, A. Santone, Explainable deep learning for pulmonary disease and coronavirus COVID-19 detection from X-rays, Comput. Methods Programs Biomed. 196 (2020) 105608. Available from: https://doi.org/10.1016/j.cmpb.2020.105608.

[17] M. Nour, Z. Cömert, K. Polat, A novel medical diagnosis model for COVID-19 infection detection based on deep features and bayesian optimization, Appl. Soft Comput. (2020) 106580. Available from: https://doi.org/10.1016/j.asoc.2020.106580.

[18] T. Tuncer, S. Dogan, F. Ozyurt, An automated Residual Exemplar Local Binary Pattern and iterative ReliefF based COVID-19 detection method using chest X-ray image, Chemom. Intell. Lab. Syst. 203 (2020) 104054. Available from: https://doi.org/10.1016/j.chemolab.2020.104054.

[19] S. Hassantabar, M. Ahmadi, A. Sharifi, Diagnosis and detection of infected tissue of COVID-19 patients based on lung x-ray image using convolutional neural network approaches, Chaos Solitons Fractals 140 (2020) 110170. Available from: https://doi.org/10.1016/j.chaos.2020.110170.

[20] T. Mahmud, M.A. Rahman, S.A. Fattah, CovXNet: a multi-dilation convolutional neural network for automatic COVID-19 and other pneumonia detection from chest X-ray images with transferable multi-receptive feature optimization, Comput. Biol. Med. 122 (2020) 103869. Available from: https://doi.org/10.1016/j.compbiomed.2020.103869.

[21] T.B. Alakus, I. Turkoglu, Comparison of deep learning approaches to predict COVID-19 infection, Chaos Solitons Fractals 140 (2020) 110120. Available from: https://doi.org/10.1016/j.chaos.2020.110120.

[22] A. Altan, S. Karasu, Recognition of COVID-19 disease from X-ray images by hybrid model consisting of 2D curvelet transform, chaotic salp swarm algorithm and deep learning technique, Chaos Solitons Fractals 140 (2020) 110071. Available from: https://doi.org/10.1016/j.chaos.2020.110071.

[23] A. Sufian, A. Ghosh, A.S. Sadiq, F. Smarandache, A survey on deep transfer learning to edge computing for mitigating the COVID-19 pandemic, J. Syst. Archit. 108 (2020) 101830. Available from: https://doi.org/10.1016/j.sysarc.2020.101830.

[24] T. Ozturk, M. Talo, E.A. Yildirim, U.B. Baloglu, O. Yildirim, U. Rajendra Acharya, Automated detection of COVID-19 cases using deep neural networks with X-ray images, Comput. Biol. Med. 121 (2020) 103792. Available from: https://doi.org/10.1016/j.compbiomed.2020.103792.

[25] A.I. Khan, J.L. Shah, M.M. Bhat, CoroNet: a deep neural network for detection and diagnosis of COVID-19 from chest x-ray images, Comput. Methods Programs Biomed. 196 (2020) 105581. Available from: https://doi.org/10.1016/j.cmpb.2020.105581.

[26] W.M. Shaban, A.H. Rabie, A.I. Saleh, M.A. Abo-Elsoud, A new COVID-19 Patients Detection Strategy (CPDS) based on hybrid feature selection and enhanced KNN classifier, Knowl.-Based Syst. 205 (2020) 106270. Available from: https://doi.org/10.1016/j.knosys.2020.106270.

[27] K. He, X. Zhang, S. Ren, J. Sun, Deep residual learning for image recognition, ArXiv151203385 Cs, Dec. 2015, <http://arxiv.org/abs/1512.03385> (accessed 08.08.20).

[28] G. Huang, Z. Liu, L. van der Maaten, K.Q. Weinberger, Densely connected convolutional networks, ArXiv160806993 Cs, Jan. 2018, <http://arxiv.org/abs/1608.06993>.

[29] A. G. Howard et al., MobileNets: Efficient Convolutional Neural Networks for Mobile Vision Applications, ArXiv170404861 Cs, Apr. 2017,<http://arxiv.org/abs/1704.04861>.

[30] F. Chollet, Xception: deep learning with depthwise separable convolutions, in: 2017 IEEE Conference on Computer Vision and Pattern Recognition (CVPR), Honolulu, HI, Jul. 2017, pp. 1800–1807, doi: 10.1109/CVPR.2017.195.

[31] K. Simonyan, A. Zisserman, Very deep convolutional networks for large-scale image recognition, ArXiv14091556 Cs, <http://arxiv.org/abs/1409.1556>, Apr. 2015 (accessed 8.8.20).

[32] C. Szegedy et al., Going deeper with convolutions, in: 2015 IEEE Conference on Computer Vision and Pattern Recognition (CVPR), Boston, MA, USA, Jun. 2015, pp. 1–9, doi: 10.1109/CVPR.2015.7298594.

[33] X. Wu, et al., Top 10 algorithms in data mining, Knowl. Inf. Syst. 14 (1) (2008) 1–37.

[34] J. Han, J. Pei, M. Kamber, Data Mining: concepts and techniques, Elsevier, 2011.

[35] L. Breiman, Random Forests, Mach. Learn. 45 (1) (2001) 5–32.

[36] M. Hall, I. Witten, E. Frank, Data Mining: Practical Machine Learning Tools and Techniques, Kaufmann Burlington, 2011.

[37] T. Chen, C. Guestrin, XGBoost: a scalable tree boosting system, in: Proc. 22nd ACM SIGKDD International Conference on Knowledge Discovery and Data Mining, pp. 785–794, Aug. 2016, doi: 10.1145/2939672.2939785.

[38] A.J. Viera, J.M. Garrett, Understanding interobserver agreement: the kappa statistic, Fam. Med. 37 (5) (2005) 360–363.

[39] C.A. Lantz, E. Nebenzahl, Behavior and interpretation of the κ statistic: Resolution of the two paradoxes, J. Clin. Epidemiol. 49 (4) (1996) 431–434.

[40] J. Cohen, A coefficient of agreement for nominal scales, Educ. Psychol. Meas. 20 (1) (1960) 37–46.

[41] Z. Yang, M. Zhou, Kappa statistic for clustered physician–patients polytomous data, Comput. Stat. Data Anal. 87 (2015) 1–17.

[42] (PDF) Classification of COVID-19 in chest X-ray images using DeTraC deep convolutional neural network (2020). https://www.researchgate.net/publication/340332332_Classification_of_COVID-19_in_chest_X-ray_images_using_DeTraC_deep_convolutional_neural_network (accessed 09.08.20).

[43] H.X. Bai, et al., AI augmentation of radiologist performance in distinguishing COVID-19 from pneumonia of other etiology on chest CT, Radiology 296 (2020). Available from: https://pubs.rsna.org/doi/full/10.1148/radiol.2020201491 (accessed 09.08.20).

[44] C. Zheng et al., Deep learning-based detection for COVID-19 from chest CT using weak label, infectious diseases (except HIV/AIDS), preprint, Mar. 2020. doi: 10.1101/2020.03.12.20027185.

[45] X. Mei, et al., Artificial intelligence-enabled rapid diagnosis of patients with COVID-19, Nat. Med. (2020) 1–5. Available from: https://doi.org/10.1038/s41591-020-0931-3.

[46] F. Ozyurt, T. Tuncer, A. Subasi, An automated COVID-19 detection based on fused dynamic exemplar pyramid feature extraction and hybrid feature selection using deep learning, Comput. Biol. Med. 132 (2021) 104356.

[47] T. Tuncer, F. Ozyurt, S. Dogan, A. Subasi, A novel Covid-19 and pneumonia classification method based on F-transform, Chemometr. Intell. Lab. Syst. 210 (2021) 104256. Available from: https://doi.org/10.1016/j.chemolab.2021.104256. 15 March 2021.

[48] A. Subasi, S.A. Qureshi, T. Brahimi, A. Serireti, COVID-19 detection from X-Ray images using artificial intelligence, Artificial Intelligence and Big Data Analytics for Smart Healthcare, Elesevier, 2021.

[49] A. Subasi, A. Mitra, F. Ozyurt, T. Tuncer, Automated Covid-19 detection from CT images using deep learning, in: V. Bajaj, G.R. Sinha (Eds.), Computer-aided Diagnosis and Design Methods for Biomedical Applications, CRC Press, Taylor & Francis, 2021.

CHAPTER

9

Detection and classification of Diabetic Retinopathy Lesions using deep learning

Siddhesh Shelke[1] *and Abdulhamit Subasi*[2,3]

[1]Indian Institute of Technology, Indore, Madhya Pradesh, India [2]Institute of Biomedicine, Faculty of Medicine, University of Turku, Turku, Finland [3]Department of Computer Science, College of Engineering, Effat University, Jeddah, Saudi Arabia

OUTLINE

9.1 Introduction

Diabetic retinopathy (DR) is a diabetic condition that causes the retina's blood vessels to enlarge and leak fluids and blood. If DR progresses to an advanced degree, it might result in visual loss. DR is responsible for 2.6% of blindness worldwide. Diabetes patients who have been sick with the condition for a long time are more likely to get DR. Regular retina screening is necessary for diabetic people to identify and treat DR early enough to avoid blindness. The presence of

Applications of Artificial Intelligence in Medical Imaging
DOI: https://doi.org/10.1016/B978-0-443-18450-5.00004-9

various sorts of lesions on a retina picture is used to identify DR. DR is one of the leading causes of blindness in the working age of developing countries. It is estimated that more than 93 million people will be affected. It is an "eye disease" that causes meningitis as a long-term consequence of diabetes, which results in progressive damage to the eye and even blindness [1]. Because diabetes is a progressive disease, doctors recommend that people with diabetes be checked at least twice a year to identify symptoms regularly.

Risks factors of DR:

- Duration of diabetes:
  - A patient diagnosed before age 30 years
  - 50% DR after 10 years
  - 90% DR after 30 years
- Poor metabolic control:
  - It is less essential but quite pertinent to the onset and progression of DR.
  - Increased HbA1c is associated with the increased risk.
- Pregnancy:
  - It is linked to a rapid progression of DR
  - Prediction factors—Poor pregnancy control of diabetes mellitus (DM), too rapid control during the early stages of pregnancy, preeclampsia, and fluid imbalance are all risk factors.
- Hypertension:
  - It is most common in patients with DM type 2.
  - Should strictly control (<140/80 mm of Hg).
- Nephropathy:
  - Associated with the worsening of DR.
  - Renal transplantation may be linked to a reduction in DR and a better response to photocoagulation.
- Others:
  - Smoking
  - Obesity
  - Hyperlipidemia
  - Anemia

DR is a chronic disease that appears only in the late stages when it is difficult and impossible to treat. Early detection by diagnosis is essential because it is often treated effectively in the early stages. The cost of this task is high, so early detection is important to reduce labor. It is needed to automatically detect defects in the eye image at low cost using digital image processing and artificial intelligence (AI) algorithms. In DR, blood vessels that help in sustaining the retina begins to leak fluid and blood onto the retina, which can result in formation of visual features known as lesions such as microaneurysms, hemorrhages, hard exudates, cotton spots, and vessels area [1]. In a medical diagnosis, an ophthalmologist examines an image of a colored background to examine the patient's condition. This diagnosis is difficult and time-consuming and introduces additional errors. Furthermore, because of the vast number of diabetics and a lack of health resources in some locations, most DR patients are unable to be detected and treated on time, suffer permanent eyesight loss, and even lack vision.

Rapid detection of DR, especially in early stages, can effectively control and delay degenerative conditions. At the same time, the impact of hand interpretation depends to a large extent on the understanding of the doctor. Medical malpractice occurs due to incompetence of doctors. Convolutional neural networks (CNNs) have surpassed all previous image analysis techniques in computer vision and image classification tasks over the past decade. Computer-assisted diagnosis is more effective because it allows screening for a large number of diseases. The conditions that cause microangiopathy can lead to the formation of microaneurysms. Hard exudates are white or creamy colors that are very bright in the retina. If they appear near the middle of the macula and show fluid in the eyeball, they are considered very dangerous. Bleeding scarring is the most common type of bleeding due to DR. It is a small hemorrhage that originates from the cervical network. Planting lesions are upper lesions on the yellow or white eyeball along the side of the hair. They describe eyelid edema as a result of ischemia. They finally recovered on their own after 3 months. In case foundational ischemic condition is not diagnosed, fresh lesions can form in a variety of locations [2].

As per the International Clinical Diabetic Retinopathy and Early Treatment Diabetic Retinopathy Study Research Group [3] severity scale, various categories of DR (Fig. 9.1) are defined as:

1. There is no visible retinopathy.
2. Diabetic with mild nonproliferative complications—retinopathy is defined as the presence of at least one microaneurysm, with or without the presence of other lesions.
3. Moderate nonproliferative DR—numerous microaneurysms and retinal hemorrhages with or without cotton wool spots are present.
4. Severe nonproliferative DR—(1) numerous hemorrhages and microaneurysms in all four quadrants of the retina, (2) cotton wool spots in two or more quadrants, and (3) intraretinal microvascular abnormalities in at least one quadrant.
5. Proliferative DR—an advanced stage in which new thin and delicate blood vessels are formed, increasing the risk of spillage and causing severe vision loss or blindness.
6. Macular edema exudates or visible stiffening within one disk diameter of the fovea could also be caused by DR, which can be vision-threatening.

Diabetes is the leading cause of blindness in many developed countries. The same applies to developing countries. People with diabetes

Retina Images	DR Levels
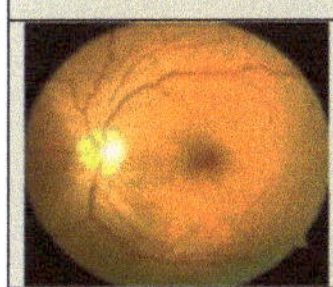	Normal (No DR)
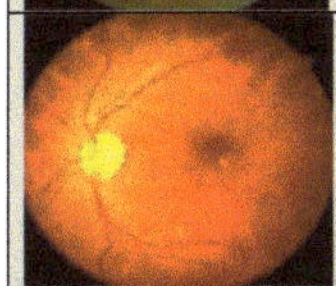	Mild DR
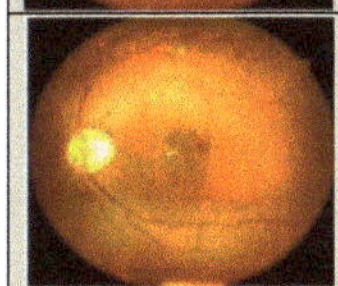	Moderate DR
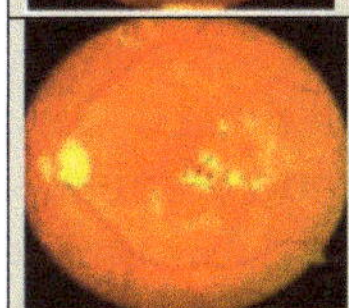	Severe DR
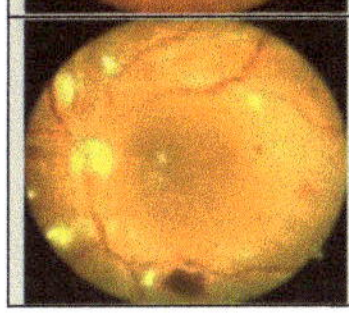	Proliferative DR

FIGURE 9.1 Diabetic retinopathy levels of the retinal images.

can be more blind than patients without diabetes. AIDS and macular edema (both essential in the hospital) can cause severe vision loss. This affects the eyeballs and can cause blindness in diabetics. DR affects many diabetics in developed countries [2].

The goal of this chapter is to use fundus image classification to enforce a direct synthesis of DR. We are working on categorizing fundus images depending on the level of DR, with the goal of achieving end-to-end actual classification from fundus images to medical status. Rather than the doctors' manual control with expertise, it helps to relieve their pressure on the diagnosing and treating of DR in a simple and accurate manner. For this task, a variety of image preprocessing and AI techniques to extract many key features is used and then classify them into their corresponding classes. We use CNN architecture to detect DR in two datasets. The precision, recall, accuracy, receiver operating characteristic (ROC), and area under curve (AUC) measures are all evaluated. We also plot the confusion matrix, which helps us to confirm the strength of the model visually.

9.2 Literature survey on diabetic retinopathy detection

Much effort has been made in DR detection. There are many ways to find a DR. Scientists have worked on a variety of techniques to treat a variety of injuries such as blood vessels, microaneurysms, secretions, and bleeding. Changes in the shape and size of blood vessels can be a positive sign of DR, as well as the presence of various types of lesions that contribute to the diagnosis of diabetes. As a result, various studies on the autonomic nervous system fall into two categories [4–6].

9.2.1 Traditional diabetic retinopathy detection approach

Chandrashekar [7] proposed a method for extracting retinal vessels from retinal fundus images using morphological approaches. Kaur and Sinha [8] suggested a blood vessel segmentation approach based on morphological filters. There is no noticeable improvement in performance when increasing the number of filter banks; instead, the convolution process, which is time-consuming task, is increased. Jaffar et al. [9] proposed a way that uses reconciling thresholding for exudate detection and eliminates artifacts from the exudates; the retinal structure area unit is utilized in classification. The projected technique failed to cowl all the DR signs; it has to be explored. Jiang and Mojon [10] proposed a way, reconciling thresholding on verification-based multithreshold inquisitory approach. With international thresholding, the blood vessels can't be divided because of the image gradients. So, image inquisitory with varied threshold values does not extract the thresholded image. Clara I Sánchez et al. [11] proposed a combination model that separates the exudates from the image background, and edge detection strategies area unit want to separate laborious and soft exudates.

Goh et al. [12] classified retinal images using various classifiers. On fundus images, segmentation was used to differentiate blood vessels, microaneurysms, and exudates. The classifiers were given the segmented region, textural data, and other information derived from Gray-Level Co-Occurrence Matrix (GLCM) to categorize the normal and abnormal images. On normal images, the detection system has a success rate of 92%, while on aberrant images, it has a success rate of 91%. Liew et al. [13] employed a statistical technique to demonstrate the relationship between retinal vascular indicators and the importance of both qualitative and quantitative evaluation of retinal vasculature. The suggested technology will require skilled assistance in identifying blood vessels. Reza et al. [14] used a marker-controlled primarily based watershed segmentation methodology that detects the exudates and points. Before that, average filtering and distinction improvement were applied to remove artifacts.

Hoover and Goldbaum [15] suggested a fuzzy-based voting system for detecting the location of the optic disk, where numerous feature elements overlap. Li and Chutatape [16] employed an active shape model to extract the vascular components based on the optic disk location, which is a model-based technique. The central macular region was discovered using the retrieved data. Gardener et al. [17] used artificial neural network to seek out completely different options such as exudates, blood vessels, and hemorrhages with 93.1%, 91.7%, and 73.8% of accuracy. Yun et al. [18] planned associate automatic DR arrangement that classifies delicate, moderate, associated severe nonproliferative DR and proliferative DR and achieved an accuracy of 72% that extracts six options. In general, the normal approach for DR detection and classification will be created, as shown in Fig. 9.2. Initially, the images are collected and preprocessed. Image phase action will be performed to segment the essential half, image options square measure extracted manually, and classification will be mistreated binary or multiclass classifiers singly.

9.2.2 Binary and multilevel classification

In DR disease detection, good number of studies are created on many different classification techniques. In binary classification, the DR illness has been classified into two categories solely. As per studies, the two categories are often DR or No-DR. In multiclass classification problem, the DR illness has been classified into several categories as No-DR, Mild, Moderate, Severe, Proliferate-DR.

9.2.2.1 *Binary classification*

Quellec et al. [19] projected an automatic detection methodology in which three CNNs

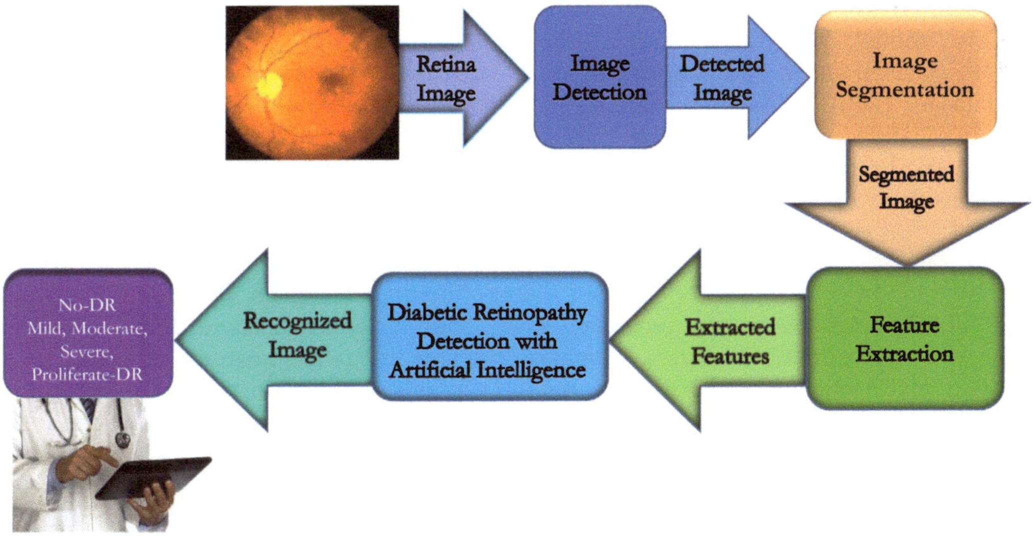

FIGURE 9.2 Diabetic retinopathy detection and classification process using traditional approach.

(AlexNet and two different networks) were used to observe microaneurysms, hemorrhages, soft and laborious exudates from three completely different datasets, Kaggle, DiaretDB1, and E-ophtha (private). The images of complex body part were resized, cropped, normalized, augmented, and therefore the morphological filter was applied within the preprocessing section. The illness was classified into two categories, attributable and nonreferable DR, and made a mythical creature price of 0.954 and 0.949 in Kaggle and E-ophtha, respectively.

Jiang et al. [20] projected a model wherever three pretrained CNNs (Inceptionv3, ResNet152, and InceptionResNetv2) do not to classify the dataset as attributable DR or nonreferable DR. Before CNN training, the images were resized, enhanced, and improved, and so the models were integrated with AdaBoost technique. Further to update the network weights, Adam optimizer was used and therefore the system achieved 88.21% accuracy and AUC of 0.946.

Zago et al. [21] projected a technique wherever two CNNs (pretrained VGG16 and a CNN) were utilized to observe DR or non-DR images supported the red lesion patches likelihood. This model was trained on the DIARETDB1 dataset, and it had been tested on the few datasets: IDRiD, Messidor, Messidor-2, DDR, DIARETDB0, and Kaggle. The model achieved the good results on the Messidor dataset with a sensitivity of 0.94 and AUC of 0.912.

9.2.2.2 *Multilevel classification*

Pratt et al. [22] planned a way wherever a CNN was used with 10 CNN layers, 8 max-pooling layers, and 3 totally connected layers, and a softmax classifier was accustomed classify the Kaggle dataset images into five categories according to the severity levels of DR. Throughout the preprocessing part, the images are color normalized and resized, and to scale back overfitting, L2 regularization and dropout techniques were used. The system created a specificity of 95%, accuracy of 75%, and a sensitivity of 30%.

Gulshan et al. [23] planned a way wherever 10 CNNs (pre-trained Inceptionv3) were trained to discover diabetic macular dropsy (DME) and DR. Eyepacs-1 and Messidor-2 datasets were accustomed to check the CNN model. The dataset images were initially normalized, resized, and fed into the CNN model to classify the images into ascribable DME, moderate/worse DR, severe/worse DR, or totally hierarchical DR. The model created a specificity of 93% in two of the datasets taken and sensitivity of 97.5% and 96.1% in yepacs-1 and Messidor-2 datasets, respectively.

9.2.3 Datasets

a) **Diabetic-Ratinopathy_Sample_Dataset_Binary**[i] - This dataset is sample data from the Diabetic Retinopathy Competition. It takes an excessive amount of time in preprocessing, so this dataset contains resized images (90,128,264) for saving a while. It contains only 526 samples 1/2 of the samples has DR, and half have not. Metadata of array are given in Binary Dataframe CSV, and images are on an exhibition on the identical index.
b) **Diabetic Retinopathy 224 × 224 Gaussian Filtered**[ii] - The given image used to identify the diabetes virus. The first dataset can be purchased at Discover Blind APTOS 2019. This image has been reshaped to 224 × 224 pixels, making it easier to use with many pretrained advanced learning models. With the dashboard, all images are stored in a folder, depending on the intensity/degree of diabetes. A CSV file is provided. There

[i] https://www.kaggle.com/sohaibanwaar1203/prepossessed-arrays-of-binary-data.

[ii] https://www.kaggle.com/sovitrath/diabetic-retinopathy-224x224-gaussian-filtered.

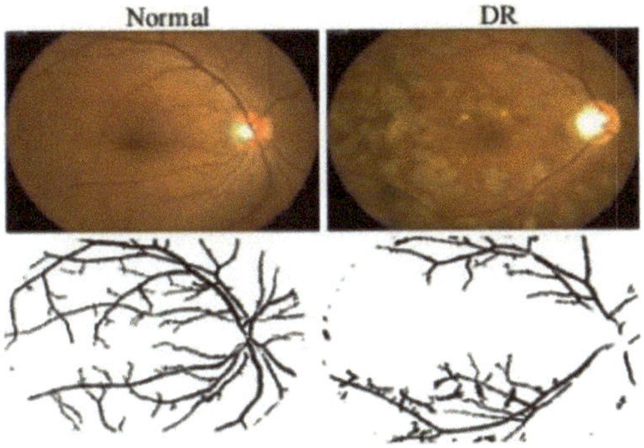

FIGURE 9.3 Comparison between normal and DR-infected retina. *DR*, Diabetic retinopathy.

are 5 different image folders: 0—DR None, 1—Easy, 2—Medium, 3—Intense, 4—lifearaDR. The database contains output. The pkl file contains a ResNet34 model that has been trained on datasets for 20 years using the FastAI library. Sample images are given in Fig. 9.3.

9.3 Deep learning methods for diabetic retinopathy detection

This study utilizes some deep learning (DL) techniques for the early detection of DR. This disease could be a real threat to the human species and be kept in check as soon as possible. With the advancement of computer vision and CNNs, finding out about diabetes has become a major topic of debate. In recent years, emerging biological networks have emerged as a major model of in-depth imaging, discovery, and dissemination activities. This type of neural network has been discussed for many years. [24,25]. Nonetheless, it gained popularity following the advent of DL [26–28] which were powered by graphic processing units [29]. However, while deep CNNs are becoming more and more popular today, strong evidence suggests that medium to medium-sized interventions do not improve performance. This idea has led to efforts to improve other aspects of CNN, such as offline gaming activities, directors, limiting methods, and regularization strategies. Another well-known area of research in this area is the integration of state-of-the-art methods for better methods, technologies, and algorithms. While several parts of the world have supported CNN's mission, many readers use their critical thinking position in neuroscience, neurobiology, and research science to explore new ways of evolving with living things. We tried to improve performance by combining and developing new discoveries to help build the system.

DL is a form of engineering learning system that uses offline workflows for learning and process experience [30]. One of the computer-aided diagnoses and detection methods is DL [31]. Image processing, detection, search, and registration are examples of DL applications in medical imaging research. Recently, DL has been used to detect and classify DR. Many different methods are combined to manage the characteristics of the installation files [32]. There are numerous DL-supported models such as restricted Boltzmann machines, CNNs, autoencoders, and sparse coding [33]. When the amount of training data increases, the performance of those methods improves [34]. This is due to an increase in the learned features, as opposed to machine learning methods. Furthermore, DL techniques eliminated the need for handcrafted feature extraction.

9.3.1 Deep neural networks

Because this system uses a mathematical "weight scale" to determine the possible output of the input data, deep neural network (DNN) analysis is more likely to handle such transactions. The "weights" are adjusted by creating a network with data from well-known databases. You can also test the web with new data to give the possibility that this data is included in a particular output. Such computer programs have been used in medicine to evaluate chest radiographs [35] and images from histopathology [36]. The use of a DNN in the screening of patients by mammography provided a better prediction of the detection of malignancy than inexperienced radiographers [37]. Moreover, DNNs are often used to medicate visual perception deficiencies in ophthalmology [38] (Fig. 9.4).

A neural network was trained in this study to recognize the features of diabetic fundus images, and its accuracy, F1 score, was tested. To see if a trained DNN can be used in the diagnosis of diabetic patients, the system's predictive power to correctly detect retinopathy in fundus images was investigated. While using this method, 10 layers of neural networks can be used by the number of neurons varying from 8 to 256 in a manner mainly in $8 \times 16 \times 32 \times 64 \times 128 \times 256 \times 128 \times 64 \times 32 \times 16 \times 8$ neuron configuration. The accuracy was not impressive since the data was imbalanced concerning the number of images in each class.

```
import tensorflow as tf
from tensorflow.keras.applications import *
from tensorflow.keras.optimizers import *
from tensorflow.keras.losses import *
from tensorflow.keras.layers import *
from tensorflow.keras.models import *
from tensorflow.keras.callbacks import *
from tensorflow.keras.preprocessing.image import *
from tensorflow.keras.utils import *

dnn_model=Sequential()
dnn_model.add(Dense(8, input_dim=2, kernel_initializer = 'uniform', activation = 'relu'))
# dnn_model.add(BatchNormalization())
# dnn_model.add(Dropout(0.2))
dnn_model.add(Dense(16, kernel_initializer = 'uniform', activation = 'relu' ))
# dnn_model.add(BatchNormalization())
# dnn_model.add(Dropout(0.2))
dnn_model.add(Dense(32, kernel_initializer = 'uniform', activation = 'relu' ))
# dnn_model.add(BatchNormalization())
# dnn_model.add(Dropout(0.2))
dnn_model.add(Dense(64, kernel_initializer = 'uniform', activation = 'relu' ))
# dnn_model.add(BatchNormalization()
# dnn_model.add(Dropout(0.2))
dnn_model.add(Dense(128, kernel_initializer = 'uniform', activation = 'relu'))
# dnn_model.add(BatchNormalization())
# dnn_model.add(Dropout(0.2))
dnn_model.add(Dense(256, kernel_initializer = 'uniform', activation = 'relu' ))
# dnn_model.add(BatchNormalization())
# dnn_model.add(Dropout(0.2))
dnn_model.add(Dense(128, kernel_initializer = 'uniform', activation = 'relu' ))
# dnn_model.add(BatchNormalization())
# dnn_model.add(Dropout(0.2))
dnn_model.add(Dense(64, kernel_initializer = 'uniform', activation = 'relu' ))
# dnn_model.add(BatchNormalization())
# dnn_model.add(Dropout(0.2))
dnn_model.add(Dense(32, kernel_initializer = 'uniform', activation = 'relu' ))
# dnn_model.add(BatchNormalization())
# dnn_model.add(Dropout(0.2))
dnn_model.add(Dense(16, kernel_initializer = 'uniform', activation = 'relu' ))
# dnn_model.add(BatchNormalization())
# dnn_model.add(Dropout(0.2))
dnn_model.add(Dense(8, kernel_initializer = 'uniform', activation = 'relu' ))
# dnn_model.add(BatchNormalization())
# dnn_model.add(Dropout(0.2))
dnn_model.add(Dense(2,activation='softmax'))
dnn_model.summary()
```

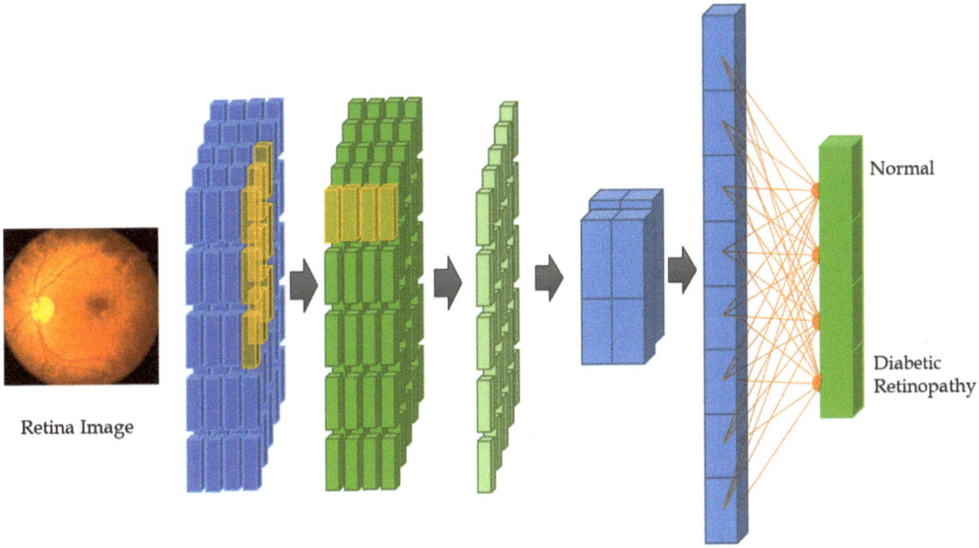

FIGURE 9.4 Deep learning model for diabetic retinopathy detection and classification.

9.3.2 Convolutional neural networks

A CNN is a process model that takes small pixels as input and routes them through a defined set of elements in the network. During learning, the network itself generates a low-level DR signal, gradually converting and integrating it into higher-resolution DR functions. These properties are automatically combined to provide a great opportunity to map the image as normal or abnormal [39].

```
model = Sequential()
model.add(Conv2D(32, (3, 3), padding='same', activation='relu', input_shape=(264,264,3)))
model.add(MaxPooling2D(pool_size=(2, 2)))model.add(Dropout(0.25))
model.add(Conv2D(64, (3, 3), padding='same', activation='relu'))
model.add(Conv2D(64, (3, 3), padding='same', activation='relu'))
model.add(MaxPooling2D(pool_size=(2, 2)))model.add(Dropout(0.25))
model.add(Conv2D(128, (3, 3), padding='same', activation='relu'))
model.add(Conv2D(128, (3, 3), padding='same', activation='relu'))
model.add(MaxPooling2D(pool_size=(2, 2)))model.add(Dropout(0.25))
model.add(Conv2D(256, (3, 3), padding='same', activation='relu'))
model.add(Conv2D(256, (3, 3), padding='same', activation='relu'))
model.add(MaxPooling2D(pool_size=(2, 2)))model.add(Dropout(0.25))
model.add(Conv2D(128, (3, 3), padding='same', activation='relu'))
model.add(Conv2D(128, (3, 3), padding='same', activation='relu'))
model.add(MaxPooling2D(pool_size=(2, 2)))model.add(Dropout(0.25))
model.add(Conv2D(64, (3, 3), padding='same', activation='relu'))
model.add(Conv2D(64, (3, 3), padding='same', activation='relu'))
model.add(MaxPooling2D(pool_size=(2, 2)))model.add(Dropout(0.25))
model.add(Conv2D(32, (3, 3), padding='same', activation='relu'))
model.add(Conv2D(32, (3, 3), padding='same', activation='relu'))
model.add(MaxPooling2D(pool_size=(2, 2)))model.add(Dropout(0.25))
model.add(Dense(16, activation='relu'))model.add(Dropout(0.5))
model.add(Flatten())model.add(Dense(16, activation='relu'))
model.add(Dropout(0.5))model.add(Dense(5, activation='softmax'))
```

9.3.2.1 Layers in convolutional neural network

In this experiment, the CNN-DRD model developed has two to eight flexible layers, schematics, and a reading system. CNN is a well-known DL model used by scientists to distribute natural images and has proven to be a successful method for medical images. For example, using monetary imaging, the CNN model plays an important role in the proper distribution of the Nonproliferative Diabetic Retinopathy (NPDR). It also increases the efficiency, availability, and cost-effectiveness of your Diabetic Retinopathy (DR) scoring system. The DR scoring system is optimized for a variety of high-quality images and different settings compared to traditional hand-crafted methods [39]. Considering the popularity of CNN architectures for DR diagnosis, they have significant limitations that are described below.

1. The current CNN model for detecting DR focuses only on the DR distribution and does not detect the status of DR lesions in the background image. They work from start to finish. From end to end of the CNN means that the input image is imported directly into the CNN and the output image provides DR intensity. Although the details of the DR injury are important to hospital staff.
2. The latest CNN models require high-quality information models for training, which is an expensive and time-consuming task. In contrast, CNN models, which can be learned from several examples, are the most effective and demanding practice in medical education today in terms of the value of DR.
3. CNN models that had been developed failed to learn the complex behavior of DR lesions. Initially, the fundus image was split into discrete segments that were used to feed into the CNN. As a result, small lesions are difficult to detect due to the incomprehensible nature of the optic nerve of the eye. Studying these vulnerabilities is essential and necessary for a real DR classification system. Moreover, this manually designed method does not cover all DR signals in the image, which is a waste of time investigating common problems. As a result, the DR solution is limited. Unlike hand-drawn images, a DL technique known as CNN has significantly increased DR values and reduced the time constraint on the need to use more data to know financial characteristics. CNN's traditional technique and functional design fall into the accessories category. The manually designed method demonstrates the process of moving an object using advanced algorithms and expert techniques. CNN is a DL process that works and is created by the human brain system. Learn key ideas from your data and understand the design process with minimal control. CNN is very motivated to practice accurate predictions, even if there are blockages, posts, and the true nature of the object is not possible. In addition, the benefits of expanding data are used to learn how to make changes and improve mapping capabilities through the CNN system. Scientists are currently adapting several modifications to the CNN system to improve their applications in physics [40].

9.3.3 Transfer learning

In order to reach high accuracy in training a deep CNN, a substantial amount of training data is typically required, which can be costly to obtain. This problem is addressed by transfer learning, which transfers knowledge learnt on a big dataset with a similar domain to the training dataset. A common method for CNNs is to train them on a big source dataset and then exploit their feature extraction skills. Furthermore, fine-tuning the resulting pre-trained model on a smaller target dataset with a comparable domain but a different goal has been shown to enhance task accuracy even further. DL techniques require many examples or a large set of data. The number of samples in our dataset are fewer as compared to modern data standards, that is, the number of MRI images for the classification function is fewer, so we can go for transfer learning. Transfer learning is about taking features learned on one issue, and using them on a different, related issue. Export learning is aimed at improving learning in the target task by incorporating information from the source task. The key facts to test before using transfer learning are that the source task should have been trained on a larger dataset than the target task, and the source and target tasks should be identical in nature. There are three common behaviors that can improve learning through a transfer. First is the first achievable output in the target task that uses only the information sent, compared to the first output of the ignorant person before further learning takes place. Second, there is a relationship between the time it takes to fully understand the target's mission and the time it takes to understand it from scratch, even though the information is being transmitted and third, the final level of output which can be reached in the target role relative to the final level without transition.

Here we used different models such as ResNet50, ResNet101, VGG16, VGG19, InceptionV3, MobileNet, MobileNetV2, DenseNet169, DenseNet121, InceptionResNetV2, and MobileNetV2. Simonyan and Zisserman introduced the VGG network architecture in the 2014 article "Very Deep Convolutional Networks for Large-Scale Image Recognition." This network is characterized by its simplicity, using only three universal layers on top of each other in detail. Maximum pooling works up to large volumes. Compared to traditional sequential network architectures such as AlexNet, Over Feat, and VGG, ResNet is a type of "exotic architecture" based on microarchitecture modules (also known as "network-in-network architectures"). The term microarchitecture refers to the collection of building blocks used to create a network. Was an original study showing it can be given? There are 3 and 5 brackets in the same network module, and these filter outputs are stacked along the dimensions of the channel before being sent to the next network layer (Fig. 9.5).

9.4 Diabetic retinopathy detection using deep learning

AI is a critical approach that enhances human expertise in the health-care industry by

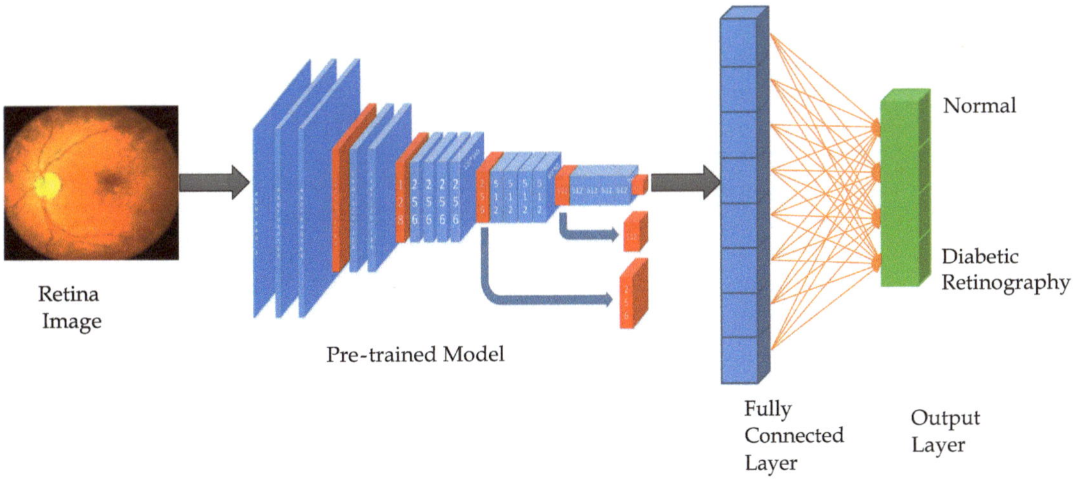

FIGURE 9.5 Transfer learning model for diabetic retinopathy detection and classification.

training complex medical data using algorithms that create advanced research data. AI and DL are applications in the field of ophthalmology because most of the data is image-based and the results are consistent with image recognition. AI applications in DR research have succeeded in providing recommendations and research presentations. The role of AI in the diagnosis of rheumatoid arthritis (RDR), expressed as moderate to severe or higher DR retinopathy (NPDR), is very consistent with or without macular degeneration (DME). This offers the right benefits. Over 90% of the studies use various AI algorithms [41]. AI algorithms require skilled professionals to obtain clear and concise images for use as input data, and ophthalmologists/ophthalmologists (informants) to provide full field accuracy for imaging. A recent study of eye imaging on smartphones and EyeArt AI software showed a high efficiency of 95.8% for detecting DR of any severity and over 99% efficiency for expertise in detecting RDR and STDR [42].

9.4.1 Prediction and classification

CNNs are similar to neural networks, with a few exceptions. CNN has three aspects: width, height, and depth. There is no absolute relationship between the inner line, that is the nerves in one layer are not connected to the nerves in the next line. The end result was optimized for a single vector. Two main steps of CNNs are to improve features and distribution. Collaboration is the process of removing tasks through a series of conflicts and integration tasks. The issue of this extraction feature will be the input for distribution. On top of extracted features, a fully connected layer will act as a classifier. Once the feature extraction is done, the classification of images will start.

This section will go through two datasets, one imbalanced and the other is balanced. The imbalanced dataset consists of Binary. npz files that have 1000 images. We split them into 600 training, 200 validation, and 200 testing images. We predict only 2 Classes (0 and 1) using this dataset. The balanced dataset consists of 3662 224 × 224 Gaussian images. We split them 2966 training images, 329 validation images, and 367 testing images. We predict both 2 Classes (0 and 1) and 5 Classes (0, 1, 2, 3, and 4) using this dataset.

9.4.1.1 For Dataset a) Diabetic-Retinopathy Sample_Dataset_Binary

In this experiment, we apply CNN with different layers and different neuron configurations. This approach did not give significant differences in the evaluation metrics.

```
import numpy as np
import matplotlib.pyplot as plt
y_test=np.argmax(y_test, axis=1)
pred=np.argmax(model.predict(x_test),axis=-1)
cm=confusion_matrix(y_test,pred)
cm_plot=plot_confusion_matrix(cm,classes=['0','1'])
```

We can observe that in some circumstances, the evaluation metrics are not so good. This occurs because the dataset is severely imbalanced, resulting in relatively poor scores. To deal with this, we could use data augmentation or Generative Adversarial Networks (GAN) to balance the dataset.

9.4.1.2 For Dataset b) Diabetic Retinopathy 224 × 224 Gaussian Filtered

This dataset is relatively balanced than the previous dataset, which was imbalanced. In this experiment, we have applied CNN with different layers and other DL. Also, we predict both 2 Classes and 5 Classes. We get good results in the 2 Classes since the number of images for Class 0 is almost equal to the number of images for Class 1.

9.4.2 Performance evaluation metrics

It is critical to estimate how precisely a classification model predicts the correct result when developing one. This estimation, however, is insufficient because it can produce deceptive results in some cases. And it is at this point where the new requirements become an important factor in determining the more significant estimations of the constructed model.

For classification models, accuracy is a critical metric. It is straightforward to understand and apply to binary and multiclass classification problems. Accuracy indicates the proportion of tangible results in the total number of records tested. The classification model, which is entirely built from balanced datasets, is accurate enough to be tested. Precision is defined as the ratio of true positives to predicted positives. Another important metric is recall, which gives information if all potential positives must be captured. Recall is the percentage of overall positive samples that were correctly predicted as positive. The recall is one if all positive samples are predicted to be positive. If an optimal combination of precision and recall is required, these two measures can be combined to form the F1 score. The F1 score is the harmonic mean of precision and recall, ranging from 0 to 1.

Many performance metrics are used to assess the classification performance of DL methods. Accuracy, recall, precision, and area under the ROC curve are some of the commonly used metrics in DL. The percentage of abnormal images classified as abnormal is referred to as sensitivity, and the percentage of normal images classified as usual is referred to as specificity [43].

AUC is area under the curve that is formed by plotting sensitivity versus specificity. The percentage of correctly classified images is referred to as accuracy. The equations for each measurement are listed below:

(1) Accuracy = TN + TP/(TN + TP + FN + FP)
(2) Precision = TP/(TP + FP)
(3) Recall = TP/(TP + FN)
(4) F1 score = (2 × Precision × Recall)/(Precision + Recall)
(5) The area under the curve is abbreviated as AUC. AUC is a more comprehensive measurement that takes into account both true negative and true positive outcomes. The higher the AUC, the better the model's performance.

```
from sklearn.metrics import classification_report
y_test = list(test_df.Label)
from sklearn import metrics
print('Accuracy:', np.round(metrics.accuracy_score(y_test, pred),5))
print('Precision:', np.round(metrics.precision_score(y_test, pred, average='weighted'),5))
print('Recall:', np.round(metrics.recall_score(y_test,pred, average='weighted'),5))
print('F1 Score:', np.round(metrics.f1_score(y_test, pred, average='weighted'),5))
print('Cohen Kappa Score:', np.round(metrics.cohen_kappa_score(y_test, pred),5))
print(classification_report(y_test, pred))

from sklearn.metrics import confusion_matrix
import seaborn as sns
cf_matrix = confusion_matrix(y_test, pred, normalize='true')
plt.figure(figsize = (10,6))
sns.heatmap(cf_matrix,annot=True, xticklabels = sorted(set(y_test)),
yticklabels = sorted(set(y_test)))
plt.title('Normalized Confusion Matrix')
plt.show()
```

The number of disease images classified as true positives (TP) is equal to the number of disease images classified as true positives (TP).

A true negative is an outcome where the model correctly predicts the negative class (TN).

The number of standard images classified as disease is referred to as the false-positive (FP) rate.

The number of false-negative (FN) disease images is equal to the number of normal disease images.

In practice, a prototype must have precision and recall of 1, resulting in an F1 score of 1, that is, 100% accuracy, which is not possible in a classification task. As a result, the classifier that is created should have higher precision and recall. In addition to the agreement observed in the confusion matrix, Cohen Kappa accounts for agreement that occurs by chance.

9.4.3 Experimental results

Different architectures were trained with pretrained weights. Apart from the popular architectures, we wished to see how the performance would change as the CNNs of higher levels were being used. We conducted experiments on the data for the same and came up with the following tables.

9.4.3.1 For Dataset a) Diabetic-Retinopathy Sample Dataset Binary

This section contains all the experimental results and observations for the abovementioned dataset (Table 9.1).

Since the given dataset might be imbalanced as a result, and the Kappa score turned out to be hostile or equal to 0. Due to the high number of layers with no residual skip in between the layers, the information could not get propagated throughout the model, and thus it gave an abysmal performance. The softmax activation function has to be added to the output to create normalized arrays that determine whether the retinal image is healthy or disabled (Table 9.2).

TABLE 9.1 The comparison of convolutional neural network (CNN) layers using Diabetic Retinopathy Detection binary dataset.

CNN layers	Training accuracy	Validation accuracy	Test accuracy	F1 score	ROC area
2 Layer CNN	0.7726	0.7556	0.7	0.668	0.96
3 Layer CNN	0.7034	0.7556	0.785	0.69	0.9407
4 Layer CNN	0.705	0.7556	0.785	0.69	0.94
5 Layer CNN	0.7285	0.7556	0.785	0.69	0.9358
6 Layer CNN	0.7172	0.7556	0.785	0.69	0.9358
7 Layer CNN	0.70	0.76	0.785	0.69	0.915
8 Layer CNN	0.69	0.7556	0.785	0.69	0.9133

```
model = Sequential()
model.add(ResNet50(input_shape=(264,264,3),include_top=True,weights=None))
model.add(Flatten())
model.add(BatchNormalization())
model.add(Dense(16,kernel_initializer='he_uniform'))
model.add(BatchNormalization())
model.add(Activation('relu'))
model.add(Dropout(0.5))
model.add(Dense(64,kernel_initializer='he_uniform'))
model.add(BatchNormalization())
model.add(Activation('relu'))
model.add(Dropout(0.5))
model.add(Dense(128,kernel_initializer='he_uniform'))
model.add(BatchNormalization())
model.add(Activation('relu'))
model.add(Dropout(0.5))
model.add(Dense(256,kernel_initializer='he_uniform'))
model.add(BatchNormalization())
model.add(Activation('relu'))
model.add(Dropout(0.5))
model.add(Dense(128,kernel_initializer='he_uniform'))
model.add(BatchNormalization())
model.add(Activation('relu'))
model.add(Dropout(0.5))
model.add(Dense(64,kernel_initializer='he_uniform'))
model.add(BatchNormalization())
model.add(Activation('relu'))
model.add(Dropout(0.5))
model.add(Dense(16,kernel_initializer='he_uniform'))
model.add(BatchNormalization())
model.add(Activation('relu'))
model.add(Dropout(0.5))
# Creating an output Layer
model.add(Dense(units= 5, activation='softmax'))
c3=tf.keras.callbacks.ReduceLROnPlateau(
monitor="val_loss",
factor=0.1,
patience=2,
mode="auto",
min_delta=0.0001,
cooldown=0,
    min_lr=0.001)
model.compile(optimizer='adam',loss='categorical_crossentropy', metrics=['accuracy','AUC'])
history=model.fit(x_train,y_train,epochs=20,batch_size=16,validation_split=0.2)
```

For the models mentioned earlier, the softmax activation function must be added to the output to create normalized arrays that determine whether the retinal image is healthy or disabled (0 or 1). We can see that the Kappa score is zero and negative in some cases. This happens because the dataset is highly unbalanced as a result the scores are relatively not good. To encounter this, we shall make the dataset balanced using data augmentation or GAN.

```
class Generator(nn.Module):
def __init__(self, ngpu):
super(Generator, self).__init__()
self.ngpu = ngpu
self.main = nn.Sequential(
# input is Z, going into a convolution

nn.ConvTranspose2d( nz, ngf * 8, 4, 1, 0, bias=False),
nn.BatchNorm2d(ngf * 8),
nn.ReLU(True),
# state size. (ngf*8) x 4 x 4

nn.ConvTranspose2d(ngf * 8, ngf * 4, 4, 2, 1, bias=False),
nn.BatchNorm2d(ngf * 4),
nn.ReLU(True),
# state size. (ngf*4) x 8 x 8

nn.ConvTranspose2d(ngf * 4, ngf * 4, 3, 1, 1, bias=False),
nn.BatchNorm2d(ngf * 4),
nn.ReLU(True),
# state size. (ngf*4) x 8 x 8

nn.ConvTranspose2d( ngf * 4, ngf * 2, 4, 2, 1, bias=False),
nn.BatchNorm2d(ngf * 2),
nn.ReLU(True),
# state size. (ngf*2) x 16 x 16

nn.ConvTranspose2d( ngf * 2, ngf, 4, 2, 1, bias=False),
nn.BatchNorm2d(ngf),
nn.ReLU(True),
# state size. (ngf) x 32 x 32

nn.ConvTranspose2d( ngf, ngf, 3, 1, 1, bias=False),
nn.BatchNorm2d(ngf),
nn.ReLU(True),
# state size. (ngf) x 32 x 32

nn.ConvTranspose2d( ngf, ngf, 4, 2, 1, bias=False),
nn.BatchNorm2d(ngf),
nn.ReLU(True),
# state size. (ngf) x 32 x 32

nn.ConvTranspose2d( ngf, ngf, 4, 2, 1, bias=False),
nn.BatchNorm2d(ngf),
nn.ReLU(True),
# state size. (ngf) x 32 x 32

nn.ConvTranspose2d( ngf, nc, 4, 2, 1, bias=False),
nn.Tanh()
# state size. (nc) x 64 x 64
)

def forward(self, input):
return self.main(input)
```

```
# Create the generator
netG = Generator(ngpu).to(device)
netG.apply(weights_init)
# Print the model
print(netG)


class Discriminator(nn.Module):
def __init__(self, ngpu):
super(Discriminator, self).__init__()
self.ngpu = ngpu
self.main = nn.Sequential(
# input is (nc) x 256 x 256
nn.Conv2d(nc, ndf , 4, 4, 1, bias=False),
nn.BatchNorm2d(ndf),
nn.LeakyReLU(0.2, inplace=True),

# input is (ndf) x128 x 128
nn.Conv2d(ndf, ndf * 2, 4, 4, 1, bias=False),
nn.BatchNorm2d(ndf * 2),
nn.LeakyReLU(0.2, inplace=True),

# input is (ndf*2)x 32 X 32
nn.Conv2d(ndf*2, ndf*4, 4, 4, 1, bias=False),
nn.LeakyReLU(0.2, inplace=True),
nn.Conv2d(ndf * 4, 1, 4, 1, 0, bias=False),
nn.Sigmoid())

def forward(self, input):
return self.main(input)

# Create the Discriminator
netD = Discriminator(ngpu).to(device)
netD.apply(weights_init)
print(netD)
```

TABLE 9.2 The comparison of other deep learning models using Diabetic Retinopathy Detection binary dataset.

CNN layers	Training accuracy	Validation accuracy	Test accuracy	F1 score	ROC area
ResNet50	0.689	0.7556	0.785	0.69	0.9255
VGG16	0.73	0.7556	0.79	0.69	0.9326
VGG19	0.731	0.7556	0.78	0.69	0.9302
Inception_v3	0.78	0.75	0.78	0.687	0.93
MobileNet	0.707	0.7556	0.79	0.69	0.9322
DenseNet169	0.694	0.7556	0.785	0.69	0.9402
DenseNet121	0.722	0.7556	0.79	0.69	0.93
InceptionResNetV2	0.7	0.7556	0.785	0.69	0.9324
MobileNetV2	0.672	0.7556	0.785	0.69	0.9286
ResNet101	0.77	0.7556	0.79	0.69	0.9405

9.4.3.2 For Dataset b) Diabetic Retinopathy 224x224 Gaussian Filtered

This section contains all the experimental results and observations for the dataset, as mentioned earlier. We can infer two different possibilities in this dataset. We can use this dataset to predict both 2 Classes and 5 Classes in diabetic retinopathy disease (Table 9.3).

The dataset, as mentioned above, is balanced with around 1700 images in both classes 0 and 1. The Cohen Kappa Score turned out to be greater than 0.8 in all the experiments, which are good signs showing an excellent strength of agreement. The softmax activation function has to be added to the output to create normalized arrays that determine whether the retinal image is healthy or disabled (0 or 1) (Table 9.4).

```
model = tf.keras.Sequential([
layers.Conv2D(16, (3,3), padding="same", input_shape=(224,224,3), activation = 'relu'),
layers.MaxPooling2D(pool_size=(2,2)),
layers.BatchNormalization(),

layers.Conv2D(32, (3,3), padding="same", activation = 'relu'),
layers.MaxPooling2D(pool_size=(2,2)),
layers.BatchNormalization(),

layers.Conv2D(64, (3,3), padding="same", activation = 'relu'),
layers.MaxPooling2D(pool_size=(2,2)),
layers.BatchNormalization(),

layers.Conv2D(64, (3,3), padding="same", activation = 'relu'),
layers.MaxPooling2D(pool_size=(2,2)),
layers.BatchNormalization(),

layers.Conv2D(128, (3,3), padding="same", activation = 'relu'),
layers.MaxPooling2D(pool_size=(2,2)),
layers.BatchNormalization(),

layers.Conv2D(128, (3,3), padding="same", activation = 'relu'),
layers.MaxPooling2D(pool_size=(2,2)),
layers.BatchNormalization(),

layers.Conv2D(256, (3,3), padding="same", activation = 'relu'),
layers.MaxPooling2D(pool_size=(2,2)),
layers.BatchNormalization(),

layers.Conv2D(256, (3,3), padding="same", activation = 'relu'),
layers.MaxPooling2D(pool_size=(1,1)),
layers.BatchNormalization(),

layers.Flatten(),
layers.Dense(32, activation = 'relu'),
layers.Dropout(0.15),
layers.Dense(2, activation = 'softmax')])
model.compile(optimizer=tf.keras.optimizers.Adam(lr = 1e-5),
loss=tf.keras.losses.BinaryCrossentropy(),
metrics=['acc','AUC',tensorflow_addons.metrics.F1Score(num_classes=2, average='weighted'),
tensorflow_addons.metrics.CohenKappa(num_classes=5)])
history = model.fit(train_batches,
epochs=12,
validation_data=val_batches)
```

TABLE 9.3 The comparison of convolutional neural network (CNN) layers using Diabetic Retinopathy Detection 224 × 224 Gaussian images for 2 Class classification.

CNN layers	Training accuracy	Validation accuracy	Test accuracy	F1 score	Kappa score	ROC area
2 Layer CNN	0.9844	0.9236	0.9527	0.9527	0.9055	0.9874
3 Layer CNN	0.9676	0.96	0.9272	0.9273	0.8546	0.9664
4 Layer CNN	0.9602	0.9327	0.9181	0.9182	0.8363	0.9763
5 Layer CNN	0.9465	0.9218	0.9127	0.9127	0.8255	0.9554
6 Layer CNN	0.9559	0.9364	0.9072	0.9073	0.8145	0.9682
7 Layer CNN	0.98	0.92	0.9272	0.9273	0.8546	0.9716
8 Layer CNN	0.9859	0.9309	0.93	0.929	0.8581	0.9724

TABLE 9.4 The comparison of DL model using Diabetic Retinopathy Detection 224 × 224 Gaussian images for 2 Class classification.

Models	Training accuracy	Validation accuracy	Test accuracy	F1 score	Kappa score	ROC area
ResNet50	0.9257	0.92	0.9183	0.9182	0.84	0.9613
VGG16	0.9556	0.9666	0.962	0.9618	0.923	0.9898
VGG19	0.9495	0.9483	0.9427	0.9428	0.8851	0.985
Inception_v3	0.9708	0.9544	0.9619	0.9618	0.9235	0.996
MobileNet	0.9795	0.9574	0.9564	0.9564	0.9128	0.9978
DenseNet169	0.9672	0.9848	0.9646	0.9646	0.9292	0.9921
DenseNet121	0.9652	0.9666	0.9782	0.9782	0.9564	0.9934
InceptionResNetV2	0.96	0.9544	0.951	0.9508	0.9015	0.9909
MobileNetV2	0.9707	0.9514	0.9427	0.9428	0.885	0.9932
ResNet101	0.8907	0.8693	0.8855	0.8857	0.77	0.9506

For the models mentioned above, the softmax activation function has to be added to the output to create normalized arrays that determine whether the retinal image is healthy or disabled (0 or 1) (Tables 9.5 and 9.6).

```
train_generator,test_generator,train_images,val_images,test_images=create_gen()
# Load the pretained model
pretrained_model = tf.keras.applications.ResNet50(
input_shape=(224, 224, 3),
include_top=False,
weights='imagenet',
pooling='avg')
pretrained_model.trainable = False

inputs = pretrained_model.input
x = tf.keras.layers.Dense(128, activation='relu')(pretrained_model.output)
x = tf.keras.layers.Dense(128, activation='relu')(x)
outputs = tf.keras.layers.Dense(2, activation='softmax')(x)
model = tf.keras.Model(inputs=inputs, outputs=outputs)
model.compile(
optimizer='adam',
loss='categorical_crossentropy',
metrics=['accuracy','AUC'])
history = model.fit(
train_images,
validation_data=val_images,
batch_size = 32,
epochs=10,
callbacks=[
tf.keras.callbacks.EarlyStopping(
monitor='val_loss',
patience=3,
restore_best_weights=True
)
])
```

TABLE 9.5 The comparison of different convolutional neural network (CNN) layers using Diabetic Retinopathy Detection 224 × 224 Gaussian images for 5 Class classification.

CNN layers	Training accuracy	Validation accuracy	Test accuracy	F1 score	Kappa score	ROC area
2 Layer CNN	0.9059	0.9025	0.9003	0.69	0.58	0.9346
3 Layer CNN	0.8953	0.8927	0.8916	0.6444	0.5	0.9152
4 Layer CNN	0.9179	0.9055	0.901	0.71	0.59	0.94
5 Layer CNN	0.8884	0.8909	0.8843	0.6142	0.47	0.88
6 Layer CNN	0.8646	0.8793	0.8687	0.62	0.41	0.85
7 Layer CNN	0.85	0.85	0.84	0.55	0.33	0.87
8 Layer CNN	0.8869	0.8775	0.8822	0.66	0.47	0.9212

TABLE 9.6 The comparison of deep learning model using Diabetic Retinopathy Detection 224 × 224 Gaussian images for 5 Class classification.

Models	Training accuracy	Validation accuracy	Test accuracy	F1 score	Kappa score	ROC area
ResNet50	0.7272	0.71	0.71	0.632	0.522	0.92
VGG16	0.7459	0.7324	0.7575	0.704	0.613	0.9364
VGG19	0.7342	0.772	0.725	0.676	0.5542	0.941
Inception_v3	0.8358	0.73	0.7438	0.7049	0.6	0.9753
MobileNet	0.8472	0.7872	0.7929	0.7755	0.67	0.9823
DenseNet169	0.8188	0.7781	0.7847	0.754	0.665	0.9714
DenseNet121	0.8235	0.7751	0.8092	0.80226	0.71	0.9697
InceptionResNetV2	0.7827	0.7781	0.8065	0.7674	0.685	0.957
MobileNetV2	0.8102	0.7568	0.7874	0.7693	0.6556	0.9668
ResNet101	0.6766	0.7325	0.703	0.6416	0.5268	0.9023

9.5 Discussion

The current study aims to detect DR in retinal images by employing a CNN deep network approach. Unlike some other authors who are using CNN to produce a label on an image's pixels, our model deals with patches and therefore can localize possible future regions of lesions, creating a strong method for a red retinal lesion specialist and resulting to further improvements in DR detection.

Our primary objective was to simplify the framework while enhancing its efficiency. To that final moment, we developed an effective framework for finding training blotches, ensuring that complicated examples obtain extra attention during the training process. Furthermore, because our goal was to achieve a rough localization of the regions rather than precise segmentation, we undersampled the images, resulting in a significant reduction in generalization processing time, which is critical in a real-world application scenario. All of the studies mentioned in this chapter used DL techniques to manipulate the DR screening system. With the increasing number of diabetics, the need for a better eye examination system has become a major problem. Using DL to detect DR and distribution solves the problem of improved ML feature selection. On the other hand, training requires a lot of data. Many researchers have used data enhancement to increase the number of images and controls during training.

A few of the drawbacks of the usage of deep gaining knowledge of withinside the clinical area is the dimensions of the datasets required to teach the DL structures, as DL calls for a huge volume of information. The consequences of DL structures are closely decided with the aid of using the quantity of the schooling facts in addition to its exceptional and sophistication equilibrium. As a result, the modern-day public dataset sizes need to be increased, while huge datasets, consisting of the general public Kaggle dataset, need to be delicate to cast off mislabeled and low-exceptional facts.

The studying defined right here differs from the usage of DL technology. They vary withinside the quantity of reads they devise their personal CNN gadget and the quantity of reads they opt to use for present structures which includes VGG, ResNet, and MobileNet, and withinside the transition to the smaller ones. Building a brand-new CNN model from scratch takes a whole lot of attempt and time. It is an awful lot simpler to apply switch learning, which hastens the layout and improvement of latest buildings. On the alternative hand, it ought to be cited that the consistency of the gadget with the CNN model itself is extra than that utilized in present structures.

Researchers must focus on this issue and need to do more research to determine both conditions. Improved DR detection system that can detect different types of lesions and stages DR results in an improved monitoring system for DR patients which avoids the risk of vision loss. Five DR levels need to be identified correctly and the gap closed in a system that can detect DR damage. This can be seen as a current challenge for researchers for future research. Use the map to show the DR level of each pixel to determine the distribution. We used a simple method to calculate the label function for the entire image. The advantage is that it does not need to be reused. In particular, our method distinguishes between injury detection time (studied by CNN) and DR tagging by CNN events. As a result, this product can be used in any live database without any additional changes.

9.6 Conclusion

DM is one of the complications of diabetes and is a cause of blindness. An effective and automatic study of the incidence of DM is of clinical significance. Early detection allows faster treatment. This is important because early detection can prevent disability. Automatic diagnosis of DM from previous images can help clinicians effectively with the diagnosis of DM, which can improve the quality of diagnosis. This chapter presents a framework to identify the diabetic-related disease, study how to disseminate images of those with the disease, and learn the function that directly removes the image's feature. Image courtesy is standardized and expanded as a product selection and model training. The maximum accuracy of the test was 97.82% using the DenseNet121 architecture, but the two-dimensional projection uses a 224 × 224 Gaussian image dataset to improve the accuracy of the diabetic image distribution. Also, unlike modern ones, the number of images required to train a small sample, which is very important due to the weight of the coverage.

The automated DR detection system reduces the time required to perform diagnostic work, saves ophthalmologists work and costs, and results in timely patient care. Automatic DR detection plays an important role in early DR detection. The DR level is supported by the type of color that appears on the eyeball. This chapter presents modern automation systems for diabetes diagnosis and classification using advanced learning techniques. Most researchers use CNN for classification, so they detect DR images from their size. This chapter also describes the main methods which could be used to classify and diagnose DR using DL.

References

[1] C. Agurto, et al., Multiscale AM-FM methods for diabetic retinopathy lesion detection, IEEE Trans. Med. Imaging 29 (2) (2010) 502–512.

[2] R. Acharya, Y.K.E. Ng, J.S. Suri, Image Modeling of the Human Eye, Artech House, 2008.

[3] Early Treatment Diabetic Retinopathy Study Research Group, Grading diabetic retinopathy from stereoscopic color fundus photographs—an extension of the modified Airlie House classification: ETDRS report number 10, Ophthalmology 98 (5) (1991) 786–806.

[4] P. Liskowski, K. Krawiec, Segmenting retinal blood vessels with deep neural networks, IEEE Trans. Med. Imaging 35 (11) (2016) 2369–2380.

[5] J. Odstrcilik, et al., Retinal vessel segmentation by improved matched filtering: evaluation on a new high-resolution fundus image database, IET Image Processing 7 (4) (2013) 373–383.

[6] J. Jan, et al., Retinal image analysis aimed at blood vessel tree segmentation and early detection of neural-layer deterioration, Comput. Med. Imaging Graph. 36 (6) (2012) 431–441.

[7] M. Chandrashekar, An approach for the detection of vascular abnormalities in diabetic retinopathy, Int. j. data min. Tech. App. 02 (2013) 246–250.

[8] J. Kaur, H.P. Sinha, Automated localization of optic disc and macula from fundus images, Int. j. adv. res. 2 (4) (2012) 242–249.

[9] H.F. Jaafar, A.K. Nandi, W. Al-Nuaimy, Detection of exudates from digital fundus images using a region based segmentation technique, in: 19th European Signal Processing Conference, Barcelona, Spain, 2011.

[10] X. Jiang, D. Mojon, Adaptive local thresholding by verification based multi-threshold probing with application to vessel detection in retinal imagesin IEEE Trans. Pattern Anal. Mach. Intell. 25 (1) (2003) 131–137.

[11] C.I. Sanchez, M. Garcia, A. Mayo, M. Lopez, R. Hornero, Retinal image analysis based on mixture models to detect hard exudates, Med. Image Anal. 13 (2009) 650–658.

[12] J. Goh, L. Tang, G. Saleh, L. Al Turk, Y. Fu, A. Browne, Filtering normal retinal images for diabetic retinopathy screening using multiple classifiers, in: International Conference on Information Technology and Applications in Biomedicine, 2009, pp. 1–4.

[13] G. Liew, T.Y. Wong, P. Mitchell, J.J. Wang, Retinal vascular imaging – a new tool in micro vascular disease research, Circ.: Cardiovasc. Imaging (2008) 156–161.

[14] A.W. Reza, C. Eswaran, K. Dimyati, Diagnosis of diabetic retinopathy: automatic extraction of optic disc and exudates from retinal images using marker-controlled watershed transformation, J Med Systems 35 (2011) 1491–1501.

[15] A. Hoover, M. Goldbaum, Locating the optic nerve in a retinal image using the fuzzy convergence of the blood vesselsin IEEE Trans. Med. Imaging 22 (8) (2003) 951–958.

[16] H. Li, O. Chutatape, Automated feature extraction in color retinal images by a model based approachin IEEE. Trans. Biomed. Eng. 51 (2) (2004) 246–254.

[17] G. Gardener, D. Keating, T. Williamson, A. Elliott, Automatic detection of diabetic retinopathy using an artificial neural network: a screening tool, Br. J. Ophthalmol. (1996).

[18] L.W. Yun, U.R. Acharya, Y.V. Venkatesh, C. Chee, L. C. Min, E.Y.K. Ng, Identification of different stages of diabetic retinopathy using retinal optical images, Inf. Sci. (Ny) 178 (2008) 106–121.

[19] G. Quellec, K. Charriere, Y. Boudi, B. Cochener, M. Lamard, Deep image mining for diabetic retinopathy screening, Med. Image Anal. 39 (2017) 178–193.

[20] H. Jiang, K. Yang, M. Gao, D. Zhang, H. Ma, W. Qian, An interpretable ensemble deep learning model for diabetic retinopathy disease classification in 41st Annual International conference of the IEEE engineering in medicine and biology society (EMBC), 2019, pp. 2045–2048.

[21] G.T. Zago, R.V. Andreao, I.B. Dorizz, E.O. Teatini Salles, Diabetic retinopathy detection using red lesion localization and convolutional neural networks, Comput. Biol. Med. (2020).

[22] H. Pratt, F. Coenen, D.M. Broadbent, S.P. Harding, Y. Zheng, Convolutional neural networks for diabetic retinopathy, Procedia Comput Sci 90 (2016) 200–205.

[23] V. Gulshan, L. Peng, M. Coram, et al., Development and validation of a deep learning algorithm for detection of diabetic retinopathy in retinal fundus photographs, JAMA 316 (22) (2016) 2402–2410.

[24] Y. LeCun, B. Boser, J.S. Denker, D. Henderson, R.E. Howard, W. Hubbard, et al., Backpropagation applied to handwritten zip code recognition, Neural Comput. 1 (4) (1989) 541–551. Available from: https://doi.org/10.1162/neco.1989.1.4.541.

[25] Y. LeCun, B. Boser, J. Denker, D. Henderson, R. Howard, W. Hubbard, et al., Handwritten digit recognition with a back-propagation network, Adv. Neural Inf. Process. Syst. 2 (1990) 396–404URL. Available from: http://citeseerx.ist.psu.edu/viewdoc/summary?doi = 10.1.1.32.5076.

[26] G.E. Hinton, S. Osindero, Y.W. Teh, A fast learning algorithm for deep 310 belief nets, Neural Comput. 18 (7) (2006) 1527–1554. Available from: https://doi.org/10.1162/neco.2006.18.7.1527.

[27] G.E. Hinton, R.R. Salakhutdinov, Reducing the dimensionality of data with neural networks, Science 313 (5786) (2006) 504–507. Available from: https://doi.org/10.1126/science.1127647.

[28] Y. Bengio, P. Lamblin, D. Popovici, H. Larochelle, Greedy layer-wise training of deep networksin Adv. Neural Inf. Process. Syst. (2007) 153–160.

[29] L. Deng, D. Yu, 7, Deep Learning: Methods and Applications, Now Publishers Inc, 2013. arXiv:1309.1501. Available from: https://doi.org/10.1561/2000000039.

[30] D. Li., A tutorial survey of architectures, algorithms, and applications for deep learning, APSIPA Trans. Signal Inf Process 3 (2) (2014) 1–29.

[31] V. Vasilakos, A. Tang, Y. Yao, Neural networks for computer-aided diagnosis in medicine: a review, Neurocomputing 216 (2016) 700–708.

[32] X.W. Chen, X. Lin, Big data deep learning: challenges and perspectives, IEEE Access 2 (2014) 514–525.

[33] Y. Guo, Y. Liu, A. Oerlemans, S. Lao, S. Wu, M.S. Lew, Deep learning for visual understanding: a review, Neurocomputing 187 (2016) 27–48.

[34] L. Deng, D. Yu, Deep learning: methods and applications, Found Trends® Signal Process 7 (3–4) (2014) 197–387.

[35] S.C. Lo, M.T. Freedman, J.S. Lin, S.K. Mun, Automatic lung nodule detection using profile matching and back-propagation neural network techniques, J. Digit. Imaging 6 (1993) 48–54.

[36] M.L. Astion, P. Wilding, The application of back-propagation neural networks to problems in pathology and laboratory medicine, Arch. Path. Lab. Med. 116 (1992) 995–1001.

[37] Y. Wu, M.L. Giger, K. Doi, C.J. Vyborny, R.A. Schmidt, C.E. Metz, Artificial neural networks in mammography: application to decision making in the diagnosis of breast cancer, Radiology, 187, 1993, pp. 81–87.

[38] S.E. Spenceley, D.B. Henson, D.R. Bull, Visual field analysis using artificial neural networks, Ophthal. Physiol. Opt. 14 (1994) 239–248.

[39] Q. Abbas, M.E.A. Ibrahim, M.A. Jaffar, Video scene analysis: an overview and challenges on deep learning algorithms, Multimed. Tools Appl. 77 (16) (2018) 20415–20453. Available from: https://doi.org/10.1007/s11042-017-5438-7.

[40] H. Greenspan, B. van Ginneken, M. Ronald, Summers Guest editorial deep learning in medical imaging: Overview and future promise of an exciting new technique, IEEE Trans. Med. Imaging 35 (5) (2016) 1153–1159. Available from: https://doi.org/10.1109/tmi.2016.2553401.

[41] R. Raman, S. Srinivasan, S. Virmani, S. Sivaprasad, C. Rao, R. Rajalakshmi, Fundus photograph-based deep learning algorithms in detecting diabetic retinopathy, Eye 33 (2019) 97–109.

[42] R. Rajalakshmi, R. Subashini, R.M. Anjana, V. Mohan, Automated diabetic retinopathy detection in smartphone-based fundus photography using artificial intelligence, Eye 32 (2018) 1138–1144.

[43] W. Zhang, et al., Automated identification and grading system of diabetic retinopathy using deep neural networks, Knowl. Base Syst. 175 (2019) 12–25.

Further reading

W.L. Alyoubi, W.M. Shalash, M.F. Abulkhair, Diabetic retinopathy detection through deep learning techniques: a review, Inform. Med. Unlocked 20 (2020) 10037.

CHAPTER

10

Automated detection of colon cancer using deep learning

Aayush Rajput[1] *and Abdulhamit Subasi*[2,3]

[1]Indian Institute of Technology, Kharagpur, West Bengal, India [2]Institute of Biomedicine, Faculty of Medicine, University of Turku, Turku, Finland [3]Department of Computer Science, College of Engineering, Effat University, Jeddah, Saudi Arabia

OUTLINE

10.1 Introduction

Colon cancer is a type of cancer that affects and starts from the large intestine end of the digestive tract. In starting phase of colon cancer, clumps of cells begin to grow on the inner side of the large intestine. As these clumps of cells called polyps grow and multiply further, they became cancerous. In the initial phase, they are not cancerous. The small polyps can be removed easily in the initial stage to avoid the threat of cancer. The average age of men and women diagnosed with colon cancer is 68 and 72, respectively [1]. Although it can occur in young persons also, most people having colon cancer are older than 50. The probability

Applications of Artificial Intelligence in Medical Imaging
DOI: https://doi.org/10.1016/B978-0-443-18450-5.00001-3

of development of colon cancer is slightly more in men than women. One man among 23 and one woman in every 25 have colon cancer in the United States [2]. Colorectal cancer is the term used for colon and rectal cancer collectively. Colorectal cancer is the third most common type of cancer, excluding the skin cancers in the United States, with 104,270 cases of colon cancer and 45,230 cases of rectal cancer in 2021 [2]. People with a family history of colon cancer have more risk of developing it, especially when any close family member below the age of 60 is diagnosed with colon cancer. Other situation can also increase the risk of developing colon cancer, such as people who do not do much physical activity, overweight, smokers, people having a history of any other type of cancer such as ovarian cancer or uterine cancer, and people having adenomas [1].

Sometimes it is possible that a person having colon cancer does not show any symptoms in the initial stage of development of cancer [3]. The symptoms of colon cancer include a frequent change in bowel habits, diarrhea, the feeling of bowel does not empty, constipation, thinner stools than usual, tiredness or fatigue, anemia, blood in stool [3]. A person should immediately consult a doctor if any of the symptoms last for a long time regardless of age because colon cancer can develop at any age.

The most common type of colon cancer is adenocarcinoma colon cancer [4]. Benign colon cancer is the initial stage of colon cancer which can be treated easily and is not life-threatening. Metastasis is when cancer develops in one part of body and spread to other parts [5]. Doctors use many types of tests for the diagnosis of colon cancer. For checking metastasis, tests are also done. The test used by the doctor depends on the various factors and symptoms seen in the patient. The first physical examination of the patient is done. Then tests such as biopsy, colonoscopy, computed tomography (CT), biomarker testing, blood test, magnetic resonance imaging (MRI) are used. In a biopsy to detect colon cancer, a small tissue from the affected region's large intestine is removed from the patient's body and examined under a microscope by a pathologist [6]. The biopsy is capable of making a definite diagnosis of colon cancer. Colonoscopy is a method performed by the colonoscopist in which a colonoscopist checks inside the colon in the patient's body. During the colonoscopy, if cancer is detected, then the tumor has to be removed surgically from the site for a complete diagnosis. In CT scans, X-rays are used to produce a 3D image of the human body, and sometimes a die is also used for more precise results [7]. The die can be given to the patient by injecting directly into the veins or can be taken as a pill.

From the resulting image of the CT scan, abnormalities can be detected. In colon cancer, internal bleeding in the large intestine occurs. The person becomes anemic, so a blood test is done to detect the blood count; a lower value of blood count shows the presence of internal bleeding and hence the colon cancer. Blood tests are also done to detect the carcinoembryonic antigen level in blood. A higher value of carcinoembryonic antigen (CEA) indicates the spread of cancer to other parts of the body. MRI works similarly to the CT scan. Only the difference is that here magnetic rays are used to make the 3D image of the body, and the size of the tumor can be detected. The treatment of colon cancer has physical, emotional, social, and financial effects on the patient. These effects can vary from person to person for the same treatment for the same type of cancer. To cope with the side effect of the treatment, various things need to be done. After the detection of a cancer, patient can have sadness, anxiety, or anger. Taking help from the counselor can relieve the patient from the emotional side effect of the treatment. The treatment of colon cancer is costly, due to which sometimes patients cannot take complete treatment and put their life at risk. Any financial problem should be consulted with the supportive team. Family and friends play an essential role in dealing with the side effects of colon cancer treatment. They also help in providing emotional support to the patient.

This study aims to use the power of artificial intelligence and deep learning (DL) to detect the type of colon cancer (benign or adenocarcinoma) from the tissue images like in biopsy. Using DL, the time can be reduced, and the speed to detect cancer from tissue images can be increased to a large extent. The manual process of doing biopsy took a lot of time of a skilled doctor, and that also can sometimes give wrong results that can have a very negative effect on the patient as the treatment has many side effects. DL will help doctors to determine colon cancer in a faster way accurately.

10.2 Literature review

DL is gaining the attention of many researchers in computer vision, where objects are detected with images. As convolutional neural networks (CNNs) are excellent in detecting objects from images, in the field of medical science, DL is beneficial in many situations. Various research have been done to detect diseases such as cancer from the images.

Kim et al. [8] used semantic gland segmentation from morphology images to detect colon cancer. They detected polyp-causing cancer from the images. They proposed a method in which features are extracted from diagnostic tests are given as input to the DL network with the semantic segmentation algorithm. They used different architecture for detecting the cancer-causing polyp. CNN, SegNet, U-Net, and FCN are used, and the U-Net architecture gave the highest intersection over union score of 92%. Various metrics evaluate predicted images with the actual label, particularly dice index, sensitivity, and Hausdorff distance. CNN, SegNet, and FCN gave the 74.80% sensitivity score, 86.36% dice score, and 62.7% mean intersection over union score.

Song et al. [9] used 411 slides from the Chinese People's Liberation Army General Hospital, of which 232 slides were of colorectal adenomas and 179 of normal mucosa. They used the deep learning model based on DeepLab v2 with ResNet-34 with the addition of skip layer fusion approach in which the higher level features extracted by the model are combined with the lower level features. The evaluation metrics on which the model's performance was measured were accuracy, sensitivity, and specificity. They compared their model results with the other DL model architectures, ResNet50, DenseNet, InceptionV3, U-Net, and DeepLab v3. These models gave an accuracy score of 89.8%, 87.8%, 90.3%, 77.7%, and 88.3%, while the improved DeepLab v2 got an accuracy score of 90.4%. The models were tested on the 194 cases which were not used for training them.

Ribeiro et al. [10] trained different models to classify the image as nonneoplastic and neoplastic. The database used in the study was an endoscopic image database with eight different imaging modalities acquired by an HD endoscope with an image size of 256×256 pixels. They first trained the CNN model from scratch and got average accuracy of 79%, which is not very good. Then the hyperparameters of the model were tuned, such as changing the size and number of filters. Five different models with different hyperparameters were used. The best result got from tuning the hyperparameter was an 89% accuracy score. In the next experiment, they trained CNN for a particular group of databases using leave one out cross-validation with a different stride; the result showed that there is not much difference in the increasing the stride for an image. In further experiments, transfer learning is used with different data augmentations, and the results showed that features extracted from CNN and VGG-16 provided better feature descriptors of colonic polyps.

Walker et al. [11] used the 307 colorectal cancer-related digital slides from St. Paul's

Hospital. Of these slides, 85 were from normal colorectal tissue, 222 slides were colorectal cancer, 275 slides were used for training, and the remaining for testing. Pretrained InceptionV3 architecture was used with the input size as 299×299 pixels. The training was done to 20 epochs with a batch size of 92 and a learning rate 3×10^{-4}. The overall classification accuracy and Receiver operating characteristic (ROC) score were 95.1% and 99%, respectively. The image segmentation was also done. The prediction performance across all the slides on the independent dataset has the mean accuracy score, specificity score, sensitivity score, and dice score of 87.8%, 90.0%, 85.2%, and 87.2%.

Hornbrook et al. [12] used machine learning (ML) tools for the early detection of colorectal cancer using gender, age, and complete blood count data. The data consist of 900 colorectal cancer cases and 9108 no cancer cases. The data was taken from the Kaiser Permanente Northwest Region's Tumour Registry. The performance of the model was evaluated using the specificity, Area under the ROC curve (AUC), and odds ratio. The model gave the area under curve score of 80% with 99% specificity. They used the ColonFlagR model for the detection of undiagnosed colorectal cancer. The study showed that ColonFlag identifies individuals 10 × higher risk of undiagnosed colon cancer and is more accurate at identifying right-sided colorectal cancers.

Ponzio et al. [13] proposed a DL technique using CNNs to detect adenocarcinoma from healthy and benign lesions. The data used was taken from a public repository of H&E stained whole-slide images available on the website of University of Leeds Virtual Pathology Project. The total data consist of 13,500 patches, from which 9000 were used for training and the rest 4500 for testing. They also used a pretrained model, VGG16, and principal component analysis (PCA) was used for the feature reduction technique. A fully trained CNN model and a pretrained VGG16 model with support vector machine (SVM) classifier and a fine-tuned pretrained model fixing the weights of lower level blocks were used. The three techniques gave an accuracy score of 90.37%, 96.46%, and 96.82%.

Li et al. [14] used DL methods to classify colorectal cancer lymph node metastasis images. Different ML techniques were used for the task, and their result was compared. Data used in the study was taken from the Harbin Medical University Cancer Hospital; total data consist of 1646 positive and 1718 negative samples. Features used for training the models are extracted by three techniques: gray-level histogram, textural features by the gray-level cooccurrence matrix, and the scale-invariant feature transform. The pretrained CNN model AlexNet is also used. CNN model LeNet and AlexNet are used with full training, classical ML models AdaBoost, Decision Tree, KNN, Logistic Regression, and multilayer perceptron, Naive Bayes (NB), Stochastic gradient descent (SGD), and SVM are used. The evaluation metrics used for comparing the model's performance are accuracy, AUC, sensitivity, specificity, positive predictive value, and negative predictive value. The pretrained AlexNet model outperformed the other model giving accuracy, AUC, sensitivity, specificity, Positive Predictive Value (PPV), and Negative Predictive Value (NPV) scores of 75.83%, 79.41%, 80.04%, 79.97%, 79.92%, and 80.09%, respectively.

10.3 Artificial intelligence for colon cancer detection

DL is a very useful tool in medical science that can significantly affect many ways, like giving accurate and fast predictions. Normally detection of colon cancer from the tissue slides takes a lot of time for a skilled doctor to detect a cancer-causing tumor, which can be avoided using DL. The important techniques used in this study are explained in the following sections.

10.3.1 Artificial neural networks

Artificial neural network (ANN) is a brain neuron-inspired model used for processing the input to give a prediction based on the input. Neural networks are based on the radial basis function. ANN can also learn to detect the features in an image. ANN consists of one input and output layer with hidden layers in between them. While training an ANN, information is feed into the forward and backward direction. In the forward propagation, the values of each node of previous layers are multiplied by its respective weight. Then the sum of these products is passed to another function known as activation function; the output of this function is the value of the node of the next element; this process is repeated for every node of the next layer. In ANN, every node of a layer is connected to every node of the next layer; this connection from one node to the node of the next layer has its value called its weight. In backpropagation, complex mathematical calculations are done, which change the weights to make more accurate predictions [15]. The backpropagation method was derived in 1960 by Seppo Linnainmaa; however, Rumelhart et al. [16] showed how it could be used to train the ANN models. Each layer also has a bias node, which helps the model better fit the training data. The activation function is associated with each layer of ANN; it also helps the model fit the data better. Activation functions keep the input value for a node in the layers after the input layer to a fixed range. It also makes the model capable of fitting into nonlinear data. An activation function should be computationally inexpensive. Otherwise, it will increase the training time and take more memory for the model to train; it should also be symmetric about zero, differentiable function, and vanishing gradient. Some of the famous activation functions are sigmoid, softmax, tanh, and ReLU. Sigmoid is computationally more expensive, so it is avoided in most cases. It is defined as $\mathrm{sig}(x) = \frac{1}{1+e^{-x}}$ [17].

It has the range of 0–1 and is not zero centered and causes a vanishing gradient problem. Softmax is defined as $S(x)i = \frac{e^{(y_i)}}{\sum_{(j=1)}^{n} e^{(y_j)}}$ It is similar to sigmoid and is used in the output layer of a multiclass classification model. It has a range from 0 to 1. Tanh is defined as $\tanh(x) = \tan^{-1}x$ and is zero centered but have other problem such as vanishing gradient and is computationally expensive. ReLU is the most widely used activation function among the other functions, it is defined as $f(x) = \max(0, x)$ [18]. It is computationally inexpensive and does not have the problem of vanishing gradient. It is mostly used in CNN models. The deeper an ANN is, the more features it can detect from the data can give better results. Still, sometimes it can lead to overfitting the training data so ANN with different hidden layers should be trained. The model giving the best results of the validation data should be selected.

10.3.2 Deep learning

DL is a subset of ML. It is used for a complex task on which classical ML algorithms cannot give good results. ANNs are the core of any DL model. The algorithms used in this study are DL algorithms. Classical ML models cannot fit the nonlinear data as the ANN can, but the data required to train DL models is more. With the increase in the amount of data, ANNs perform better than classical ML models. The only disadvantage of using ANN is that they are computationally expensive. Medical science data used to be more complex, so classical ML models cannot get good results. DL also gives the ability to combine two or more models to get better results [19]. Transfer learning is a way through which the huge DL models trained for weeks by the people having best resources can be used directly by any other person for a different problem. These pretrained models give very good results in a much shorter time because they already have the best weights and do not need to be changed [20].

The results using transfer learning are generally better than training a DL model from scratch. For the computer vision tasks, CNNs outperform the other algorithms. In this study, CNN models are trained from scratch, pretrained models and pretrained models with classical ML algorithms are used, and their results are compared. Besides computer vision, DL is used in many other fields such as natural language processing, visual recognition, recommendation systems, self-driving cars, etc. [21]. Today, we have enough data resources to train these DL models that take more data than other ML models. DL has a direct effect on people's lives. DL in medical science has made it easier for people to get accurate and faster results. From getting recommendations on social media to robotics everywhere, DL is used. Many research are going on DL to make it better.

10.3.3 Convolutional neural networks

CNN is a type of ANN which can take a 3D array as input and give the required results. Every image is a 3D array with each pixel as the value so that images can be given as input in CNN. CNN can detect the features in an image using the filters. During the training, the weights of these filters are changed to get a minimum possible error in the true and predicted values. Like ANN, in CNN, there is an input layer, an output layer, and in between input and output layers, there are hidden layers. In CNN, each neuron or unit is not connected to every neuron of the next layers. Only a small portion of the layer is used for getting the value of one neuron of the next layer. In CNN, no feature engineering is done to get good results CNN takes care of the features itself while using classical ML; a lot of work has to be done to get the important features. The initially hidden layers of CNN are responsible for detecting the low-level features in the image, and high-level features are detected using the farther hidden layers. The feature detection in CNN is done by the filters associated with each layer; a layer can have many filters [22]. A filter is a matrix that moves over a CNN layer, and the value of its weights has multiplied the value of the neurons to get an output value for the next layer's neuron. This operation is called the convolution operation. The pooling layer is used for reducing the size of a layer to reduce the computation cost. There are no weights associated with filters in these layers. Max pooling and average pooling are the two types of pooling mainly used [23]. The maximum value from a portion of layer on which filter is applied is taken in max pooling. In average pooling takes the average of the portion on which kernel is applied is taken. During the convolution operation, the pixel in the middle will affect more neurons of the output layer than the corners. It can lead to loss of information present in the corner pixels; to avoid this problem, padding is done. Padding is adding zeros to the convolution layer [24]. This can keep the shape of the next layers the same as the previous layer—this kind of padding is called the same padding. Valid padding means no padding at all. A filter of CNN has the property of stride; it defines the movement of a filter after doing one convolution operation. If the stride of a filter is n, it means that after doing one convolution operation, it will move n pixels to reach its new position. These all factors are to be given by the user designing the CNN.

```
from keras.layers import Dense, Flatten, Conv2D, MaxPooling2D, Dropout,
                                                                  Input
from keras.layers.normalization import BatchNormalization
from keras.models import Model, Sequential

inp = Input(shape = (128, 128, 3))
model = BatchNormalization()(inp)
model = Conv2D(filters = 64, kernel_size = (3, 3), padding = 'same',
                                          activation='relu')(model)
model = BatchNormalization()(model)
model = Conv2D(filters = 128, kernel_size = (3, 3), padding = 'same',
                                          activation='relu')(model)
model = MaxPooling2D()(model)
model = Dropout(0.2)(model)
model = Conv2D(filters = 128, kernel_size = (3, 3), padding = 'same',
                                          activation='relu')(model)
model = MaxPooling2D()(model)
model = Dropout(0.2)(model)
model = Conv2D(filters = 64, kernel_size = (3, 3), padding = 'same',
                                          activation='relu')(model)
model = MaxPooling2D()(model)
model = Dropout(0.2)(model)
model = Flatten()(model)
output = Dense(units = len(train_it.class_indices), activation =
                                               'softmax')(model)

model = Model(inputs=inp, outputs=output)
```

10.4 Disease detection using artificial intelligence

10.4.1 Feature extraction using deep learning

Feature extraction is a technique of converting the original data to a more useful and lower size. Before developing DL, various techniques were used to extract important features from the original data, which was very time-consuming and required expertise. Redundant and unnecessary features can decrease the performance model. But with the development of DL and CNN, the feature extraction is done automatically by the CNN. In a CNN, filters are used to extract the features from a layer and pass them to the next layers. The filters of the beginning layers detect lower level features while the filters of ending layers extract the high-level features. Better the weights of filters of the model, better the results it will give. A large CNN model is computationally very expensive to get the best weights of the complex data. To avoid this computation cost, pretrained models are used for extracting the features from the data. The pretrained models already have the best weights, which can detect important features in the data; there is no need to change the weights of the pretrained model. In this study pretrained CNN architectures VGG16 [25], VGG19 [25], ResNet50 [26], ResNet101 [26], MobileNetV2 [27], MobileNet [27], InceptionV3 [28], InceptionResNetV2 [29], DenseNet169 [30], DenseNet121 [30], and Xception [31] are used as feature extractor and then feature extracted using these models are then flattened and used for training the classical ML model. The time taken to train models by the extracted features is very less than training the model from scratch. The performance of pretrained models is also very good.

```
def get_features(base_model, train, validate):
    X_train = base_model.predict(train)
    y_train = train.classes
    X_val = base_model.predict(validate)
    y_val = validate.classes

    X_val, X_test, y_val, y_test = train_test_split(X_val,y_val,test_size=0.5,shuffle=True)
    return (X_train, X_val, X_test, y_train, y_val, y_test)
```

10.4.2 Dimension reduction

Dimension reduction reduces the size of data without losing much important information of the data. In ML, various algorithms are used for reducing the size of data. PCA [32], linear discriminant analysis (LDA) [33], neural autoencoder [34], and t-distributed stochastic neighbor embedding (t-SNE) [35] are some dimension reduction algorithms. In this study, PCA is used for reducing the dimension of the features extracted by the pretrained models. In PCA, data is projected on a plane on which the variance of the data is nearest to the original data. As PCA is a distance-based algorithm, data should be normalized before performing the PCA for reducing the size. After applying PCA, data can be converted to original dimensions, but that will not be the same as the original data. PCA will project the data on a hyperplane, and the value after the project will be the value of data. This is the most widely used and older dimension reduction algorithm. In LDA, linear combinations of input columns are calculated in a way that the calculated features of a particular class are extracted from another class of data. Neural autoencoders convert the data to a lower size by removing the noisy and redundancy in it. T-SNE is a recent dimension reduction technique. It is a nonlinear technique which maps the data to a lower dimensional space where the similarity of data point to its neighbors is maintained. This is also used for visualizing the data

```
from sklearn.decomposition import PCA
pca = PCA(n_components=7000)
X_train = pca.fit_transform(X_train)
X_val = pca.transform(X_val)
model.fit(X_train, y_train)
val_pred = model.predict(X_val)
```

10.4.3 Prediction and classification

Every ML task can be classified into two types: one is prediction and another is classification. In the prediction task, continuous-valued functions are used to assign a value to an instance of data. In contrast, classification is the task of assigning a particular instance of data to a class label. In prediction, only one neuron is present in the output layer of CNN. The activation function used in the prediction task is linear or ReLU. The activation function used for predicting the class in a multiclass classification is the softmax activation function. It predicts the probability of a sample to belong a particular class. The sum of the values predicted by the softmax layer is equal to one. It can also be applied for a binary classification task. The task here is binary classification. The number of units in the last or output layer will be equal to the number of classes in data. The output value given by the model will be an array having n values. Classical ML algorithms KNN [36], SVM [37], Random Forest [38], AdaBoost [39], and XGBoost [40] can be used for multiclass classification.

```
from keras.models import Sequential
from sklearn.neighbors import KNeighborsClassifier
from sklearn.svm import SVC
from sklearn.ensemble import RandomForestClassifier, AdaBoostClassifier
from xgboost import XGBClassifier


def get_models():
      ANN = Sequential()
      ANN.add(Dense(128, input_dim = X_train.shape[1], activation = 'relu'))
      ANN.add(BatchNormalization())
      ANN.add(Dropout(0.2))
      ANN.add(Dense(64, activation='relu'))
      ANN.add(Dense(32, activation='relu'))
      ANN.add(Dense(16, activation='relu'))
      ANN.add(Dense(8, activation='relu'))
      ANN.add(Dense(len(train_it.class_indices), activation='softmax'))

      ANN.compile(loss='sparse_categorical_crossentropy', optimizer='adam',
                                                     metrics=['accuracy'])

      KNN = KNeighborsClassifier()

      SVM = SVC(kernel = 'linear')

      RF = RandomForestClassifier(n_estimators = 50)

      ADB = AdaBoostClassifier()

      XGB = XGBClassifier(n_estimators = 50, use_label_encoder=False)

      return (ANN, KNN, SVM, RF, ADB, XGB)

def fit_model(model, X_train, y_train):
      model.fit(X_train, y_train)
      return model
```

10.4.4 Experimental data

The data used in this study is taken from the images generated from the original sample of Health Insurance Portability and Accountability Act (HIPAA) compliant and validated sources, consisting of 500 total images of colon tissue and augmented to 10,000 using the augmenter package [41]. There are two classes present in the data: colon adenocarcinoma and benign colon tissue. Each has 5000 images. This dataset is available publicly on Kaggle as colon cancer histopathological images [42]. Each image is in jpeg file format, and the size of each image is 768×768 pixels. The images were reshaped to 128×128 pixels to reduce the computational cost and training time.

10.4.5 Performance evaluation measures

For measuring the model's performance on the data, various metrics can be used as accuracy scores can sometimes be misleading. High accuracy scores do not always imply that the model is accurate. Sometimes, the imbalanced model can predict a certain class and get a high accuracy score. Thus, to avoid this situation, training accuracy score, validation accuracy score, test accuracy score, F1 score, Cohen kappa score, ROC AUC score, recall, and precision score are calculated. Based on these scores, best model is selected. 70% of the data is used for training, and for each validation and testing, 15% of data is used.

10.4.6 Experimental results

First, the CNNs with different layers are trained from scratch, and then the pretrained CNN architectures are used for the task. The weights of pretrained models are fixed and fully connected layer is the attached as the last layer of pretrained CNN, which takes the flattened output from the CNN as its input and two neurons at last with softmax activation function for predicting the class an instance belongs (Fig. 10.1). The results of CNN models with different layers are given in (Table 10.1).

The results obtained by training the whole CNNs are not very good. They are not able to classify the images correctly and giving just random results. This is quite expected because the images are complex and simple CNNs do not have the ability to give the results.

Table 10.2 clearly shows that the performance of the pretrained models is far better than the own designed and trained CNN models. This is because the data is very complex, and it needs a huge network and good weights to classify image classes correctly. In the above table, ResNet50 is giving the best results, followed by VGG16, ResNet101, and VGG19 with an F1 score of 99.8%, 99.33%, 99.26%, and 99.06%, respectively.

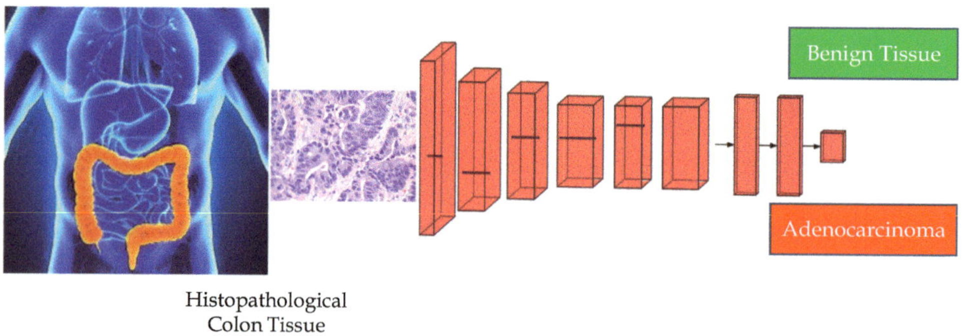

FIGURE 10.1 The general framework for Histopathological image classification using a CNN model. *CNN*, Convolutional neural network.

TABLE 10.1 Results of different convolutional neural network (CNN) models.

Classifier	Training accuracy	Validation accuracy	Test accuracy	F1 score	KAPPA	ROC score
CNN 2 Layer	0.67343	0.67533	0.64267	0.60878	0.28869	0.78553
CNN 3 Layer	0.5	0.502	0.498	0.33111	0	0.5
CNN 4 Layer	0.5	0.498	0.502	0.33556	0	0.5
CNN 5 Layer	0.5	0.496	0.50333	0.33749	− 0.00133	0.49868
CNN 6 Layer	0.91386	0.90667	0.91133	0.91115	0.82262	0.97803
CNN 7 Layer	0.50	0.50	0.50267	0.3363	0	0.5
CNN 8 Layer	0.5	0.504	0.496	0.3289	0	0.65413

TABLE 10.2 Results of different transfer learning models.

Classifier	Training accuracy	Validation accuracy	Test accuracy	F1 score	KAPPA	ROC score
ResNet50	0.99929	0.998	0.998	0.998	0.996	0.99933
VGG16	0.99957	0.99067	0.99333	0.99333	0.98666	0.99656
VGG19	0.99757	0.988	0.99067	0.99067	0.98131	0.99516
Inception_v3	0.67986	0.70267	0.66867	0.63419	0.35282	0.69598
MobileNet	0.98671	0.934	0.932	0.93183	0.86407	0.9883
DenseNet169	0.86757	0.85067	0.876	0.87402	0.75183	0.96002
DenseNet121	0.98157	0.966	0.96467	0.96467	0.92936	0.99618
InceptionResNetV2	0.62929	0.62267	0.64667	0.64667	0.29329	0.66052
MobileNetV2	0.88443	0.86667	0.85933	0.85691	0.71828	0.96966
ResNet101	1	0.992	0.99267	0.99267	0.98532	0.99923

Then pretrained models are used with classical ML algorithms. In this method, pretrained models are used as a feature extractor, and the output is flattened and fed to the classical ML models as input (Fig. 10.2).

Table 10.3 shows that VGG16 is giving very good results on the dataset. The best result is given by the VGG16 as feature extractor followed by SVM with test accuracy of 99.73% and the worst result is given by the KNN with test accuracy of 84.8%.

VGG19 achieved very closely similar results as VGG16 model so the results are also nearly same as the VGG16. The best result is obtained by the VGG19 is with the SVM model with test accuracy of 99.66% and worst result are obtained with the KNN model with test accuracy of 88.26% (Table 10.4).

The results obtained by using ResNet50 are very accurate, with any ML model the test accuracy is around 96%. The best result with ResNet50 is given by the SVM with test accuracy of 99.86% and worst is given by KNN with test accuracy of 95.93%, which is similar to the previous case (Table 10.5).

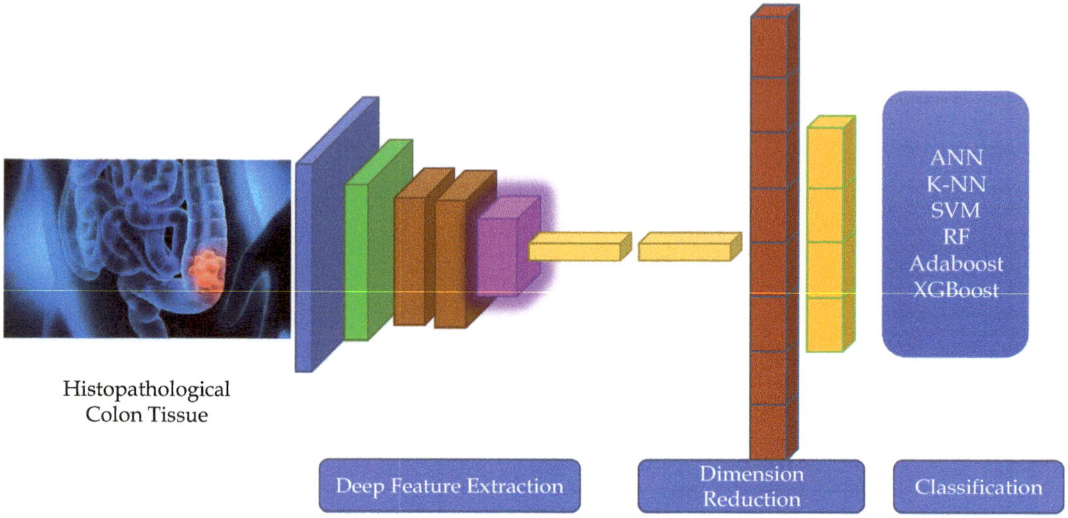

FIGURE 10.2 The general framework for Histopathological image classification using deep feature extraction.

TABLE 10.3 Performance of VGG16 deep features extraction with different machine learning models.

Model	Training accuracy	Validation accuracy	Test accuracy	F1 score	Kappa	Recall	Precision
ANN	0.99943	0.99533	0.99667	0.99667	0.99333	0.99667	0.99667
KNN	0.896	0.848	0.848	0.8469	0.69517	0.848	0.87796
SVM	1	0.99467	0.99733	0.99733	0.99467	0.99733	0.99733
Random Forest	1	0.988	0.986	0.986	0.97199	0.986	0.98607
AdaBoost	0.99286	0.97933	0.984	0.984	0.968	0.984	0.984
XGBoost	1	0.99333	0.98733	0.98733	0.97466	0.98733	0.98737

TABLE 10.4 Performance of VGG19 deep features extraction with different machine learning models.

Model	Training accuracy	Validation accuracy	Test accuracy	F1 score	Kappa	Recall	Precision
ANN	**0.99957**	**0.99333**	**0.99267**	**0.99267**	**0.98533**	**0.99267**	**0.99274**
KNN	**0.904**	**0.894**	**0.88267**	**0.88105**	**0.76482**	**0.88267**	**0.90271**
SVM	**1**	**0.99467**	**0.99667**	**0.99667**	**0.99333**	**0.99667**	**0.99667**
Random Forest	**1**	**0.98267**	**0.97333**	**0.97332**	**0.94665**	**0.97333**	**0.97373**
AdaBoost	**0.98857**	**0.97867**	**0.97733**	**0.97733**	**0.95466**	**0.97733**	**0.97733**
XGBoost	**1**	**0.99**	**0.97933**	**0.97933**	**0.95865**	**0.97933**	**0.97958**

TABLE 10.5 Performance of ResNet50 deep features extraction with different machine learning models.

Model	Training accuracy	Validation accuracy	Test accuracy	F1 score	Kappa	Recall	Precision
ANN	0.99929	0.99667	0.99533	0.99533	0.99067	0.99533	0.99536
KNN	0.979	0.96867	0.95933	0.95933	0.91867	0.95933	0.95952
SVM	1	0.998	0.99867	0.99867	0.99733	0.99867	0.99867
Random Forest	1	0.98267	0.976	0.976	0.95199	0.976	0.97603
AdaBoost	0.99686	0.98267	0.97867	0.97867	0.95733	0.97867	0.97867
XGBoost	1	0.99267	0.98533	0.98533	0.97066	0.98533	0.98536

TABLE 10.6 Performance of ResNet101 deep features extraction with different machine learning models.

Model	Training accuracy	Validation accuracy	Test accuracy	F1 score	Kappa	Recall	Precision
ANN	0.99957	0.99067	0.99067	0.99067	0.98133	0.99067	0.9907
KNN	0.977	0.95533	0.95067	0.95066	0.90135	0.95067	0.95107
SVM	1	0.99267	0.997	0.994	0.988	0.994	0.994
Random Forest	1	0.97133	0.97133	0.97133	0.94265	0.97133	0.97147
AdaBoost	0.99057	0.96867	0.968	0.968	0.936	0.968	0.96801
XGBoost	1	0.984	0.98733	0.98733	0.97466	0.98733	0.98733

TABLE 10.7 Performance of MobileNetV2 deep features extraction with different machine learning models.

Model	Training accuracy	Validation accuracy	Test accuracy	F1 score	Kappa	Recall	Precision
ANN	0.98929	0.93667	0.944	0.94397	0.88795	0.944	0.94452
KNN	0.929	0.87533	0.89267	0.8925	0.78517	0.89267	0.89457
SVM	1	0.92267	0.92267	0.92267	0.84532	0.92267	0.92267
Random Forest	0.99986	0.87267	0.87267	0.87265	0.7454	0.87267	0.87319
AdaBoost	0.892	0.84067	0.87333	0.87334	0.74667	0.87333	0.87339
XGBoost	1	0.91	0.90933	0.90933	0.81866	0.90933	0.90935

ResNet101 achieved similar results as the ResNet50 model with deeper network. The results are also similar to the results of ResNet50 model. Here best test accuracy is given by the SVM with a value of 99.7% and worst test accuracy is given by KNN model with test accuracy of 95.06% (Table 10.6).

The performance of MobileNetV2 model is not as good as VGG or ResNet. The best accuracy given by this is with ANN model with test accuracy value of 94.4% and worst is given with AdaBoost with a value of 87.33% which is quite low (Table 10.7).

MobileNet is the earlier version of MobileNetV2 model and is similar to the V2 of MobileNet. Therefore the results are also similar of this model's best test accuracy score is 95.9% with ANN model and worst accuracy

TABLE 10.8 Performance of MobileNet deep features extraction with different machine learning models.

Model	Training accuracy	Val accuracy	Test accuracy	F1 score	Kappa	Recall	Precision
ANN	0.99357	0.95467	0.95933	0.95933	0.91865	0.95933	0.95939
KNN	0.911	0.86	0.858	0.85668	0.71651	0.858	0.87334
SVM	1	0.948	0.95667	0.95667	0.91334	0.95667	0.95677
Random Forest	1	0.87533	0.874	0.87386	0.74816	0.874	0.87626
AdaBoost	0.913	0.866	0.886	0.886	0.77202	0.886	0.88618
XGBoost	1	0.91933	0.92333	0.92333	0.84669	0.92333	0.92369

TABLE 10.9 Performance of InceptionV3 deep features extraction with different machine learning models.

Model	Training accuracy	Validation accuracy	Test accuracy	F1 score	Kappa	Recall	Precision
ANN	0.90143	0.882	0.88667	0.88648	0.77316	0.88667	0.88861
KNN	0.91	0.852	0.86533	0.86515	0.73048	0.86533	0.86676
SVM	1	0.91067	0.88867	0.88867	0.77733	0.88869	0.88869
Random Forest	1	0.86667	0.86667	0.86664	0.73341	0.86667	0.8673
AdaBoost	0.85271	0.826	0.82333	0.82334	0.64666	0.82333	0.82336
XGBoost	0.90143	0.882	0.88667	0.88648	0.77316	0.88667	0.88861

TABLE 10.10 Performance of InceptionResNetV2 deep features extraction with different machine learning models.

Model	Training accuracy	Validation accuracy	Test accuracy	F1 score	Kappa	Recall	Precision
ANN	0.59686	0.602	0.584	0.51663	0.17377	0.584	0.70005
KNN	0.791	0.674	0.69667	0.69563	0.39267	0.69667	0.69868
SVM	0.82086	0.76867	0.76	0.76	0.51995	0.76	0.76
Random Forest	0.99986	0.734	0.74	0.73999	0.48007	0.74	0.7402
AdaBoost	0.74829	0.70467	0.69733	0.69731	0.39455	0.69733	0.69733
XGBoost	0.99129	0.75267	0.72467	0.72467	0.44929	0.72467	0.72467

score is with AdaBoost with test accuracy score of 88.6% which is similar to the MobileNetV2 (Table 10.8).

InceptionV3 is developed by the Google and is upgraded version of InceptionV1. This is not suitable for the colon dataset as the results are not very good. The best test accuracy is given by InceptionV3 is with SVM with value of 88.86% and worst is AdaBoost with a value of 82.33% (Table 10.9).

This model is inspired by the Inception and ResNet architecture. It has 164 layers and can classify up to 1000 different classes on a single dataset. But the result of this model is not good highest test accuracy is reached with the SVM with value of 76%, while the worst value is given with KNN with test accuracy value of 69.66% which are very low (Table 10.10).

The results obtained with DenseNet169 are impressive. This model is able to classify 99.13% of test images correctly with SVM. The worst test accuracy is given with the AdaBoost with value of 93.66% (Table 10.11).

DenseNet121 is a simpler version of DenseNet169. The results obtained by this model are worse than DenseNet169. The best accuracy is given with the SVM model with value of 98.86% and worst accuracy is given with AdaBoost with value of 92.46% (Table 10.12).

TABLE 10.11 Performance of DenseNet169 deep features extraction with different machine learning models.

Model	Training accuracy	Validation accuracy	Test accuracy	F1 score	Kappa	Recall	Precision
ANN	0.93514	0.92267	0.92067	0.92014	0.84109	0.92067	0.93071
KNN	0.969	0.94733	0.96133	0.96133	0.92266	0.96133	0.96134
SVM	1	0.98933	0.99133	0.99133	0.98267	0.99133	0.99138
Random Forest	1	0.972	0.954	0.954	0.90802	0.954	0.95444
AdaBoost	0.97314	0.95133	0.93667	0.93667	0.87332	0.93667	0.93667
XGBoost	1	0.98067	0.974	0.974	0.948	0.974	0.97415

TABLE 10.12 Performance of DenseNet121 deep features extraction with different machine learning models.

Model	Training accuracy	Validation accuracy	Test accuracy	F1 score	Kappa	Recall	Precision
ANN	0.98914	0.97267	0.97067	0.97066	0.94135	0.97067	0.97154
KNN	0.956	0.93667	0.93733	0.93731	0.87471	0.93733	0.93849
SVM	1	0.98867	0.98867	0.98867	0.97733	0.98867	0.98869
Random Forest	1	0.94533	0.93933	0.93933	0.87868	0.93933	0.93952
AdaBoost	0.94871	0.92933	0.92467	0.92467	0.84933	0.92467	0.92469
XGBoost	1	0.97133	0.97333	0.97333	0.94667	0.97333	0.9735

TABLE 10.13 Performance of Xception deep features extraction with different machine learning models.

Model	Training accuracy	Validation accuracy	Test accuracy	F1 score	Kappa	Recall	Precision
ANN	0.87357	0.83933	0.83133	0.8287	0.66188	0.83133	0.85104
KNN	0.874	0.76133	0.76733	0.76671	0.53421	0.76733	0.76955
SVM	1	0.89	0.87067	0.87066	0.74138	0.87067	0.87096
Random Forest	1	0.81867	0.81733	0.81734	0.63466	0.81733	0.81737
AdaBoost	0.83786	0.79867	0.77333	0.77333	0.54671	0.77333	0.77347
XGBoost	1	0.85533	0.85333	0.85328	0.70667	0.85333	0.85421

Xception model has 73 layers which are lesser than the above used models, so the results are also worse than many of the previously used models. Here the best accuracy is given with the SVM model with a test accuracy value of 87.06% and the worst accuracy is given with 77.33% with AdaBoost (Table 10.13).

10.5 Discussion

From the above tables of the second method, it can be seen that ANN, SVM, and XGBoost are giving the best results with a particular model. Here also, ResNet and VGG models are outperforming other pretrained models. The best results in given by the ResNet50 with the SVM model with test accuracy score, F1 score, kappa score, recall score, and a precision score of 99.867%, 99.867%, 99.733%, 99.869%, and 99.867%, respectively. While using the classical ML models, the extracted features from the pretrained models were reduced to lower dimensions, still there are no significant differences in the results obtained by both methods. Using pretrained models drastically increased the performance of models.

10.6 Conclusion

Colon cancer is a very common disease that can lead to loss of life if not treated early. The cancer is very hard to treat in the last stages. All the steps should be taken to avoid the development of colon cancer in the body, and immediate action should be taken if any of the cancer symptoms are detected. The detection of cancer through biopsy is one of the most trusted methods by a doctor. Artificial intelligence can help the doctor reduce the time taken for detection of cancer and help get accurate results as the incorrect results can lead to very bad results and sometimes even the loss of life. The false-positive result will also be very bad for the patient as the treatment evolves many side effects and huge cost.

References

[1] Cancer.Net, Colorectal cancer - risk factors and prevention, <https://www.cancer.net/cancer-types/colorectal-cancer/risk-factors-and-prevention>, Jun. 25, 2012 (accessed 16.05.21).

[2] Colorectal Cancer Statistics | How common is colorectal cancer? <https://www.cancer.org/cancer/colon-rectal-cancer/about/key-statistics.html> (accessed 16.05.21).

[3] Mayo Clinic, Colon cancer - symptoms and causes, <https://www.mayoclinic.org / diseases-conditions / colon-cancer/symptoms-causes/syc-20353669> (accessed 25.05.21).

[4] Stanford Health Care, Types of colorectal cancer. https://stanfordhealthcare.org/medical-conditions/cancer/colorectal-cancer/types.html (accessed 25.05.21).

[5] Memorial Sloan Kettering Cancer Center, Treatment for metastatic colon cancer, <https://www.mskcc.org/cancer-care/types/colon/treatment/metastases> (accessed 25.05.21).

[6] Testing for Colorectal Cancer | How is colorectal cancer diagnosed? <https://www.cancer.org/cancer/colon-rectal-cancer/detection-diagnosis-staging/how-diagnosed.html> (accessed 25.05.21).

[7] Cancer.Net, Colorectal cancer - diagnosis, <https://www.cancer.net/cancer-types/colorectal-cancer/diagnosis>, Jun. 25, 2012 (accessed 25.05.21).

[8] T.-H. Kim, D. Bhattacharyya, M. Mahanty, D. Midhunchakkaravarthy, Detection of colorectal cancer by deep learning an extensive review, Int. J. Curr. Res. Rev. 12 (2020). Available from: https://doi.org/10.31782/IJCRR.2020.122234.

[9] Z. Song, et al., Automatic deep learning-based colorectal adenoma detection system and its similarities with pathologists, BMJ Open 10 (9) (2020) e036423.

[10] E. Ribeiro, A. Uhl, G. Wimmer, M. Häfner, Exploring deep learning and transfer learning for colonic polyp classification, Comput. Math. Methods Med. (2016) e6584725. Available from: https://doi.org/10.1155/2016/6584725. Oct. 2016.

[11] L. Xu, et al., Colorectal cancer detection based on deep learning, J. Pathol. Inform. 11 (2020).

[12] M.C. Hornbrook, et al., Early colorectal cancer detected by machine learning model using gender, age, and complete blood count data, Dig. Dis. Sci. 62 (10) (2017) 2719–2727.

[13] F. Ponzio, E. Macii, E. Ficarra, S. Di Cataldo, Colorectal cancer classification using deep convolutional networks, in: Proc. 11th International Joint Conference on Biomedical Engineering Systems and Technologies, 2018, 2, pp. 58–66.

[14] J. Li, P. Wang, Y. Zhou, H. Liang, K. Luan, Different machine learning and deep learning methods for classification of colorectal cancer lymph node metastasis images, Front. Bioeng. Biotechnol. 8 (2020) 1521.

[15] Scribd, Artificial neural network a study | PDF | Artificial neural network | Nervous system,. <https://www.scribd.com/document/525467375/artificial-neural-network-a-study> (accessed 04.04.22).
[16] D.E. Rumelhart, G.E. Hinton, R.J. Williams, Learning representations by back-propagating errors, Nature 323 (6088) (1986) 533–536.
[17] M. Manavazhahan, A study of activation functions for neural networks, 2017.
[18] S. Sharma, Activation functions in neural networks, medium, <https://towardsdatascience.com/activation-functions-neural-networks-1cbd9f8d91d6>, Jul. 04, 2021 (accessed 04.04.22).
[19] N. Ganatra, A. Patel, A comprehensive study of deep learning architectures, applications and tools, Int. J. Comput. Sci. Eng. 6 (12) (2018) 701–705.
[20] F. Zhuang, et al., A comprehensive survey on transfer learning, Proc. IEEE 109 (1) (2020) 43–76.
[21] MathWorks, What is deep learning? | How it works, techniques & applications. <https://in.mathworks.com/discovery/deep-learning.html> (accessed 04.04.22).
[22] S. Indolia, A.K. Goswami, S.P. Mishra, P. Asopa, Conceptual understanding of convolutional neural network-a deep learning approach, Procedia Comput. Sci. 132 (2018) 679–688.
[23] J. Brownlee, A Gentle Introduction to Pooling Layers for Convolutional Neural Networks, Machine Learning Mastery, Apr. 21, 2019. https://machinelearningmastery.com/pooling-layers-for-convolutional-neural-networks/ (accessed Apr. 04, 2022).
[24] Padding (Machine Learning), DeepAI, May 17, 2019. https://deepai.org/machine-learning-glossary-and-terms/padding (accessed Apr. 04, 2022).
[25] K. Simonyan, A. Zisserman, Very deep convolutional networks for large-scale image recognition, ArXiv Prepr. ArXiv14091556, 2014.
[26] K. He, X. Zhang, S. Ren, J. Sun, Deep residual learning for image recognition, in: Proc. the IEEE Conference on Computer Vision and Pattern Recognition, 2016, pp. 770–778.
[27] A.G. Howard et al., MobileNets: efficient convolutional neural networks for mobile vision applications, ArXiv170404861 Cs, [Online]. <http://arxiv.org/abs/1704.04861>, Apr. 2017 (accessed 14.05.21).
[28] C. Szegedy, V. Vanhoucke, S. Ioffe, J. Shlens, Z. Wojna, Rethinking the inception architecture for computer vision, in: Proc. IEEE Conference on Computer Vision and Pattern Recognition, 2016, pp. 2818–2826.
[29] C. Szegedy, S. Ioffe, V. Vanhoucke, A. Alemi, Inception-v4, inception-resnet and the impact of residual connections on learning, in: Proc. AAAI Conference on Artificial Intelligence, 2017, 31, 1.
[30] G. Huang, Z. Liu, L. Van Der Maaten, K.Q. Weinberger, Densely connected convolutional networks, in: Proc. IEEE Conference on Computer Vision and Pattern Recognition, 2017, pp. 4700–4708.
[31] F. Chollet, Xception: deep learning with depthwise separable convolutions, in: Proc. IEEE Conference on Computer Vision and Pattern Recognition, 2017, pp. 1251–1258.
[32] I.T. Jolliffe, J. Cadima, Principal component analysis: a review and recent developments, Philos. Trans. R. Soc. Math. Phys. Eng. Sci. 374 (2065) (2016) 20150202.
[33] A. Tharwat, T. Gaber, A. Ibrahim, A.E. Hassanien, Linear discriminant analysis: a detailed tutorial, AI Commun. 30 (2) (2017) 169–190.
[34] F.-N. Yuan, L. Zhang, J.T. Shi, X. Xia, G. Li, Theories and applications of auto-encoder neural networks: a literature survey, Chin. J. Comput. 42 (01) (2019) 203–230.
[35] G. Hinton, S.T. Roweis, Stochastic neighbor embedding, in NIPS, 2002, 15, pp. 833–840.
[36] G. Guo, H. Wang, D. Bell, Y. Bi, K. Greer, KNN model-based approach in classification, in: OTM Confederated International Conferences On the Move to Meaningful Internet Systems, 2003, pp. 986–996.
[37] S.R. Gunn, Support vector machines for classification and regression, ISIS Tech. Rep. 14 (1) (1998) 5–16.
[38] L. Breiman, Random forests, Mach. Learn. 45 (1) (2001) 5–32.
[39] R.E. Schapire, Explaining adaboost, Empirical Inference, Springer, 2013, pp. 37–52.
[40] T. Chen, C. Guestrin, Xgboost: a scalable tree boosting system, in: Proc. 22nd ACM SIGKDD International Conference on Knowledge Discovery and Data Mining, 2016, pp. 785–794.
[41] A.A. Borkowski, M.M. Bui, L.B. Thomas, C.P. Wilson, L.A. DeLand, S.M. Mastorides, Lung and Colon Cancer Histopathological Image Dataset (LC25000), ArXiv Prepr. ArXiv191212142, 2019.
[42] Kaggle, Lung and colon cancer histopathological images. <https://kaggle.com/andrewmvd/lung-and-colon-cancer-histopathological-images> (accessed 14.05.21).

CHAPTER

11

Brain hemorrhage detection using computed tomography images and deep learning

Abdullah Elen[1], *Aykut Diker*[1] *and Abdulhamit Subasi*[2,3]

[1]Department of Software Engineering, Faculty of Engineering and Natural Sciences, Bandirma Onyedi Eylul University, Bandirma, Balikesir, Turkey [2]Institute of Biomedicine, Faculty of Medicine, University of Turku, Turku, Finland [3]Department of Computer Science, College of Engineering, Effat University, Jeddah, Saudi Arabia

OUTLINE

11.1 Introduction

Intracranial hemorrhage (ICH) is a fatal lesion that occurs in the brain, and it has a high rate of mortality as reported in Ref. [1]. ICH is a type of stroke, which is caused by an artery in the brain. If the stroke bursts and bleeds, it could harm the surrounding brain tissues. Hence, the continuous bleeding kills brain cells [2]. ICH is known to be clinically hazardous since it has a high risk for causing brain injuries that may lead to paralysis and even death. Since ICH is fatal, opportune diagnosis of ICH is significant for taking appropriate actions such as surgical intervention [3]. Depending on where it is located in the brain, ICH can be divided into five distinct categories: intraventricular (IVH), intraparenchymal (IPH),

Applications of Artificial Intelligence in Medical Imaging
DOI: https://doi.org/10.1016/B978-0-443-18450-5.00014-1

subarachnoid (SAH), epidural (EDH), and subdural (SDH) [4]. Studies conducted on populations indicate that incidence of ICH is estimated to be about 10–30 per 100 000 people [5,6]. Furthermore, 3 months later, more than one-third of survivors have significant disabilities [7].

Computed tomography (CT) can be used to determine the source of hemorrhage and its localization. CT uses consecutive 2D slices and stacks them to generate 3D image as an output [8]. The types of ICH can be diagnosed by an expert with the help of their properties in the CT images such as lesion shape, size, etc. and therefore manual diagnosis is a tedious procedure [9]. Hence, automated diagnosis and detection of hemorrhage has gained attention in the last decades [10–12]. Deep learning-based methods have shown good performance in medical image classification [12–14], medical image segmentation [15–17], and disease diagnosis [18–21]. Convolutional neural network (CNN) is employed for various classification tasks related with the medical images [19–21]. CNNs are capable of extracting features from images and learning automatically. They are devised such that they can understand the image content [22]. CNNs are optimized for images. There are concepts such as weight sharing, pooling, and convolution that can be used for maintaining essential operations of CNNs [23]. CNNs can extract meaningful features and classify them further, for the diagnosis of hemorrhage from CT scan images. Sample normal brain CT image and hemorrhage CT image are given in Fig. 11.1.

Since CNNs are popular in the field of computer vision, researchers proposed more advanced deep neural networks to obtain more accuracy. Deep neural network models such as residual networks (ResNet) [24], AlexNet [25], VGGNet [26], and other different models are proposed, which are usually large and specialized networks relative to the classical CNNs. Each network has its own pros and cons, as mentioned in their original study. Variants or competitors of these deep neural network models are usually employed in transfer learning [27–29]. In transfer learning, the model is trained on a dataset, and weights of the trained model are saved for further usage. Then, most of the layers are frozen during the training phase, and only a few layers are trained. It can be seen that these networks are accepted as state-of-the-art methods in the literature and preferred by most of the researchers in the field of computer vision and medical imaging.

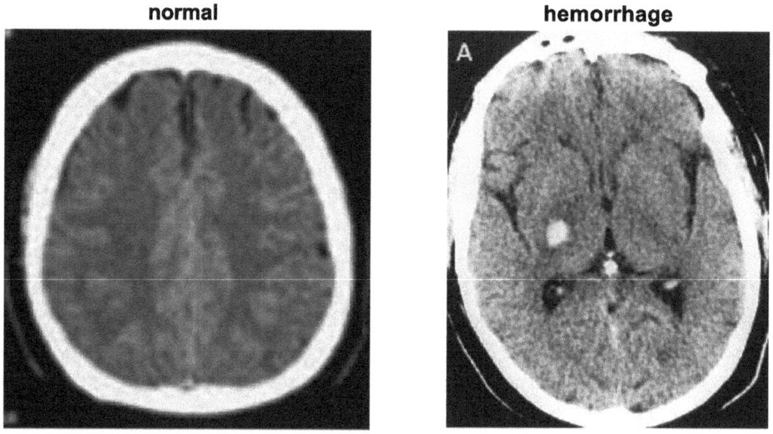

FIGURE 11.1 Normal and hemorrhage image samples.

Starting from this point, in this chapter, some of the popular deep learning models are employed for hemorrhage detection using brain CT images. Moreover, the brain hemorrhage CT image dataset is exploited for hemorrhage detection.

The rest of this chapter is organized as follows: some of the methods proposed for brain hemorrhage detection are reviewed and presented in Section 11.2. The deep learning models and methods employed in this chapter are explained in Section 11.3. The obtained experimental results are provided in Section 11.4. Discussions and drawn conclusions are included in Section 11.5 and Section 11.6, respectively.

11.2 Literature survey in brain hemorrhage detection

In the recent decades, many studies have been conducted for ICH detection from the brain CT scan images. In this section, some of these studies are reviewed in an elaborative manner. The methods proposed in the literature that use various CT image datasets are examined.

In Ref. [30], authors observed the performance of different frameworks. They used AlexNet, a modified AlexNet-support vector machine (AlexNet-SVM), and AlexNet-SVM with principal component analysis (PCA), separately, for classifying brain CT images. Using the AlexNet-SVM model, on the original images and preprocessed images, highest accuracy rates of 99.86% and 99.60% are obtained, respectively. Hence, AlexNet-SVM framework outperformed AlexNet-SVM with PCA framework and AlexNet. In Ref. [31], researches employed summation images of the brain CT images in order to detect hemorrhage. Six distinct subdivisions are created according to the height of images and CT images belong to the same subdivision are summed. Authors used various artificial neural network (ANN) architectures for classifying these images and gradually increased the size of input image layer and number of hidden layers. The best accuracy rate, which is 83.30%, is obtained for subdivision 41%–50%. In Ref. [32], a deep learning-based system is proposed. The proposed system uses multiple ImageNet [33] pretrained CNNs, some preprocessing techniques, atlas creation module, and prediction-based selection module. The preprocessing provides high performance for small and imbalanced datasets. The proposed system reached an area under the curve (AUC) score of 99.30%. In Ref. [34], a deep learning-based method is presented for achieving accurate ICH detection and classification. Authors employed 2D slices of 3D CT scans, by processing DICOM files. In the first step, the method exploits CNN to extract features from the slices and classify them. Then, in the second step, feature vectors generated from the slices are supplied to the first sequence model for providing spatially coherent results. In the final stage, classification results of CNN and the first sequence model through a second sequence model is implemented. The second sequence model is employed for adaptive model averaging, therefore the final decision is made by the second sequence model. In Ref. [35], brain hemorrhage classification based on neural network (BHCNet) is proposed. Authors employed data augmentation and combined this augmentation method with the CNN model. They also evaluated the performance of CNN + long short-term memory and CNN + gated recurrent unit. Their method has reached and accuracy rate of 100.00% on the imbalanced data they employed for benchmarking. In Ref. [36], a joint model is proposed which combined CNN with recurrent neural network (RNN). The CNN-RNN model is employed for both five-type classification and two-type classification. The two-type classification is simply made for detecting the presence of hemorrhage and the five-type classification is made for ICH subtype detection. Their models named Sub-Lab and

Sub-Lab reached accuracy, sensitivity, and specificity scores that are greater than or equal to 0.98. In Ref. [37], authors proposed a CNN model for the detection of ICH. They reached an AUC score of 0.846. Their ICH detection method is also speed efficient, which processes images in 2.3 seconds on average. In Ref. [38], the authors presented a new classification technique, which is based on minimalist machine learning (MML) [39]. They also employed a feature selection algorithm, namely, dMeans. Their best model has reached an accuracy rate of 86.50%. In Ref. [40], authors proposed computer-aided diagnosis (CAD) method for SAH (which is subtype of ICH) detection. Their method applies techniques such as morphological operations, noise reduction, and segmentation, on the images. The CAD system outputs the final image which indicates hemorrhage and classification result together. They reached a classification accuracy of 88.89%. In Ref. [41], authors proposed an artificial intelligence-based tool for ICH detection. The tool makes prediction about the presence of ICH by localizing it using heat map. An atlas is created from the training images by obtaining class activation maps and visual evidence of later layers belong to the CNN model. They reached an AUC score of 0.994.

11.3 Deep learning methods

In this chapter, a model based on CNN has been proposed to classify it as hemorrhage and normal, which consists of 2 classes in total. The block diagram of the model is given in Fig. 11.2.

11.3.1 ResNet-18

Residual Network, or ResNet in short, is a form of neural network first presented by He et al. in 2015 in their study "Deep Residual Learning for Image Recognition" [24]. Additional layers have been added to deep neural networks to handle a mostly complex problem, resulting in better accuracy and performance. The idea behind adding more layers is that they would learn increasingly complicated features. For example, in image recognition and classification processes, the first layer can learn to recognize edges, the second layer can learn to identify textures, and the third layer can learn to recognize objects. The standard CNN model has a maximum depth threshold. With the introduction of a new neural network layer, the Residual Block, the challenge of training very deep networks was relieved shown in Fig. 11.3.

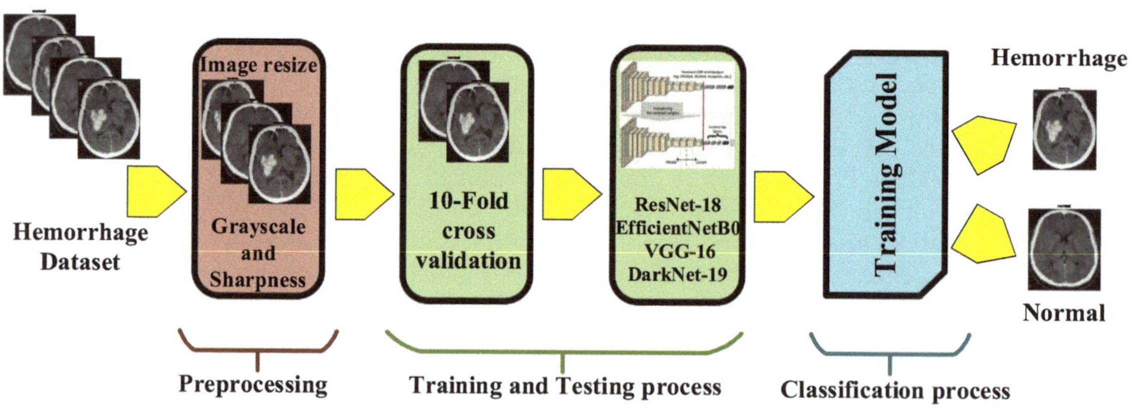

FIGURE 11.2 Block diagram of the deep learning architecture.

11.3.2 EfficientNet-B0

The EfficientNet model was presented by Tan and Le from the Google research team. These researchers examined CNN model scaling and determined that balancing in scaling the depth, width, and resolution of the network affects network performance. They proposed a new scaling method that equally scales all dimensions of the network's depth, width, and resolution based on this observation. EfficientNet consists of eight models from B0 to B7. As the model grows, the number of parameters used and the success rate increase. According to the study, the basic model, EfficientNet-B0, is created first. The base model can then be scaled up with composite scaling from EfficientNet-B1 to EfficientNet-B7 to increase accuracy and model size. With each increase, the processing power nearly doubles. In previous scaling studies, only one of the parameters among depth, width, and resolution is usually scaled. The depth of the network corresponds to the number of layers in a network. The width is related to the number of filters in a convolutional layer. Resolution is the height and width of the input image. While it is possible to arbitrarily scale two or three parameters, arbitrary scaling requires laborious manual adjustment and still often does not achieve the desired efficiency and accuracy. The EfficientNet model not only provides higher accuracy, but also increases the efficiency of the models by reducing parameters compared to cutting-edge models. Contrary to the traditional practice of randomly scaling factors with the EfficientNet model, it is proposed to evenly scale the network width, depth, and resolution with a set of fixed scaling coefficients with composite scaling. The proposed composite scaling method in the EfficientNet model is given in Fig. 11.4.

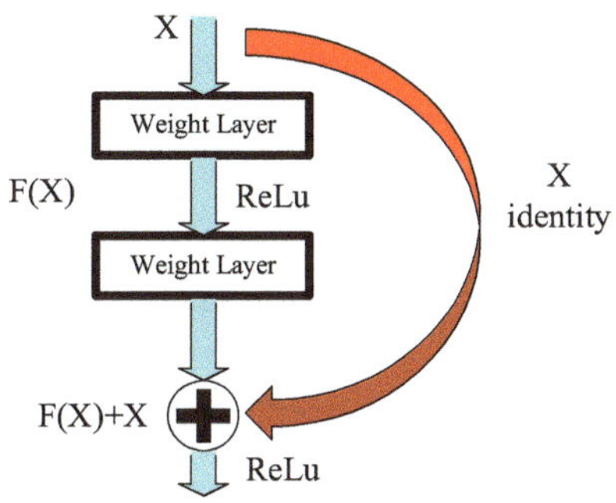

FIGURE 11.3 ResNet's architecture.

The researchers conceived by executing a baseline network using a technique called neural architecture search, which automates the

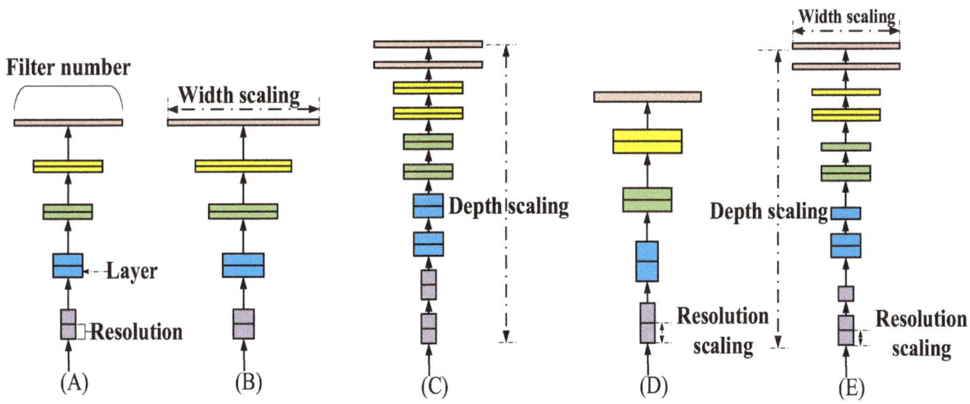

FIGURE 11.4 (A) Basic network example. (B–D) Traditional scaling that increases only one dimension of network width, depth, or resolution. (E) EfficientNet's composite scaling method.

structure of neural networks. It optimizes both the accuracy and efficiency as measured on the floating-point operations per second basis. The improved inverted bottleneck convolution is used in this architecture (MBConv). The researchers then scaled up this baseline network to create the EfficientNets family of deep learning models [42,43]. Its architecture is given in the below diagram in Fig. 11.5.

11.3.3 VGG-16

VGG-16 is a 16-layered CNN architecture that employs a 3 × 3 filter for all convolution layers, making it the smallest size filter available. RGB samples with a resolution of 224 × 224 pixels are input to the model. The image features are extracted using a collection of convolutional layers. The stride of convolution is one. The convolutional layer input must be spatially padded after each convolution process. For nonlinear activation, the convolution output is employed. Spatial pooling is handled via five pooling layers. A 2 × 2 size filter and stride 2 are used for max-pooling. Following a succession of convolutional and max-pooling layers, three fully connected layers are built. The final layer [44] is the softmax layer. The architecture of VGG-16 is depicted in Fig. 11.6.

11.3.4 DarkNet-19

A network can be improved to be small and effective at the same time. Many earlier ideas, such as the Darknet reference, Network in Network, Inception, and Batch Normalization, have been incorporated. Convolutional layers are used instead of completely connected layers in Darknet-19. It is composed of 19 convolutional and 5 max-pooling layers. To decrease the

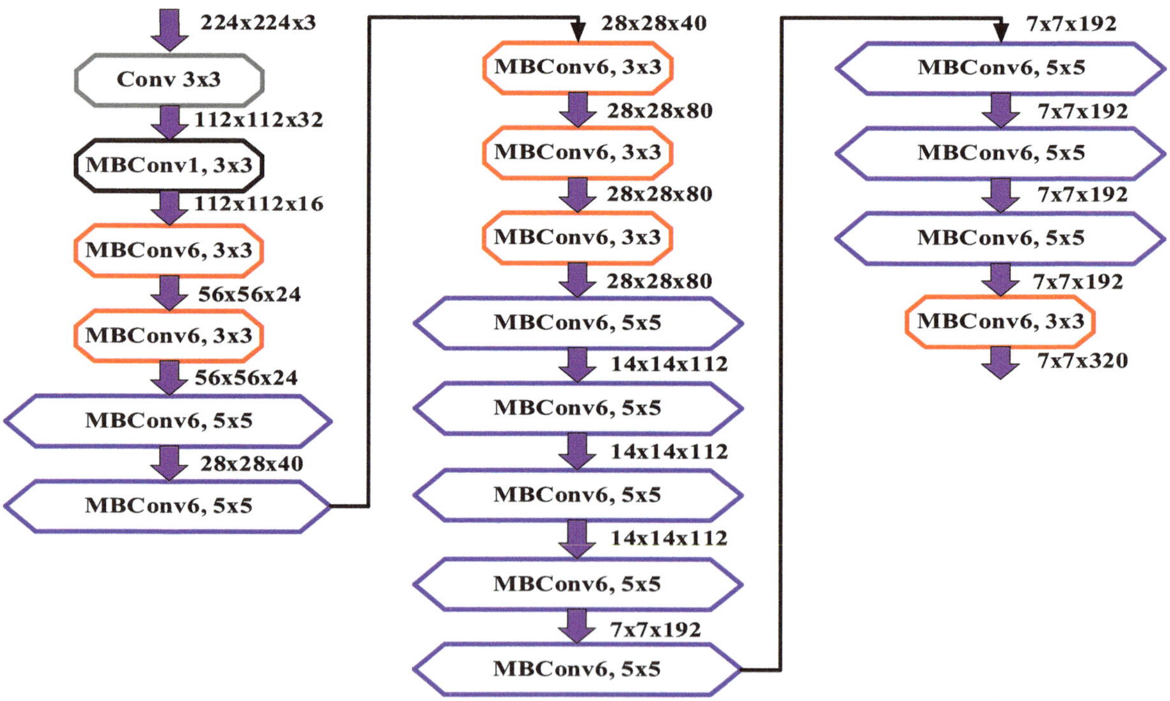

FIGURE 11.5 EfficientNet-B0's architecture.

number of parameters, it only employs 3×3 convolutional kernels and numerous 1×1 convolutional kernels. In YOLO algorithms such as YOLOv2, DarkNet-19 is frequently used [45]. The architecture of DarkNet-19 is shown in Fig. 11.7.

11.4 Experimental results

The experiments were carried out on Intel (R) Core (TM) i7 2.90 GHz CPU, 8GB RAM, and 2 GB GeForce GT 730 graphic card hardware in the computer. The environment for code development and experimental work was MATLAB (2020b).

To evaluate the proposed model, we have used the confusion matrix and commonly used performance parameters, which are obtained from this matrix such as accuracy (ACC), sensitivity (SEN), specificity (SPE), F-score, Matthews correlation coefficient (MCC), and receiver operating characteristic (ROC). It consists of four components: true positive (TP), true negative (TN), false

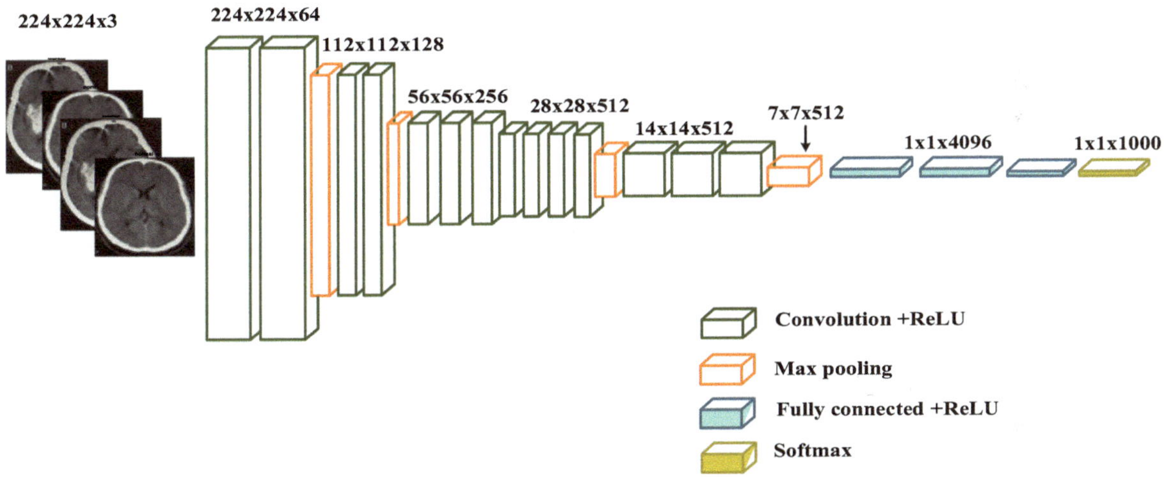

FIGURE 11.6 VGG-16's architecture.

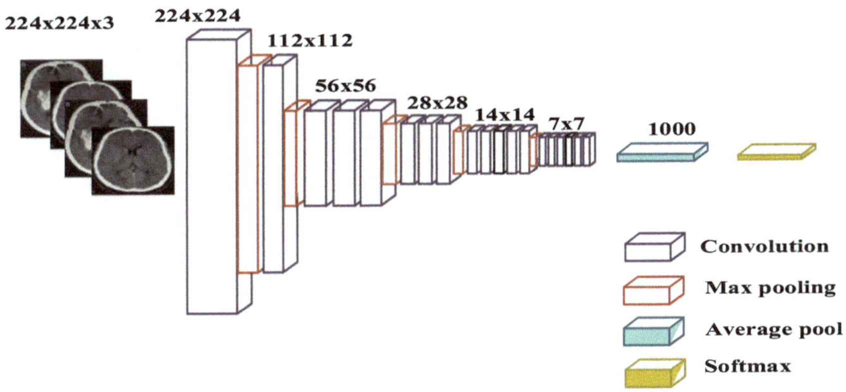

FIGURE 11.7 DarkNet-19's architecture.

positive (FP), and false negative [46]. The performance measures are given in Eqs. (11.2–11.6).

$$ACC = \frac{TP + TN}{TP + TN + FP + FN} \quad (11.2)$$

$$SEN = \frac{TP}{TP + FN} \quad (11.3)$$

$$SPE = \frac{TN}{TN + FP} \quad (11.4)$$

$$F - score = \frac{2 \times TP}{2 \times TP + FP + FN} \quad (11.5)$$

$$MCC = \frac{TPxTN - FPxFN}{\sqrt{(TP + FP)(TP + FN)(TN + FP)(TN + FN)}} \quad (11.6)$$

Hereby, TP and TN represent the number of correctly predicted positive and negative samples, whereas FP and FN correspond to the number of incorrectly predicted positive and negative samples. Additionally, ROC has been considered to evaluate the model performance. The ROC is a 2D graph that is drawn the true-positive rate against the false-negative rate. The training of the CNN models was realized in 50 epochs, the mini-batch size was 16 as can be seen from the given MATLAB code of each CNN models in Section 11.3.

11.4.1 Dataset

The Head CT Hemorrhage Image Dataset (HCTHID) contains a total of 200 CT images of 100 normal (healthy) and hemorrhage [47]. The images in the dataset have different sizes and are all in PNG format. In this study, we used all images by converting them to 32×32 pixels. Sample images of the HCTHID are shown in Fig. 11.8, the first row of the figure depicts

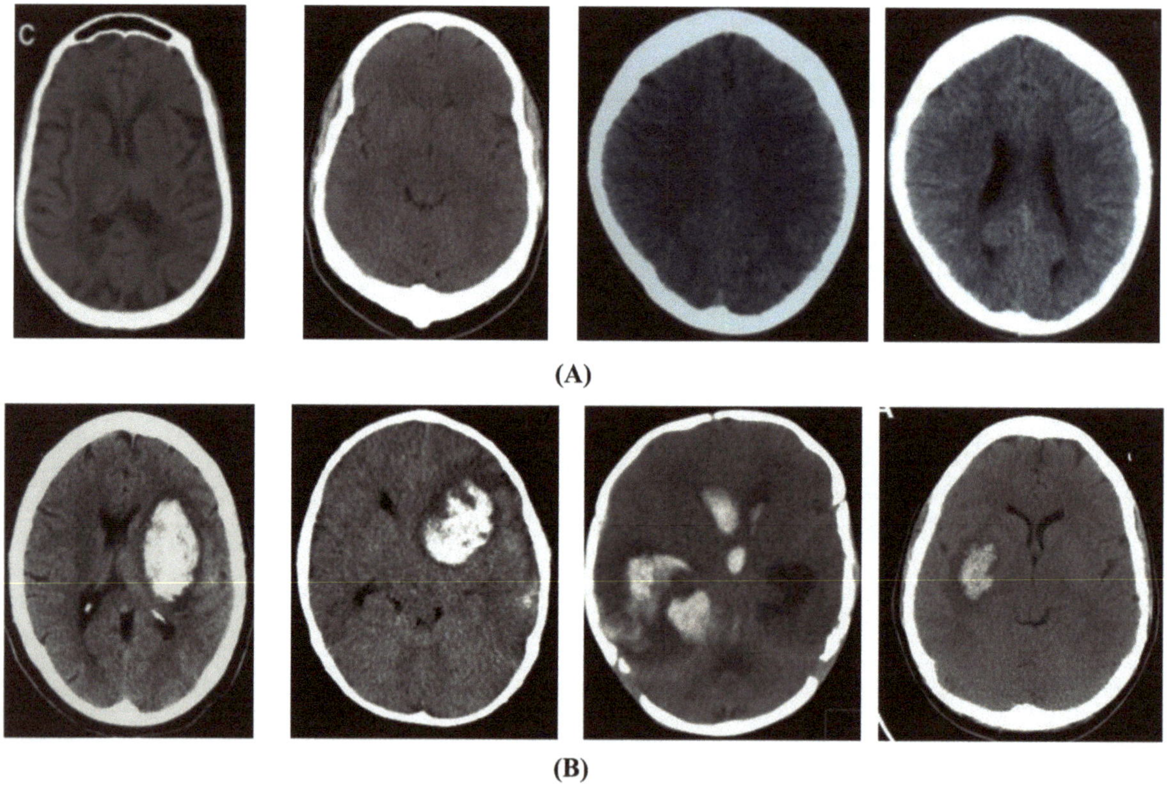

FIGURE 11.8 Sample CT images of (A) normal and (B) hemorrhage brain. *CT*, Computed tomography.

images of cases which do not have any problem, that is, normal and the images of hemorrhages are depicted in the second row.

The performance of deep learning models was investigated in different experiments for classification of the HCTHID. The confusion matrix, ROC curves, performance metrics tables of the CNN models, and classification results obtained by using the 10-fold cross-validation technique of the HCTHID are given in Figs. 11.9–11.12 and Tables 11.1–11.5.

The ACC, SEN, SPE, F-score, and MCC values were obtained by using the 10-fold cross validation of the DarkNet-19 CNN model. As seen in Fig. 11.9, the accuracy rate of the confusion matrix and mean score of the ROC curve, respectively 83.5%, and 91.50%

```
% Datastore for the image dataset.
dataStore = imageDatastore('Repository\BrainHemorrhage', ...
    'LabelSource', 'foldernames', 'IncludeSubfolders', true);
% Partition data for cross-validation.
cv = cvpartition(dataStore.Labels, 'Kfold', 10);
% Create cell array for train and test data.
trainSet = cell(10, 1); testSet = cell(10, 1);
for i = 1 : 10 % Create subset of datastore.
    % Train cases for current fold.
    trainSet{i} = subset(dataStore, find(cv.training(i)));
    % Test cases for current fold.
    testSet{i} = subset(dataStore, find(cv.test(i)));
end
% Options for training deep learning neural networks.
% Set the options to the default settings for the stochastic gradient
descent with momentum. Set the maximum number of epochs at 50,
and start the training with an initial learning rate of 0.0001.
options = trainingOptions('sgdm', ...
    'MiniBatchSize', 16, ...
    'MaxEpochs', 50, ...
    'InitialLearnRate', 1e-4);
% Load DarkNet-19 convolutional neural network.
deepNet = darknet19;
layers = layerGraph(deepNet.Layers);
for i = 1 : 10 % 10-fold cross validation.
    % Train neural network for deep learning.
    [net, info] = trainNetwork(trainSet{i}, layers, options);
    % Test the performance of the network by evaluating the prediction
accuracy of the test data.
    [predictions, ~] = classify(net, testSet{i});
    % Evaluate test accuracy of the model.
    % The accuracy is the ratio of the number of true labels in the
testdata matching the classifications from classify to the number
of images in the test data.
    ACC = mean(predictions == testSet{i}.Labels);
    disp(['Test accuracy is ', num2str(ACC)]);
end
```

The ACC, SEN, SPE, F-score, and MCC values were obtained by using the 10-fold cross validation of the EfficientNet-B0 CNN model.

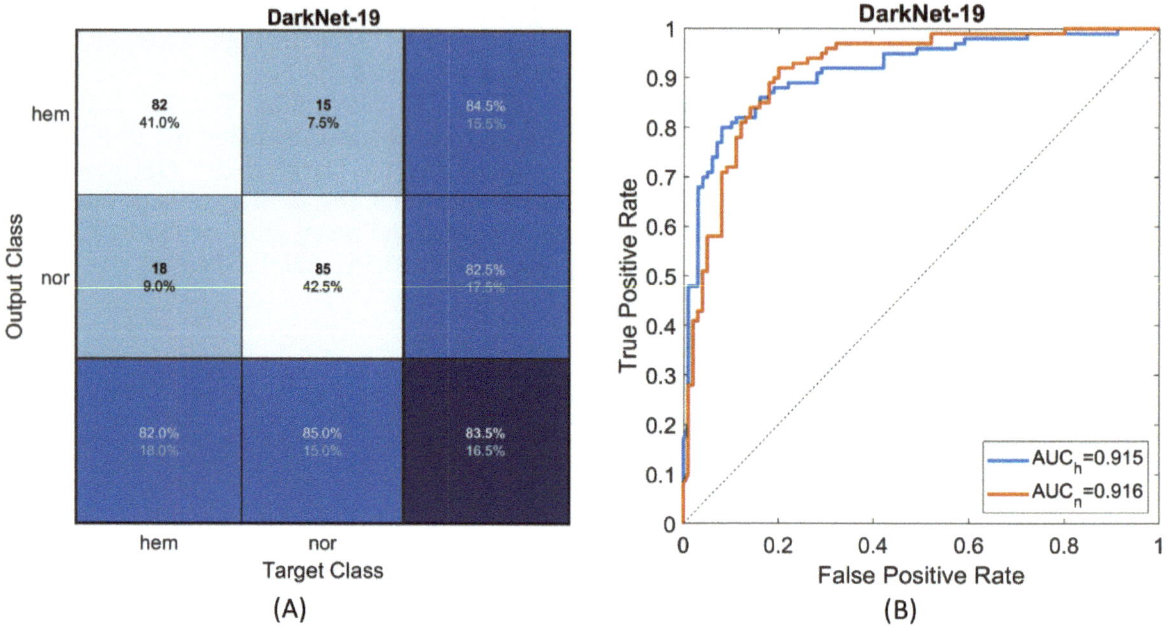

FIGURE 11.9 Test result of the DarkNet-19; (A) confusion matrix and (B) ROC curve. *ROC*, Receiver operating characteristic.

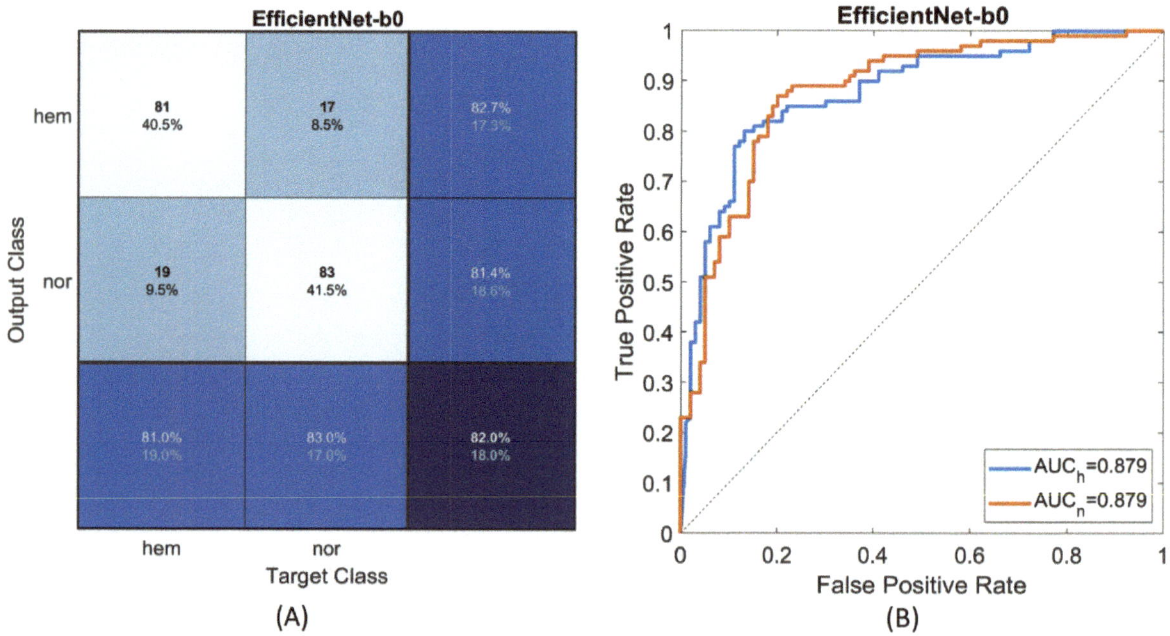

FIGURE 11.10 Test result of the EfficientNet-B0; (A) confusion matrix and (B) ROC curve. *ROC*, Receiver operating characteristic.

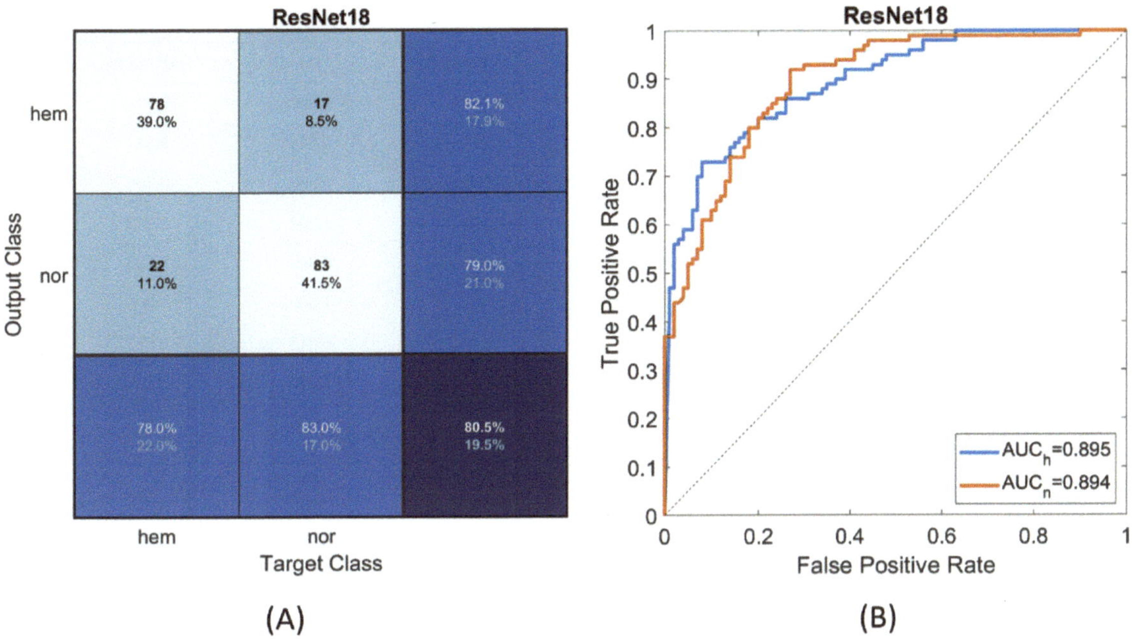

FIGURE 11.11 Test result of the ResNet-18; (A) confusion matrix and (B) ROC curve. *ROC*, Receiver operating characteristic.

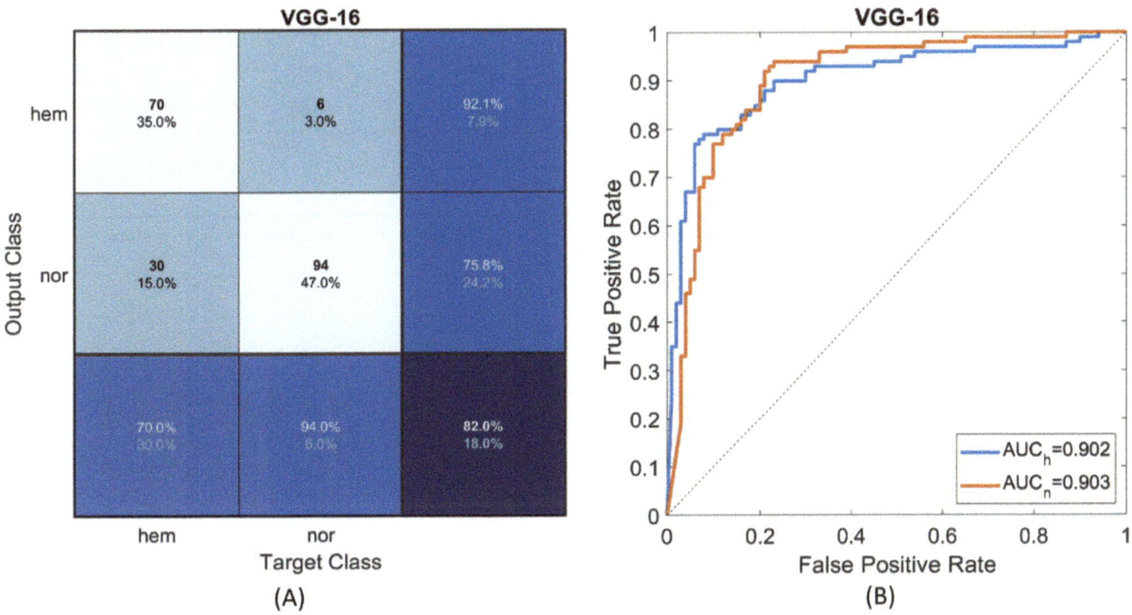

FIGURE 11.12 Test result of the VGG-16; (A) confusion matrix and (B) ROC curve. *ROC*, Receiver operating characteristic.

TABLE 11.1 Test result of the DarkNet-19.

	Fold	Accuracy	Sensitivity	Specificity	Precision	F-score	MCC
Hemorrhage	1	0.650	0.400	0.900	0.800	0.533	0.346
	2	0.850	0.800	0.900	0.889	0.842	0.704
	3	0.800	0.800	0.800	0.800	0.800	0.600
	4	0.950	0.900	1.000	1.000	0.947	0.905
	5	0.800	0.800	0.800	0.800	0.800	0.600
	6	1.000	1.000	1.000	1.000	1.000	1.000
	7	0.850	0.900	0.800	0.818	0.857	0.704
	8	0.900	0.800	1.000	1.000	0.889	0.817
	9	1.000	1.000	1.000	1.000	1.000	1.000
	10	0.550	0.800	0.300	0.533	0.640	0.116
Normal	1	0.650	0.900	0.400	0.600	0.720	0.346
	2	0.850	0.900	0.800	0.818	0.857	0.704
	3	0.800	0.800	0.800	0.800	0.800	0.600
	4	0.950	1.000	0.900	0.909	0.952	0.905
	5	0.800	0.800	0.800	0.800	0.800	0.600
	6	1.000	1.000	1.000	1.000	1.000	1.000
	7	0.850	0.800	0.900	0.889	0.842	0.704
	8	0.900	1.000	0.800	0.833	0.909	0.817
	9	1.000	1.000	1.000	1.000	1.000	1.000
	10	0.550	0.300	0.800	0.600	0.400	0.116

TABLE 11.2 Test result of the EfficientNet-B0.

	Fold	Accuracy	Sensitivity	Specificity	Precision	F-score	MCC
Hemorrhage	1	0.600	0.600	0.600	0.600	0.600	0.200
	2	0.750	0.800	0.700	0.727	0.762	0.503
	3	0.900	0.900	0.900	0.900	0.900	0.800
	4	0.850	0.900	0.800	0.818	0.857	0.704
	5	0.700	0.700	0.700	0.700	0.700	0.400
	6	1.000	1.000	1.000	1.000	1.000	1.000
	7	0.950	0.900	1.000	1.000	0.947	0.905
	8	0.700	0.600	0.800	0.750	0.667	0.408
	9	0.900	0.800	1.000	1.000	0.889	0.817
	10	0.850	0.900	0.800	0.818	0.857	0.704

(Continued)

TABLE 11.2 (Continued)

	Fold	Accuracy	Sensitivity	Specificity	Precision	F-score	MCC
Normal	1	0.600	0.600	0.600	0.600	0.600	0.200
	2	0.750	0.700	0.800	0.778	0.737	0.503
	3	0.900	0.900	0.900	0.900	0.900	0.800
	4	0.850	0.800	0.900	0.889	0.842	0.704
	5	0.700	0.700	0.700	0.700	0.700	0.400
	6	1.000	1.000	1.000	1.000	1.000	1.000
	7	0.950	1.000	0.900	0.909	0.952	0.905
	8	0.700	0.800	0.600	0.667	0.727	0.408
	9	0.900	1.000	0.800	0.833	0.909	0.817
	10	0.850	0.800	0.900	0.889	0.842	0.704

TABLE 11.3 Test result of the ResNet-18.

	Fold	Accuracy	Sensitivity	Specificity	Precision	F-score	MCC
Hemorrhage	1	0.800	0.700	0.900	0.875	0.778	0.612
	2	0.700	0.900	0.500	0.643	0.750	0.436
	3	0.850	0.700	1.000	1.000	0.824	0.734
	4	0.600	0.500	0.700	0.625	0.556	0.204
	5	0.800	0.800	0.800	0.800	0.800	0.600
	6	0.900	0.800	1.000	1.000	0.889	0.817
	7	0.900	1.000	0.800	0.833	0.909	0.817
	8	0.650	0.500	0.800	0.714	0.588	0.315
	9	1.000	1.000	1.000	1.000	1.000	1.000
	10	0.850	0.900	0.800	0.818	0.857	0.704
Normal	1	0.800	0.900	0.700	0.750	0.818	0.612
	2	0.700	0.500	0.900	0.833	0.625	0.436
	3	0.850	1.000	0.700	0.769	0.870	0.734
	4	0.600	0.700	0.500	0.583	0.636	0.204
	5	0.800	0.800	0.800	0.800	0.800	0.600
	6	0.900	1.000	0.800	0.833	0.909	0.817
	7	0.900	0.800	1.000	1.000	0.889	0.817
	8	0.650	0.800	0.500	0.615	0.696	0.315
	9	1.000	1.000	1.000	1.000	1.000	1.000
	10	0.850	0.800	0.900	0.889	0.842	0.704

TABLE 11.4 Test result of the VGG-16.

	Fold	Accuracy	Sensitivity	Specificity	Precision	F-score	MCC
Hemorrhage	1	0.850	0.700	1.000	1.000	0.824	0.734
	2	0.950	0.900	1.000	1.000	0.947	0.905
	3	0.500	0.000	1.000	–	0.000	0.000
	4	0.850	0.900	0.800	0.818	0.857	0.704
	5	0.900	0.900	0.900	0.900	0.900	0.800
	6	1.000	1.000	1.000	1.000	1.000	1.000
	7	1.000	1.000	1.000	1.000	1.000	1.000
	8	0.800	0.600	1.000	1.000	0.750	0.655
	9	0.500	0.000	1.000	–	0.000	0.000
	10	0.850	1.000	0.700	0.769	0.870	0.734
Normal	1	0.850	1.000	0.700	0.769	0.870	0.734
	2	0.950	1.000	0.900	0.909	0.952	0.905
	3	0.500	1.000	0.000	0.500	0.667	0.000
	4	0.850	0.800	0.900	0.889	0.842	0.704
	5	0.900	0.900	0.900	0.900	0.900	0.800
	6	1.000	1.000	1.000	1.000	1.000	1.000
	7	1.000	1.000	1.000	1.000	1.000	1.000
	8	0.800	1.000	0.600	0.714	0.833	0.655
	9	0.500	1.000	0.000	0.500	0.667	0.000
	10	0.850	0.700	1.000	1.000	0.824	0.734

TABLE 11.5 Mean test scores of deep learning methods.

	Model	Accuracy	Sensitivity	Specificity	Precision	F-score	MCC
Hemorrhage	DarkNet-19	0.835	0.820	0.850	0.864	0.831	0.679
	EfficientNet-B0	0.820	0.810	0.830	0.831	0.818	0.644
	ResNet-18	0.805	0.780	0.830	0.831	0.795	0.624
	VGG-16	0.820	0.700	0.940	0.936	0.715	0.653
Normal	DarkNet-19	0.835	0.850	0.820	0.825	0.828	0.679
	EfficientNet-B0	0.820	0.830	0.810	0.816	0.821	0.644
	ResNet-18	0.805	0.830	0.780	0.807	0.809	0.624
	VGG-16	0.820	0.940	0.700	0.818	0.855	0.653

As seen in Fig. 11.10, the accuracy rate of the confusion matrix and mean score of the ROC curve, respectively, 82.00%, and 87.90%.

The ACC, SEN, SPE, F-score, and MCC values were obtained by using the 10-fold cross validation of the ResNet-18 CNN model. As

```
% Datastore for the image dataset.
dataStore = imageDatastore('Repository\BrainHemorrhage', ...
    'LabelSource', 'foldernames', 'IncludeSubfolders', true);
% Partition data for cross-validation.
cv = cvpartition(dataStore.Labels, 'Kfold', 10);
% Create cell array for train and test data.
trainSet = cell(10, 1); testSet = cell(10, 1);
for i = 1 : 10 % Create subset of datastore.
    % Train cases for current fold.
    trainSet{i} = subset(dataStore, find(cv.training(i)));
    % Test cases for current fold.
    testSet{i} = subset(dataStore, find(cv.test(i)));
end
% Options for training deep learning neural networks.
% Set the options to the default settings for the stochastic gradient descent
with momentum. Set the maximum number of epochs at 50, and start the training
with an initial learning rate of 0.0001.
options = trainingOptions('sgdm', ...
    'MiniBatchSize', 16, ...
    'MaxEpochs', 50, ...
    'InitialLearnRate', 1e-4);
% Load EfficientNet-B0 convolutional neural network.
deepNet = efficientnetb0;
layers = layerGraph(deepNet.Layers);
for i = 1 : 10 % 10-fold cross validation.
    % Train neural network for deep learning.
    [net, info] = trainNetwork(trainSet{i}, layers, options);
    % Test the performance of the network by evaluating the prediction accuracy
of the test data.
    [predictions, ~] = classify(net, testSet{i});
    % Evaluate test accuracy of the model.
    % The accuracy is the ratio of the number of true labels in the test data
matching the classifications from classify to the number of images in the test data.
    ACC = mean(predictions == testSet{i}.Labels);
    disp(['Test accuracy is ', num2str(ACC)]);
end
```

seen in Fig. 11.11, the accuracy rate of the confusion matrix and mean score of the ROC curve, respectively 80.50%, and 89.50%.

The ACC, SEN, SPE, F-score, and MCC values were obtained by using the 10-fold cross validation of the VGG-16 CNN model. As seen

```
% Datastore for the image dataset.
dataStore = imageDatastore('Repository\BrainHemorrhage', ...
    'LabelSource', 'foldernames', 'IncludeSubfolders', true);
% Partition data for cross-validation.
cv = cvpartition(dataStore.Labels, 'Kfold', 10);
% Create cell array for train and test data.
trainSet = cell(10, 1); testSet = cell(10, 1);
for i = 1 : 10 % Create subset of datastore.
    % Train cases for current fold.
    trainSet{i} = subset(dataStore, find(cv.training(i)));
    % Test cases for current fold.
    testSet{i} = subset(dataStore, find(cv.test(i)));
end
% Options for training deep learning neural networks.
% Set the options to the default settings for the stochastic gradient descent
with momentum. Set the maximum number of epochs at 50, and start the training
with an initial learning rate of 0.0001.
options = trainingOptions('sgdm', ...
    'MiniBatchSize', 16, ...
    'MaxEpochs', 50, ...
    'InitialLearnRate', 1e-4);
% Load ResNet-18 convolutional neural network.
deepNet = resnet18;
layers = layerGraph(deepNet.Layers);
for i = 1 : 10 % 10-fold cross validation.
    % Train neural network for deep learning.
    [net, info] = trainNetwork(trainSet{i}, layers, options);
    % Test the performance of the network by evaluating the prediction accuracy
of the test data.
    [predictions, ~] = classify(net, testSet{i});
    % Evaluate test accuracy of the model.
    % The accuracy is the ratio of the number of true labels in the test data
matching the classifications from classify to the number of images in the test data.
    ACC = mean(predictions == testSet{i}.Labels);
    disp(['Test accuracy is ', num2str(ACC)]);
end
```

in Fig. 11.12, the accuracy rate of the confusion matrix and mean score of the ROC curve, respectively, 82.00%, and 90.20%.

```
% Datastore for the image dataset.
dataStore = imageDatastore('Repository\BrainHemorrhage', ...
    'LabelSource', 'foldernames', 'IncludeSubfolders', true);
% Partition data for cross-validation.
cv = cvpartition(dataStore.Labels, 'Kfold', 10);
% Create cell array for train and test data.
trainSet = cell(10, 1); testSet = cell(10, 1);
for i = 1 : 10 % Create subset of datastore.
    % Train cases for current fold.
    trainSet{i} = subset(dataStore, find(cv.training(i)));
    % Test cases for current fold.
    testSet{i} = subset(dataStore, find(cv.test(i)));
end
% Options for training deep learning neural networks.
% Set the options to the default settings for the stochastic gradient descent
with momentum. Set the maximum number of epochs at 50, and start the training
with an initial learning rate of 0.0001.
options = trainingOptions('sgdm', ...
    'MiniBatchSize', 16, ...
    'MaxEpochs', 50, ...
    'InitialLearnRate', 1e-4);
% Load VGG-16 convolutional neural network.
deepNet = vgg16;
layers = layerGraph(deepNet.Layers);
for i = 1 : 10 % 10-fold cross validation.
    % Train neural network for deep learning.
    [net, info] = trainNetwork(trainSet{i}, layers, options);
    % Test the performance of the network by evaluating the prediction accuracy
of the test data.
    [predictions, ~] = classify(net, testSet{i});
    % Evaluate test accuracy of the model.
    % The accuracy is the ratio of the number of true labels in the test data matching
the classifications from classify to the number of images in the test data.
    ACC = mean(predictions == testSet{i}.Labels);
    disp(['Test accuracy is ', num2str(ACC)]);
end
```

All scores of the classifiers CNN architectures are reported in Table 11.5. According to Table 11.5, the best hemorrhage classification accuracy was obtained 83.50% with DarkNet-19 CNN model.

11.5 Discussions

ICH is one of the most serious risks to human healthcare. Head trauma, excessive blood pressure, and intracranial tumors are all possible causes of brain hemorrhages. Deep learning models can never replace doctors and radiologists. But with its image processing analysis, it can make a big impact. Computer-aided, especially deep learning-based medical image methods have increased in recent years. In this chapter, four deep CNN approaches are considered in hemorrhage classification. Several machine learning models, CNN models and our method are compared according to performance criteria, which consist of accuracy, sensitivity, specificity and F-score in Table 11.6.

TABLE 11.6 Classification performance of the proposed method and comparison with other studies.

References	Methods	Image Datasets	Accuracy
Dawud et al. (2019) [48]	Deep learning and machine learning (AlexNet + SVM)	CT image dataset	93%
Shahangian et al. [49]	K-NN and MLP	CT scan image dataset	93.3%
Lee [50]	ResNet-50	CT scan image dataset with 674258 samples	F1 score 88%
Lee et al. [31]	Deep learning algorithm for ANN	250 cases with 9085 CT images samples	91.7%
Chilamkurthy et al. [51]	Deep Learning	CT scan images from 4304 samples	AUC 94.19%
This study	ResNet-18 EfficientNet-B0 DarkNet-19 VGG-16	CT image dataset with 200 samples	ACC: 83.50% SEN: 82% SPE: 85% F-score 83.20%

In Ref. [31] ICH classification and its subtype classification were made. The aim of this study was to evaluate if the method could be used to identify ICH and classify its subtypes without requiring a CNN. CT images were split into 10 subdivisions based on the intracranial height. For the classification of the ICH to subtypes, the accuracy success for subarachnoid hemorrhage was substantially impeccable at 91.7%.

Ref. [48] addressed the problem of detecting cerebral hemorrhage in the early stages of hemorrhage, which is a challenging task for radiologists. AlexNet, which is the popular CNN architecture, is used to solve this problem. Besides, the modified AlexNet is supported by the SVM classifier. With the Alexnet + SVM classifier structure used in the study, the classification of CT image accuracy rate of 93% was obtained.

11.6 Conclusion

In this chapter, a pretrained CNN model that can distinguish between hemorrhage and normal on brain CT images has emerged. The dataset used consists of two classes. The classification accuracy of DarkNet-19 CNN model has been superior compared to the other CNN models. The proposed model was tested by operating the Brain CT Image database. As a result of the best classification, the values of ACC 83.50%, SEN 82%, SPE 85%, F-Score 83.20%, and MCC 65% with DarkNet-19 CNN model were obtained. Considering the classification accuracies of the study, it can be said that it has a low accuracy rate as a disadvantage. The reason for this is that the dataset used has less data than the sources in the discussion section. Nevertheless, it can be said that promising results were obtained.

References

[1] C.J. van Asch, M.J. Luitse, G.J. Rinkel, I. van der Tweel, A. Algra, C.J. Klijn, Incidence, case fatality, and functional outcome of intracerebral haemorrhage over time, according to age, sex, and ethnic origin: a systematic review and *meta*-analysis, Lancet Neurol. 9 (2) (2010).

Available from: https://doi.org/10.1016/S1474-4422(09)70340-0.

[2] H. Chen, S. Khan, B. Kou, S. Nazir, W. Liu, A. Hussain, A smart machine learning model for the detection of brain hemorrhage diagnosis based Internet of Things in smart cities, Complexity 2020 (2020). Available from: https://doi.org/10.1155/2020/3047869.

[3] W. Kuo, C. Häne, P. Mukherjee, J. Malik, E.L. Yuh, Expert-level detection of acute intracranial hemorrhage on head computed tomography using deep learning, Proc. Natl. Acad. Sci. U. S. A. 116 (45) (2019). Available from: https://doi.org/10.1073/pnas.1908021116.

[4] S. Santhoshkumar, V. Varadarajan, S. Gavaskar, J. Jegathesh Amalraj, A. Sumathi, Machine learning model for intracranial hemorrhage diagnosis and classification, Electron. 10 (21) (2021). Available from: https://doi.org/10.3390/electronics10212574.

[5] D.L. Labovitz, A. Halim, B. Boden-Albala, W.A. Hauser, R.L. Sacco, The incidence of deep and lobar intracerebral hemorrhage in whites, blacks, and Hispanics, Neurology 65 (4) (2005) 518–522. Available from: https://doi.org/10.1212/01.wnl.0000172915.71933.00.

[6] J. Liu, et al., Prediction of hematoma expansion in spontaneous intracerebral hemorrhage using support vector machine, EBioMedicine 43 (2019) 454–459. Available from: https://doi.org/10.1016/j.ebiom.2019.04.040.

[7] L.R. Øie, et al., Functional outcome and survival following spontaneous intracerebral hemorrhage: a retrospective population-based study, Brain Behav. 8 (10) (2018) e01113. Available from: https://doi.org/10.1002/brb3.1113.

[8] A. Sage, P. Badura, Intracranial hemorrhage detection in head CT using double-branch convolutional neural network, support vector machine, and random forest, Appl. Sci. 10 (21) (2020) 1–13. Available from: https://doi.org/10.3390/app10217577.

[9] T. Lewick, M. Kumar, R. Hong, W. Wu, Intracranial hemorrhage detection in CT scans using deep learning, in: Proc. - 2020 IEEE 6th Int. Conf. Big Data Comput. Serv. Applications, BigDataService 2020, 2020, pp. 169–172. Available from: https://doi.org/10.1109/BigDataService49289.2020.00033.

[10] M.A. Al-masni, W.-R. Kim, E.Y. Kim, Y. Noh, D.-H. Kim, Automated detection of cerebral microbleeds in MR images: a two-stage deep learning approach, NeuroImage Clin. 28 (2020) 102464. Available from: https://doi.org/10.1016/j.nicl.2020.102464.

[11] M. Grewal, M.M. Srivastava, P. Kumar, S. Varadarajan, RADnet: radiologist level accuracy using deep learning for hemorrhage detection in CT scans, in: Proc. - Int. Symposium Biomed. Imaging, 2018, vol. 2018–April, pp. 281–284. Available from: https://doi.org/10.1109/ISBI.2018.8363574.

[12] C.S.S. Anupama, M. Sivaram, E.L. Lydia, D. Gupta, K. Shankar, Synergic deep learning model–based automated detection and classification of brain intracranial hemorrhage images in wearable networks, Pers. Ubiquitous Comput. 26 (1) (2022) 1–10. Available from: https://doi.org/10.1007/s00779-020-01492-2.

[13] K. Kowsari, et al., HMIC: Hierarchical Medical Image Classification, a deep learning approach, Information 11 (6) (2020) 318. Available from: https://doi.org/10.3390/info11060318.

[14] L. Cai, J. Gao, D. Zhao, A review of the application of deep learning in medical image classification and segmentation, Ann. Transl. Med. 8 (11) (2020) 713. Available from: https://doi.org/10.21037/atm.2020.02.44.

[15] X. Liu, L. Song, S. Liu, Y. Zhang, A review of deep-learning-based medical image segmentation methods, Sustain 13 (3) (2021) 1–29. Available from: https://doi.org/10.3390/su13031224.

[16] F. Renard, S. Guedria, N. De, et al., Variability and reproducibility in deep learning for medical image segmentation, Sci. Rep. 10 (1) (2020) 13724. Available from: https://doi.org/10.1038/s41598-020-69920-0.

[17] M.H. Hesamian, W. Jia, X. He, P. Kennedy, Deep learning techniques for medical image segmentation: achievements and challenges, J. Digit. Imaging 32 (4) (2019) 582–596. Available from: https://doi.org/10.1007/s10278-019-00227-x.

[18] V. Shah, R. Keniya, A. Shridharani, M. Punjabi, J. Shah, N. Mehendale, Diagnosis of COVID-19 using CT scan images and deep learning techniques, Emerg. Radiol. 28 (3) (2021) 497–505. Available from: https://doi.org/10.1007/s10140-020-01886-y.

[19] H. Li, Y. Pan, J. Zhao, L. Zhang, Skin disease diagnosis with deep learning: A review, Neurocomputing 464 (2021) 364–393. Available from: https://doi.org/10.1016/j.neucom.2021.08.096.

[20] V. Thanikachalam, S. Shanthi, K. Kalirajan, S. Abdel-Khalek, M. Omri, L.M. Ladhar, Intelligent deep learning based disease diagnosis using biomedical tongue images, Comput. Mater. Contin. 70 (3) (2022) 5667–5681. Available from: https://doi.org/10.32604/cmc.2022.020965.

[21] P. Khan, et al., Machine learning and deep learning approaches for brain disease diagnosis: principles and recent advances, IEEE Access. 9 (2021) 37622–37655. Available from: https://doi.org/10.1109/ACCESS.2021.3062484.

[22] N. Alsharman, I. Jawarneh, GoogleNet CNN neural network towards chest CT-coronavirus medical image classification, J. Comput. Sci. 16 (5) (2020) 620–625. Available from: https://doi.org/10.3844/JCSSP.2020.620.625.

[23] O.K. Oyedotun, K. Dimililer, Pattern recognition: invariance learning in convolutional auto encoder

network, Int. J. Image, Graph. Signal. Process. 8 (3) (2016) 19–27. Available from: https://doi.org/10.5815/ijigsp.2016.03.03.

[24] K. He, X. Zhang, S. Ren, and J. Sun, "Deep residual learning for image recognition," in Proc. IEEE Computer Soc. Conf. Computer Vis. Pattern Recognit., 2016, vol. 2016–Decem, pp. 770–778. Available from: https://doi.org/10.1109/CVPR.2016.90.

[25] A. Krizhevsky, I. Sutskever, G.E. Hinton, ImageNet classification with deep convolutional neural networks, Commun. ACM 60 (6) (2017). Available from: https://doi.org/10.1145/3065386.

[26] K. Simonyan, A. Zisserman, Very deep convolutional network. Large-scale image Recognition, 2015.

[27] L. Wen, X. Li, X. Li, L. Gao, A new transfer learning based on VGG-19 network for fault diagnosis, in: Proc. 2019 IEEE 23rd International Conference on Computer Supported Cooperative Work in Design, CSCWD 2019, 2019, pp. 205–209. Available from: https://doi.org/10.1109/CSCWD.2019.8791884.

[28] M.A. Fayemiwo, et al., Modeling a deep transfer learning framework for the classification of COVID-19 radiology dataset, PeerJ Comput. Sci. 7 (2021) e614. Available from: https://doi.org/10.7717/peerj-cs.614.

[29] S. Lu, Z. Lu, Y.-D. Zhang, Pathological brain detection based on AlexNet and transfer learning, J. Comput. Sci. 30 (2019) 41–47. Available from: https://doi.org/10.1016/j.jocs.2018.11.008.

[30] P. Kumaravel, S. Mohan, J. Arivudaiyanambi, N. Shajil, H.N. Venkatakrishnan, A simplified framework for the detection of intracranial hemorrhage in CT brain images using deep learning, Curr. Med. Imaging Former. Curr. Med. Imaging Rev. 17 (10) (2021) 1226–1236. Available from: https://doi.org/10.2174/1573405617666210218100641.

[31] J.Y. Lee, J.S. Kim, T.Y. Kim, Y.S. Kim, Detection and classification of intracranial haemorrhage on CT images using a novel deep-learning algorithm, Sci. Rep. 10 (1) (2020). Available from: https://doi.org/10.1038/s41598-020-77441-z.

[32] H. Lee, et al., An explainable deep-learning algorithm for the detection of acute intracranial haemorrhage from small datasets, Nat. Biomed. Eng. 3 (3) (2019) 173–182. Available from: https://doi.org/10.1038/s41551-018-0324-9.

[33] J. Deng, W. Dong, R. Socher, L.-J. Li, K. Li, Li Fei-Fei, ImageNet: A large-scale hierarchical image database, in: IEEE Conference on Computer Vision and Pattern Recognition 2010, pp. 248–255. Available from: https://doi.org/10.1109/cvpr.2009.5206848.

[34] X. Wang, et al., A deep learning algorithm for automatic detection and classification of acute intracranial hemorrhages in head CT scans, NeuroImage Clin. 32 (2021) 102785. Available from: https://doi.org/10.1016/j.nicl.2021.102785.

[35] M.F. Mushtaq, et al., BHCNet: Neural Network-Based Brain Hemorrhage Classification Using Head CT Scan, IEEE Access. 9 (2021) 113901–113916. Available from: https://doi.org/10.1109/ACCESS.2021.3102740.

[36] H. Ye, et al., Precise diagnosis of intracranial hemorrhage and subtypes using a three-dimensional joint convolutional and recurrent neural network, Eur. Radiol. 29 (11) (2019) 6191–6201. Available from: https://doi.org/10.1007/s00330-019-06163-2.

[37] M.R. Arbabshirani, et al., Advanced machine learning in action: identification of intracranial hemorrhage on computed tomography scans of the head with clinical workflow integration, NPJ Digit. Med. 1 (1) (2018) 9. Available from: https://doi.org/10.1038/s41746-017-0015-z.

[38] J.-L. Solorio-Ramírez, M. Saldana-Perez, M.D. Lytras, M.-A. Moreno-Ibarra, C. Yáñez-Márquez, Brain Hemorrhage Classification in CT scan images using minimalist machine learning, Diagnostics 11 (8) (2021) 1449. Available from: https://doi.org/10.3390/diagnostics11081449.

[39] C. Yanez-Marquez, Toward the bleaching of the black boxes: minimalist machine learning, IT Prof. 22 (4) (2020) 51–56. Available from: https://doi.org/10.1109/MITP.2020.2994188.

[40] E.L. Yuh, A.D. Gean, G.T. Manley, A.L. Callen, M. Wintermark, Computer-aided assessment of head computed tomography (CT) studies in patients with suspected traumatic brain injury, J. Neurotrauma 25 (10) (2008) 1163–1172. Available from: https://doi.org/10.1089/neu.2008.0590.

[41] S. Yune, H. Lee, S. Do, D. Ting, Case-based learning based on artificial intelli-gence radiology atlas: Example of intracranial hemorrhage and urinary stone detection, J. Gen. Intern. Med. 33 (Supplement 1) (2018).

[42] K. Ali, Z. Shaikh, A. Khan, A. Laghari, Multiclass skin cancer classification using. EfficientNets a first step towards preventing skin cancer, Neurosci. Inform. 2 (4) (2022).

[43] V. Kumar, Implementing EfficientNet: a powerful convolutional neural network. https://analyticsindiamag.com/implementing-efficientnet-a-powerful-convolutional-neural-network/, June 19, 2020 (accessed 10.02.22).

[44] P. Saha, M.S. Sadi, O.F.M.R.R. Aranya, S. Jahan, F.-A. Islam, COV-VGX: an automated COVID-19 detection system using X-ray images and transfer learning, Inform. Med. Unlocked 26 (2021) 100741.

[45] A. Benali Amjoud, M. Amrouch, Convolutional neural networks backbones for object detection, Lecture Notes Computer Sci. (2020) 282–289.

[46] A. Diker, Sıtma Hastalığının Sınıflandırılmasında Evrişimsel Sinir Ağlarının Performanslarının Karşılaştırılması, BEÜ Fen. Bilim. Derg. 9 (4) (2020) 1825–1835.

[47] Kaggle, Head CT Hemorrhage Image Dataset, 2022, https://www.kaggle.com/mrdvolk/head-ct-hemorrhage-detection-with-keras (accessed 07.01.2022).

[48] A.M. Dawud, K. Yurtkan, H. Oztoprak, Application of deep learning in neuroradiology: Brain Haemorrhage Classification Using Transfer Learning, Computational Intell. Neurosci. 2019 (2019) 1–12.

[49] B. Shahangian, H. Pourghassem, Automatic brain hemorrhage segmentation and classification in CT Scan Images, in: 2013 8th Iranian Conference on Machine Vision and Image Processing (MVIP), 2013.

[50] H.J. Lee, Intracranial Hemorrhage Classification using CNN, https://cs230.stanford.edu/projects_fall_2019/reports/26248009.pdf, Autumn 2019 (accessed 01.02.22.)

[51] S. Chilamkurthy, R. Ghosh, S. Tanamala, M. Biviji, N.G. Campeau, V.K. Venugopal, et al., Deep learning algorithms for detection of critical findings in head CT scans: a retrospective study, Lancet 392 (10162) (2018) 2388–2396.

CHAPTER

12

Artificial intelligence-based retinal disease classification using optical coherence tomography images

Sohan Patnaik[1] *and Abdulhamit Subasi*[2,3]

[1]Department of Mechanical Engineering, Indian Institute of Technology, Kharagpur, West Bengal, India [2]Institute of Biomedicine, Faculty of Medicine, University of Turku, Turku, Finland [3]Department of Computer Science, College of Engineering, Effat University, Jeddah, Saudi Arabia

OUTLINE

12.1 Introduction

Starting from the work of Adolf Fercher and associates on low/partial coherence or white-light interferometry for in vivo ocular eye measurements [1,2] in Vienna during the 1980s, imaging of natural tissue, particularly of the eye, was examined in parallel by various research groups around the world. Initially, two-dimensional in vivo portrayal of a human eye fundus along a horizontal meridian was found to be dependent on white light and interferometric depth scans were introduced at the ICO-15 SAT meeting in 1990 [3]. Further in

Applications of Artificial Intelligence in Medical Imaging
DOI: https://doi.org/10.1016/B978-0-443-18450-5.00009-8

1990 Tanno et al. [4,5], at Yamagata University, created a technology which was referred to as heterodyne reflectance tomography, and specifically in 1991, Huang et al., at Massachusetts Institute of Technology, effectively coined the term "optical rational tomography." From that point of time, OCT with micrometer resolution and cross-sectional imaging capacities has become an extraordinary biomedical tissue-imaging procedure that continuously got new specialized abilities starting from early electronic signal detection, via utilization of broadband lasers and linear pixel arrays to ultrafast tunable lasers to expand its performance and sensitivity envelope. Ocular (or ophthalmic) OCT is used heavily by ophthalmologists and optometrists to obtain high-resolution images of the retina and anterior segment.

With regard to OCT's capability to show cross-sections of tissue layers with micrometer resolution, OCT provides a straightforward and effective method of assessing cellular organization, photoreceptor integrity [6–9], axonal thickness in glaucoma [10], macular degeneration [11], diabetic macular edema [12], multiple sclerosis, and other eye diseases or systemic pathologies which have ocular signs [13]. Additionally, ophthalmologists leverage OCT to assess the vascular health of the retina via a technique called OCT angiography [14]. In ophthalmological surgery, especially retinal surgery, an OCT can be mounted on the microscope. Such a system is called an intraoperative OCT (iOCT) and provides support during the surgery with clinical benefits [15].

Diagnosing the diseases from the retinal cross-sectional images obtained using OCT is still a challenge when the number of images is very high. Here comes the need of an intelligent agent with better computational skills than a normal human being. Deep learning has made remarkable progress in the field of image classification. Keeping that in mind, we propose a deep learning-based diagnosis of three types of retinal diseases—drusen, diabetic macular edema (DME), and choroidal neovascularization (CNV). For the basic framework, we use a convolutional architecture followed by either a machine learning-based classification model or a fully connected neural network with softmax activation, which gives the probability of each of the diseases present in the retina. It is obvious to keep in mind that the retinal image might not have a disease too. So, our framework addresses that also and captures if there is no retinal disease.

12.2 Related work

The execution of autograding framework algorithms for optical coherence tomography (OCT) images has undergone a long turn of events. With the emergence of the artificial intelligence, specialists have investigated the diagnostic instrument with various strategies. The use of an autograding framework has also encountered a long turn of events. Prior algorithms started with an image segmentation model. Similar to the methodology that human experts use, the segmentation algorithms detected the edge of the features and made diagnoses by a binary classification algorithm [16]. As the convolutional neural network (CNN) came into picture, it was gradually implemented in the classification model. One study endeavored to utilize CNN to perceive the features and make classifications [17]. In recent years, some CNN models have been modified to achieve higher accuracy [18,19].

Since the start of the 21st century, OCT technology is more frequently used in detecting the features of Age-related macular degeneration (AMD) and Diabetic Macular Edema (DME) [20–22], with the increasing desire for OCT image autograding, many research communities have advanced assets in this field to attempt to accomplish more accurate models. Improvement of automated image classification/grading frameworks started with an automated segmentation algorithm [16]. In 2014 Ehlers et al. [16] proposed an automated classification framework to perceive AMD and DME. They used image segmentation to detect the particular features (RNFL and drusen) of AME and DME and made it the identifier of the classification.

In 2017 Rogers et al. [23] introduced an automatic segmentation framework based on CNNs and graph search strategies (CNN-GS). They segmented the OCT images of nonexudative AMD patients to detect nine retinal layer units. The algorithm first generated likelihood maps by the CNN model. Then they used the CNN probability map to accomplish layer segmentation. This was a novel use of a CNN that customized the CNN and integrated with a segmentation algorithm. Fang et al. [24] proposed surrogate-assisted retinal OCT image classification based on CNN for AMD and DME. In 2018 Kermany et al. [18] collected and processed a vast number of OCT images and built a CNN model dependent on the denoised images. The dataset they processed has been published and is publicly available, and we have used that dataset for our research.

12.3 Dataset

The dataset used for our research was prepared by Kermany et al. [25,26] in 2018. The dataset is organized into three folders (train, test, val) and contains subfolders for each image category (NORMAL, CNV, DME, DRUSEN). There are 84,495 X-Ray images (JPEG) and 4 categories (NORMAL, CNV, DME, DRUSEN). Images are labeled as (disease)−(randomized patient ID)−(image number by this patient) and split into four directories: CNV, DME, DRUSEN, and NORMAL.

Before the data was published by Kermany et al. [25], each image was taken through a tiered grading framework consisting of different layers of trained graders of increasing expertise for checking and verifying image labels. Each image brought into the dataset began with a label matching the most recent diagnosis of the patient. The first tier of graders consisted of undergraduate and medical students who had taken and passed an OCT interpretation course review. This first tier of graders conducted initial quality control and excluded OCT images containing severe artefacts or significant image resolution reductions. The second tier of graders consisted of four ophthalmologists who independently graded each image that had passed the first tier. The presence or absence of CNV (active or in the form of subretinal fibrosis), macular edema, drusen, and other pathologies visible on the OCT scan were recorded. Finally, a third level of two trained professionals and retinal specialists, each with over 20 years of medical retina experience, verified the true labels for each image. To account for human error in grading, a validation subset of 993 scans was graded separately by two ophthalmologist graders, with disagreement in clinical labels arbitrated by a senior retinal specialist. Some images are shown in Fig. 12.1.

12.4 Implementation details

In this section, we elaborate the three approaches that we incorporated to accomplish the task of classifying the image whether it contains the three diseases mentioned earlier or it is normal. The codebase is written in python and tensorflow was the library used to design the neural network architecture. For the machine learning-based classification, sklearn package was used. All the neural network-based models were trained for 10 epochs which took from 45 minutes to 1 hour 20 minutes depending upon the size of the model. The random seed for training purposes was set to 0.

12.4.1 Convolutional neural network-based classification

CNN stands for convolutional neural networks. These are a class of neural networks that are extensively used to deal with tasks involving images. Let's say we have an image. In order to capture certain interpretable information, we first need a feature representation of the image. This is accomplished by using the convolution operator between the image and some lower dimensional kernels. After several layers of convolution, we arrive at a small feature map that indeed captures the features of the image such as edges, colors, spatial layout, etc. Moreover, a typical CNN architecture also incorporates certain pooling layers such as max pooling and average pooling which just reduce the dimension of the feature maps without any learnable parameters.

For our research, we experimented with networks with 2- to 8-layered CNNs along with some max-pooling layers embedded in between the convolutional layers. The best accuracy and F1 score on the test set was obtained using the 7-layered CNN, the architecture of which is shown in Fig. 12.2. The input image was mapped to $150 \times 150 \times 3$, where 3 is the number of channels (RGB) and 150×150 is the spatial layout of the image. For the classification stage, the output feature map of the convolutional network was flattened and then passed to a fully connected layer in order to get a vector representation of the image. Finally, a softmax layer with four neurons was employed to obtain the probabilities of the four classes (three diseases and normal).

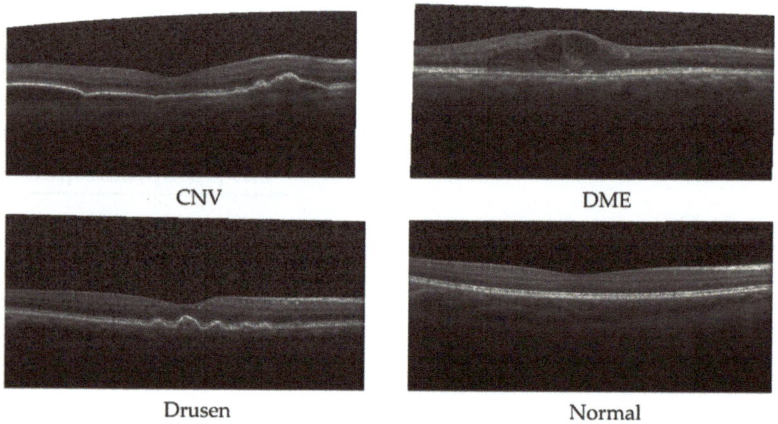

FIGURE 12.1 Retinal OCT images. *OCT*, Optical coherence tomography.

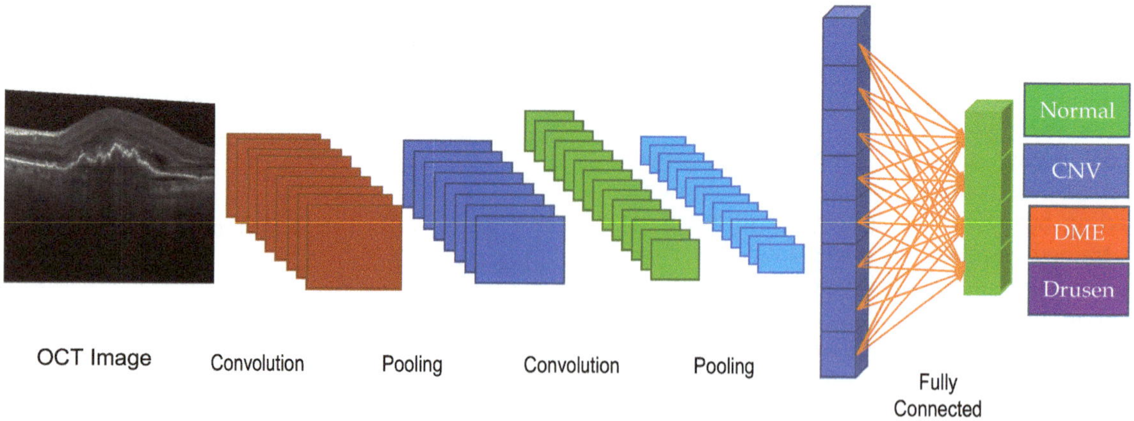

FIGURE 12.2 CNN's architecture. *CNN*, Convolutional neural network.

```
#Create the Model
model = tf.keras.models.Sequential[

    tf.keras.layers.Conv2D(16, (3, 3), activation = 'relu',
                                  input_shape = INPUT_SHAPE),
    tf.keras.layers.MaxPooling2D(2, 2),

    tf.keras.layers.Conv2D(16, (3, 3), activation = 'relu'),
    tf.keras.layers.Conv2D(32, (3, 3), activation = 'relu'),
    tf.keras.layers.MaxPooling2D(2, 2),

    tf.keras.layers.Conv2D(64, (3, 3), activation = 'relu'),
    tf.keras.layers.Conv2D(64, (3, 3), activation = 'relu'),
    tf.keras.layers.MaxPooling2D(2, 2),

    tf.keras.layers.Conv2D(32, (3, 3), activation = 'relu'),
    tf.keras.layers.MaxPooling2D(2, 2),

    tf.keras.layers.Conv2D(16, (3, 3), activation = 'relu'),
    tf.keras.layers.MaxPooling2D(2, 2),

    tf.keras.layers.Flatten(),

    tf.keras.layers.Dense(4, activation = 'softmax')
]

# Define the metrics list
metrics_list = ['accuracy',
                tf.keras.metrics.AUC(),
                tfa.metrics.CohenKappa(num_classes = 4),
                tfa.metrics.F1Score(num_classes = 4)]

# Compile the model
model.compile(loss = 'categorical_crossentropy', optimizer = 'adam',
                                          metrics = metrics_list)
# Fit the model and store the history
history = model.fit_generator(
    train_generator,
    steps_per_epoch = (83484/100),
    epochs = 10,
    validation_data = validation_generator,
    validation_steps = (32/16),
    verbose = 1)

# Obtain the test predictions
y_pred = model.predict(test_generator)
```

12.4.2 Transfer learning-based classification

The basic premise of transfer learning is simple: take a model trained on a large dataset and transfer its knowledge to a smaller dataset. For object recognition with a CNN, we freeze the early convolutional layers of the network and only train the last few layers which make a prediction. The idea is that the convolutional layers extract general, low-level features that are applicable across images such as edges, patterns, and gradients and the later layers identify specific features within an image such as eyes or wheels. Thus we can use a network trained on unrelated categories in a massive dataset (usually ImageNet) and apply it to our own problem because there are universal, low-level features shared between images.

In our experiment, we used 10 pretrained CNN-based models which were trained on ImageNet dataset. The models used were ResNet50, VGG16, VGG19, Inceptionv3, MobileNet, DenseNet169, DenseNet121, InceptionResNetV2, MobileNetV2, and ResNet101. The feature maps obtained from these pretrained models were passed through some fully connected layers before the softmax layer. A pretrained VGG19 model is given below.

```
# Import libraries
import tensorflow as tf
import tensorflow_addons as tfa
# Download VGG19 Model
vgg19 = tf.keras.applications.VGG19(
    include_top = False,
    weights = 'imagenet',
    input_tensor = None,
    input_shape = INPUT_SHAPE,
    pooling = None,
    classes = 1000
)

# Freeze the early convolutional layers of the VGG19 Pretrained Model
vgg19.trainable = False
# Create a Transfer Learning Model by Adding New Layers
model = tf.keras.models.Sequential[
    vgg19,
    tf.keras.layers.Conv2D(64, (3, 3), activation = 'relu'),
    tf.keras.layers.Flatten(),
    tf.keras.layers.Dense(100, activation = 'relu'),
    tf.keras.layers.Dense(4, activation = 'softmax')
]

# Define the Metrics
metrics_list = ['accuracy',
                tf.keras.metrics.AUC(),
                tfa.metrics.CohenKappa(num_classes = 4),
                tfa.metrics.F1Score(num_classes = 4)]

# Train the model
model.compile(loss = 'categorical_crossentropy', optimizer = 'adam',metrics=metrics_list)
history = model.fit_generator(
    train_generator,
    steps_per_epoch = (83484/100),
    epochs = 10,
    validation_data = validation_generator,
    validation_steps = (32/16),
    verbose = 1)
```

```
# Validate and test the model
model.predict(test_generator, steps = int(968/44))
model.evaluate(test_generator)
# Plot the training and validation performance curves
acc = history.history['accuracy']
val_acc = history.history['val_accuracy']
loss = history.history['loss']
val_loss = history.history['val_loss']

epochs = range(len(acc))
plt.figure(figsize=(7,7))
plt.plot(epochs, acc, 'r', label = 'Training accuracy')
plt.plot(epochs, val_acc, 'b', label = 'Validation accuracy')
plt.title('Training and validation accuracy')
plt.legend()
plt.figure(figsize = (7,7))
plt.plot(epochs, loss, 'r', label = 'Training Loss')
plt.plot(epochs, val_loss, 'b', label = 'Validation Loss')
plt.title('Training and validation loss')
plt.legend()
plt.show()

#Print the Confusion Matrix and Classification Results
from sklearn.metrics import confusion_matrix, classification_report
import pandas as pd


 Y_pred = model.predict(test_generator, int(968/44))
 y_pred = np.argmax(Y_pred, axis = 1)
cm = confusion_matrix(test_generator.classes, y_pred)
df_cm = pd.DataFrame(cm, list(test_generator.class_indices.keys()),
                         list(test_generator.class_indices.keys()))

fig, ax = plt.subplots(figsize = (10,8))
sns.set(font_scale = 1.4) # for label size
sns.heatmap(df_cm, annot = True, annot_kws = {"size": 16}, cmap = plt.cm.Blues)
plt.title('Confusion Matrix\n')
plt.savefig('confusion_matrix.png', transparent = False, bbox_inches = 'tight', dpi = 400)
plt.show()

print('Classification Report\n')
target_names = list(test_generator.class_indices.keys())
print(classification_report(test_generator.classes, y_pred, target_names = target_names))
```

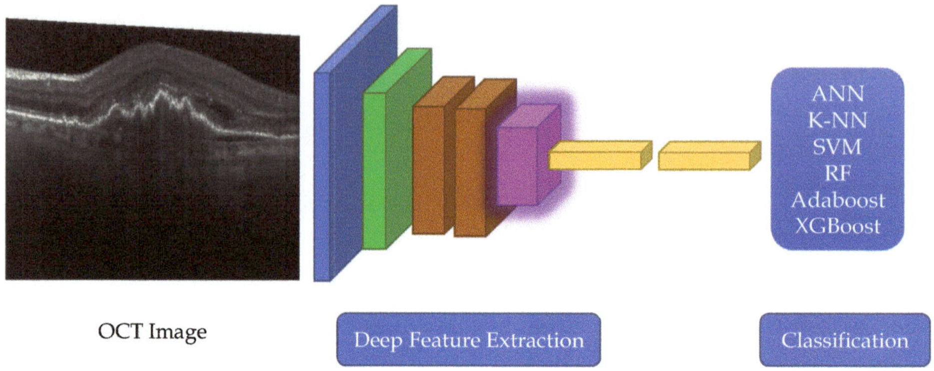

FIGURE 12.3 Deep feature extraction's architecture.

12.4.3 Deep feature extraction and machine learning

It is quite intuitive that the pretrained convolutional models indeed capture the low-level features in the image. Keeping that in mind, we first extracted the feature maps from the pretrained CNN models and flattened them to obtain a vector representation of the image. After this, some machine learning-based classification models were trained on the feature vectors obtained to get the correct class among the three diseases and normal retina. We used six machine learning models, namely, artificial neural network, K-nearest neighbors, support vector machines, random forest classifier, AdaBoost classifier, and XGBoost classifier. A basic picture of the feature extraction model on top of which the machine learning-based classification model was employed is shown in Fig. 12.3.

```
# Import the required libraries
import numpy as np
import tensorflow as tf
import tensorflow_addons as tfa
from sklearn.ensemble import AdaBoostClassifier
from keras.applications.vgg16 import VGG16

#Create the Model
model = VGG16(weights = "imagenet", include_top = False)

# Obtain the train and test features
x_train = model.predict(train_generator)
x_test = model.predict(test_generator)

# Fit the classifier
# Default parameters, change to obtain required accuracy
classifier = AdaBoostClassifier()
classifier = classifier.fit(x_train, y_train)
y_preds = classifier.predict(x_test)
```

12.5 Results and discussions

The results for the test set to classify the retinal disease using the simple CNN-based architecture is shown in Table 12.1. It can be clearly seen that the accuracy for a 7-layered CNN is equal to 0.9948 and so is the F1 score. After proper hyperparameter tuning, we arrived at the best CNN-based architecture. The insight behind this impressive high accuracy accounts for the ability of CNNs which capture the intricate features in the image with high precision. Moreover, proper hyperparameter tuning to decide the size of filters, strides and the number of max-pooling layers to be included in the architecture are also important to achieve promising results.

The accuracy plot for the train, validation, and test set is shown in Fig. 12.4.

From the plot, we can see that the training accuracy decreases as we increase the number of convolutional layers but the test accuracy increases. The reason for this can be attributed to the fact that more the convolutional layers, more high-level information is being captured by the model and less is the overfitting. Owing to this fact, the accuracy on the test set is also increasing. It is also a matter of fact that as the number of layers is increasing, we may require to train the model for a greater number of epochs owing to the reason that the number of parameters also increases, so more training iterations would be needed.

TABLE 12.1 Results for convolutional neural network (CNN) models.

Classifier	Training accuracy	Validation accuracy	Test accuracy	F1 score	Kappa	ROC area
2 Layer	0.9824	0.9062	0.9298	0.9309	0.9105	1.0
3 Layer	0.949	1	0.9638	0.9639	0.9518	0.9973
4 Layer	0.9487	0.9688	0.9742	0.9743	0.9656	0.999
5 Layer	0.9543	1	0.9897	0.9897	0.9862	0.9999
6 Layer	0.9393	1	0.9804	0.9804	0.9738	0.9995
7 Layer	0.95	1.00	0.9948	0.9948	1	0.9931
8 Layer	0.9529	1	0.9897	0.9897	0.9862	0.9999

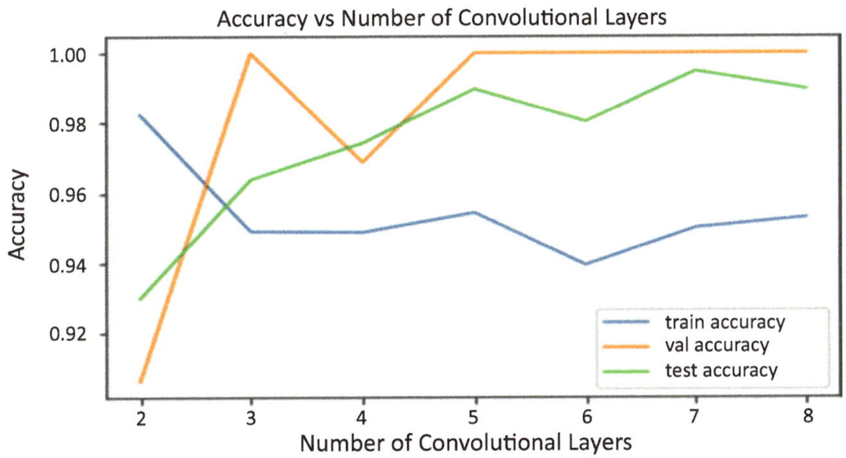

FIGURE 12.4 Train, validation, and test accuracy.

The results for the test set on pretrained image recognition models which have been trained previously on ImageNet dataset are shown in Table 12.2. Here, we can see that the maximum accuracy and F1 score, that is, 0.9628 is obtained for VGG19. Good results can be attributed to the fact that pretraining helps the models to understand and capture certain low-level features, such as edges, gradients of color, shapes, etc.

From Fig. 12.5, we can observe that almost all the pretrained image recognition models

TABLE 12.2 Results for pretrained models.

Classifier	Training accuracy	Validation accuracy	Test accuracy	F1 score	Kappa	ROC area
ResNet50	0.7551	0.6875	0.75	0.7299	0.6667	0.9338
VGG16	0.9346	0.875	0.939	0.9388	0.9187	0.9944
VGG19	0.9197	0.9688	0.9628	0.9628	0.9504	0.9962
InceptionV3	0.915	0.9375	0.8988	0.8999	0.865	0.9854
MobileNet	0.971	0.875	0.9473	0.9478	0.9298	0.9931
DenseNet169	0.9386	0.9375	0.9421	0.9423	0.9229	0.9935
DenseNet121	0.929	0.9062	0.9525	0.9527	0.9366	0.9942
InceptionResNetV2	0.9159	0.9375	0.9329	0.932	0.9105	0.9937
MobileNetV2	0.9471	0.9375	0.9576	0.9574	0.9435	0.9956
ResNet101	0.7514	0.75	0.6849	0.6541	0.5799	0.9133

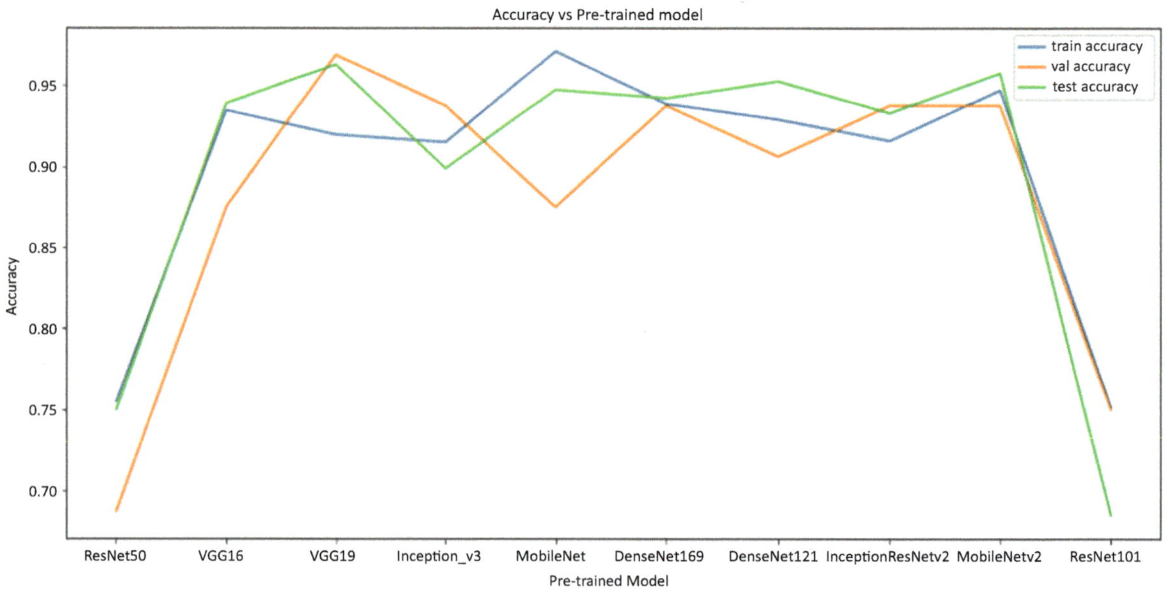

FIGURE 12.5 Accuracy of pretrained models.

perform to the same extent except for both the variants of the ResNet model. As the pre-trained models have a huge number of parameters, training for more epochs might fetch better results.

Now, the results for the feature extraction and machine learning model based classification are shown in Table 12.3–12.13. The different feature extractors in this case are VGG16, VGG19, ResNet50, ResNet101, MobileNetv2, MobileNet, Inceptionv3, InceptionResNetv2, DenseNet169, DenseNet121, and Xception. The typical machine learning-based classification models are artificial neural networks, K-nearest

TABLE 12.3 Results for features extracted from VGG16.

Model	Training accuracy	Validation accuracy	Test accuracy	F1 score	Kappa	Recall	Precision
ANN	0.8528	0.7812	0.8023	0.8104	0.8327	0.8524	0.7724
KNN	0.8324	0.75	0.7926	0.8034	0.8524	0.8254	0.7826
SVM	0.8879	0.8437	0.8726	0.8636	0.9025	0.8445	0.8836
Random Forest	0.9524	0.9375	0.9334	0.9539	0.9542	0.9722	0.9364
AdaBoost	0.9437	0.9375	0.9216	0.9133	0.9345	0.9417	0.8867
XGBoost	0.8765	0.8437	0.8236	0.8565	0.8621	0.7912	0.8336

TABLE 12.4 Results for features extracted from VGG19.

Model	Training accuracy	Validation accuracy	Test accuracy	F1 score	Kappa	Recall	Precision
ANN	0.8645	0.8125	0.8224	0.8175	0.8346	0.8005	0.8354
KNN	0.8445	0.7812	0.8356	0.8344	0.8412	0.8065	0.8644
SVM	0.8769	0.875	0.8456	0.8414	0.8521	0.8612	0.8225
Random Forest	0.9412	0.9375	0.9399	0.9353	0.9465	0.9264	0.9444
AdaBoost	0.9566	0.96875	0.9216	0.9207	0.9366	0.9625	0.8824
XGBoost	0.9026	0.9375	0.8845	0.8777	0.8934	0.8916	0.8643

TABLE 12.5 Results for features extracted from ResNet50.

Model	Training accuracy	Validation accuracy	Test accuracy	F1 score	Kappa	Recall	Precision
ANN	0.7845	0.7812	0.7656	0.7511	0.7622	0.7411	0.7614
KNN	0.8065	0.8125	0.7923	0.7894	0.8041	0.7922	0.7867
SVM	0.8366	0.8125	0.8244	0.8238	0.8491	0.8014	0.8475
Random Forest	0.8627	0.875	0.8465	0.8371	0.8538	0.8477	0.8269
AdaBoost	0.8689	0.875	0.8544	0.8541	0.8512	0.8421	0.8666
XGBoost	0.8412	0.8437	0.8214	0.818	0.8265	0.8355	0.8014

TABLE 12.6 Results for features extracted from ResNet101.

Model	Training accuracy	Validation accuracy	Test accuracy	F1 score	Kappa	Recall	Precision
ANN	0.8016	0.7812	0.7841	0.7799	0.7955	0.8005	0.7605
KNN	0.8245	0.8125	0.7944	0.7885	0.8143	0.8142	0.7644
SVM	0.8411	0.875	0.8321	0.8312	0.8314	0.8332	0.8294
Random Forest	0.8665	0.8465	0.8511	0.8514	0.8444	0.8741	0.83
AdaBoost	0.8545	0.8465	0.8512	0.8522	0.8693	0.8765	0.8294
XGBoost	0.8744	0.875	0.8497	0.8539	0.85	0.8742	0.8346

TABLE 12.7 Results for features extracted from MobileNetV2.

Model	Training accuracy	Validation accuracy	Test accuracy	F1 score	Kappa	Recall	Precision
ANN	0.9014	0.875	0.8821	0.8822	0.9147	0.8641	0.9011
KNN	0.8944	0.875	0.8912	0.8944	0.8825	0.8814	0.9078
SVM	0.9246	0.9375	0.9102	0.9034	0.9176	0.8925	0.9146
Random Forest	0.9564	0.9375	0.9234	0.9204	0.9264	0.8947	0.9478
AdaBoost	0.9677	0.9687	0.9541	0.9499	0.9645	0.9388	0.9614
XGBoost	0.9412	0.9375	0.9368	0.9339	0.9489	0.9266	0.9414

TABLE 12.8 Results for features extracted from MobileNet.

Model	Training accuracy	Validation accuracy	Test accuracy	F1 score	Kappa	Recall	Precision
ANN	0.8528	0.7812	0.8023	0.8104	0.8327	0.8524	0.7724
KNN	0.8065	0.8125	0.7923	0.7894	0.8041	0.7922	0.7867
SVM	0.8769	0.875	0.8456	0.8414	0.8521	0.8612	0.8225
Random Forest	0.8627	0.875	0.8465	0.8371	0.8538	0.8477	0.8269
AdaBoost	0.9026	0.9375	0.8845	0.8777	0.8934	0.8916	0.8643
XGBoost	0.8879	0.8437	0.8726	0.8636	0.9025	0.8445	0.8836

TABLE 12.9 Results for features extracted from InceptionV3.

Model	Training accuracy	Validation accuracy	Test accuracy	F1_score	Kappa	Recall	Precision
ANN	0.9144	0.9375	0.8971	0.8965	0.8964	0.9058	0.8874
KNN	0.8926	0.75	0.8533	0.8523	0.8935	0.8743	0.8315
SVM	0.9137	0.9375	0.8546	0.8597	0.8744	0.8744	0.8456
Random Forest	0.9564	0.9375	0.8947	0.9025	0.9348	0.9137	0.8916
AdaBoost	0.9677	0.9687	0.9244	0.9254	0.9513	0.9369	0.9142
XGBoost	0.9456	0.9375	0.9173	0.9137	0.9577	0.9315	0.8966

TABLE 12.10 Results for features extracted from InceptionResNetV2.

Model	Training accuracy	Validation accuracy	Test accuracy	F1 score	Kappa	Recall	Precision
ANN	0.8665	0.8465	0.8511	0.8514	0.8444	0.8741	0.83
KNN	0.8689	0.875	0.8544	0.8541	0.8512	0.8421	0.8666
SVM	0.8769	0.875	0.8456	0.8414	0.8521	0.8612	0.8225
Random Forest	0.9246	0.9375	0.9102	0.9034	0.9176	0.8925	0.9146
AdaBoost	0.9026	0.9375	0.8845	0.8777	0.8934	0.8916	0.8643
XGBoost	0.9026	0.9375	0.8845	0.8777	0.8934	0.8916	0.8643

TABLE 12.11 Results for features extracted from DenseNet169.

Model	Training accuracy	Validation accuracy	Test accuracy	F1 score	Kappa	Recall	Precision
ANN	0.9014	0.875	0.8821	0.8822	0.9147	0.8641	0.9011
KNN	0.8769	0.875	0.8456	0.8566	0.8521	0.8612	0.8225
SVM	0.8944	0.875	0.8912	0.8944	0.8825	0.8814	0.9078
Random Forest	0.9437	0.9375	0.9216	0.9133	0.9345	0.9417	0.8867
AdaBoost	0.9564	0.9375	0.9234	0.9204	0.9264	0.8947	0.9478
XGBoost	0.9026	0.9375	0.8845	0.8777	0.8934	0.8916	0.8643

neighbors, support vector machines, Random Forest classifier, AdaBoost classifier, and XGBoost classifier. From the tables, we can conclude that the 7-layered CNN architecture proposed by us is the best model when employed on the test set.

TABLE 12.12 Results for features extracted from DenseNet121.

Model	Training accuracy	Validation accuracy	Test accuracy	F1 score	Kappa	Recall	Precision
ANN	0.9014	0.875	0.8821	0.8822	0.9147	0.8641	0.9011
KNN	0.8944	0.875	0.8912	0.8944	0.8825	0.8814	0.9078
SVM	0.9246	0.9375	0.9102	0.9034	0.9176	0.8925	0.9146
Random Forest	0.9524	0.9375	0.9334	0.9539	0.9542	0.9722	0.9364
AdaBoost	0.9437	0.9375	0.9216	0.9133	0.9345	0.9417	0.8867
XGBoost	0.8765	0.8437	0.8236	0.8118	0.8621	0.7912	0.8336

TABLE 12.13 Results for features extracted from XceptionNet.

Model	Training accuracy	Validation accuracy	Test accuracy	F1 score	Kappa	Recall	Precision
ANN	0.9146	0.875	0.8821	0.8839	0.8885	0.9045	0.8644
KNN	0.8963	0.875	0.8912	0.9006	0.8736	0.9073	0.8941
SVM	0.9244	0.9375	0.9102	0.9137	0.9002	0.9452	0.8844
Random Forest	0.9752	0.9687	0.9234	0.9302	0.9146	0.9241	0.9364
AdaBoost	0.9644	0.9687	0.9541	0.9531	0.9436	0.9647	0.9418
XGBoost	0.9348	0.9375	0.9368	0.9364	0.9389	0.9536	0.92

12.6 Discussion

An electronic medical record (EMR) is a system composed of the patient's medical treatment files such as the records of text, symbols, charts, images, and data slices. The impressive data in the EMR system is obtained, sorted, and analyzed by medical staff through outpatient, physical examination, auxiliary examination, diagnoses, treatment, nursing, and other medical activities. It provides the most practical and abundant data for health management, medical diagnoses, treatment, and scientific research. EMR significantly improved medical quality, management level, and academic ability. Besides, it is a cost-saving approach not only for paper and folders but also for labor and storage space.

This autograding OCT image classification system is a brief diagnostic tool that can be used in an EMR system. In the treatment of AMD and DME, patients take OCT images frequently to monitor the changes in their disease. The integration of the autograding OCT image classification system and EMR system can greatly improve the efficiency of retinopathy treatments. This system could provide a better quality of healthcare. With better access to test results and automatic diagnoses, the time that doctors spend on recognizing the test result can be considerably reduced, and patients could be aware of their test results and treatment method.

12.7 Conclusion

In this chapter, we proposed robust architectures to detect the retinal diseases such as drusen, DME, and CNV from normal retina. Our work obtained an astounding high accuracy of 0.9948 and an F1 score of 0.9948 on the test set as prepared by Kermany et al. [26]. As CNNs are gaining importance in medical imaging and diagnosis tasks, we also provided an intuitive understanding why this is so. For the future, we would like to work upon zero shot learning and semisupervised learning where we will only be given normal retinal images and the model would try to identify whether an image is defective or not.

References

[1] A.F. Fercher, E. Roth, G.J. Mueller (Ed.). Ophthalmic laser interferometry, in: Proc. SPIE. Optical Instrumentation for Biomedical Laser Applications, 15 September 1986, 658: 4851.

[2] A.F. Fercher, K. Mengedoht, W. Werner, Eye-length measurement by interferometry with partially coherent light, Opt. Letters. 13 (3) (1988) 1868. March.

[3] A.F. Fercher. Ophthalmic interferometry. in: G. von Bally, S. Khanna (Eds.), Proc. International Conference on Optics in Life Sciences. Garmisch-Partenkirchen, Germany, 12–16 August 1990, pp. 221228.

[4] N. Tanno, T. Ichikawa, A. Saeki Lightwave reflection measurement, Japanese Patent # 2010042, 1990 (Japanese Language).

[5] S. Chiba, N. Tanno, Backscattering optical heterodyne tomography, in: 14th Laser Sensing Symposium (in Japanese), 1991.

[6] J. Sherman, D. Epshtein, The ABCs of OCT, Review of Optometry, 2012.

[7] J. Sherman, Photoreceptor integrity line joins the nerve fiber layer as key to clinical diagnosis, Optometry. 80 (6) (2009) 2778. June.

[8] M.A. Bonini Filho, A.J. Witkin, Outer retinal layers as predictors of vision loss, Review of Ophthalmology, 2015.

[9] N. Cuenca, I. Ortuo-Lizarn, I. Pinilla, Cellular characterization of OCT and outer retinal bands using specific immunohistochemistry markers and clinical implications, Ophthalmology (2018). March.

[10] D.S. Grewal, A.P. Tanna, Diagnosis of glaucoma and detection of glaucoma progression using spectral domain optical coherence tomography, Curr. Opin. Ophthalmol. 24 (2) (2013) 15061. March.

[11] P.A. Keane, P.J. Patel, S. Liakopoulos, F.M. Heussen, S. R. Sadda, A. Tufail, Evaluation of age-related macular degeneration with optical coherence tomography, Surv. Ophthalmol. 57 (5) (2012) 389414. September.

[12] G. Virgili, F. Menchini, G. Casazza, R. Hogg, R.R. Das, X. Wang, et al., Optical coherence tomography (OCT) for detection of macular oedema in patients with diabetic retinopathy, Cochrane Database Syst. Rev. (2015). January.

[13] J. Drr, K.D. Wernecke, M. Bock, G. Gaede, J.T. Wuerfel, C.F. Pfueller, et al., Association of Retinal and Macular Damage with Brain Atrophy in Multiple Sclerosis, PLoS ONE 6 (4) (2011) e18132. April.

[14] T. Aik Kah, CuRRL syndrome: a case series, Acta Sci. Ophthalmol. 1 (3) (2018).

[15] A.H. Kashani, C.L. Chen, J.K. Gahm, F. Zheng, G.M. Richter, P.J. Rosenfeld, et al., Optical coherence tomography angiography: a comprehensive review of current methods and clinical applications, Prog. Retinal Eye Res. 60 (2017). September.

[16] J.P. Ehlers, Y.K. Tao, S.K. Srivastava, The value of intraoperative OCT imaging in vitreoretinal surgery, Curr. Opin. Ophthalmol. 25 (2014). May.

[17] J. Sugmk, S. Kiattisin, A. Leelasantitham, Automated classification between age-related macular degeneration and Diabetic macular edema in OCT image using image segmentation, in: The 7th 2014 Biomedical Engineering International Conference, 2014.

[18] D.S. Kermany, et al., Identifying medical diagnoses and treatable diseases by image-based deep learning, Cell 172 (5) (2018) 1122–1131. e9.

[19] J. Wang, et al., Deep learning for quality assessment of retinal OCT images, Biomed. Opt. Express 10 (12) (2019) 60576072.

[20] T. Tsuji, et al., Classification of optical coherence tomography images using a capsule network, BMC Ophthalmol 20 (1) (2020) 114.

[21] S.W. Kang, C.Y. Park, D.-I. Ham, The correlation between fluorescein angiographic and optical coherence tomographic features in clinically significant diabetic macular edema, Am. J. Ophthalmol. 137 (2) (2004) 313322.

[22] M.R. Hee, et al., Optical coherence tomography of age-related macular degeneration and choroidal neovascularization, Ophthalmology 103 (8) (1996) 12601270.

[23] A.H. Rogers, A. Martidis, P.B. Greenberg, C.A. Puliafito, Optical coherence tomography findings following photodynamic therapy of choroidal neovascularization, Am. J. Ophthalmol. 134 (4) (2002) 566576.

[24] L. Fang, D. Cunefare, C. Wang, R.H. Guymer, S. Li, S. Farsiu, Automatic segmentation of nine retinal layer boundaries in OCT images of non-exudative AMD patients using deep learning and graph search, Biomed. Opt. Express 8 (5) (2017) 27322744.

[25] D.S. Kermany, M. Goldbaum, W. Cai, C.C. Valentim, H. Liang, S.L. Baxter, et al., Identifying medical diagnoses and treatable diseases by image-based deep learning, Cell 172 (5) (2018) 1122–1131.

[26] D. Kermany, K. Zhang, M. Goldbaum, Labeled optical coherence tomography and chest X-Ray images for classification, Mendeley Data v2 (2018).

CHAPTER

13

Diagnosis of breast cancer from histopathological images with deep learning architectures

Emrah Hancer[1] *and Abdulhamit Subasi*[2,3]

[1]Department of Software Engineering, Mehmet Akif Ersoy University, Burdur, Turkey [2]Institute of Biomedicine, Faculty of Medicine, University of Turku, Turku, Finland [3]Department of Computer Science, College of Engineering, Effat University, Jeddah, Saudi Arabia

OUTLINE

13.1 Introduction

Being one of the most dangerous types of cancer, breast cancer has a significant impact on the death of women between 20 and 59 years old. According to the American Cancer Society, approximately 268,600 new cases was reported for breast cancer and 41,760 people died because of it in 2019. Despite the fact, the diagnosis of breast cancer in its earlier stages can significantly increase the survival rate [1]. The diagnosis process first starts with clinical analysis through visual imaging technologies, such as ultrasound, mammography, or magnetic resonance imaging. A needle tissue biopsy is then applied to patients with high probability of breast malignancy by using the hematoxylin and eosin (H&E) stain protocol. While hematoxylin has a dark purple- or blue-colored visual effect on nucleus structures, eosin stains other structures into red, pink, and orange shades. Histopathological breast images

Applications of Artificial Intelligence in Medical Imaging
DOI: https://doi.org/10.1016/B978-0-443-18450-5.00002-5

are considered at different magnifications to investigate the cellular and tissue level variations. For instance, the tissue patterns are investigated at 100× magnification, while the size and shape of nuclei structures are investigated at 400× magnification. After obtaining information from these features, pathologists can determine a tumor slide as benign and malignant. In case of malignancy, a further analysis is carried out to grade the tumor and based on the grade a treatment is advised to the corresponding patient. Breast cancer may be in the form of different types and each type has its own microscopic features [2].

Pathologists examine morphological features of H&E stained tissue samples from related breast regions under a microscope to establish a definitive diagnosis. Any differences observed in any features of the interested region is regarded as abnormal and then a confirmation process is carried out to verify it as a malignant tumor. Pathologists also need to grade the tumor to examine the degree of cancer in some cases [3]. Unfortunately, the visual analysis manually carried out by the pathologists is an error-prone, tiresome, and subjective task, causing inevitable errors in decisions. To ease the workload on experts and/or pathologists and improve the efficiency of the diagnosis performance, researchers have focused on automating the diagnosis process through computer-aided diagnosis (CAD) systems in recent years. The basic steps of CAD systems are as follows: (1) preprocessing, (2) segmentation, (3) feature extraction, and (4) classification. The CAD systems wrapped around traditional machine learning methods use specified classifiers over a set of handcrafted features obtained from histopathology images to predict the output labels. However, the classification performance of them is not competitive and even far from need. Moreover, extracting handcrafted features is a computationally intensive and complex process due to the requirement of extensive prior-domain knowledge. Fortunately, the recent advancements from the perspective of machine learning and image processing have resulted in the emergence of deep learning discipline, which has widely attracted the interest of researchers from different fields. Deep learning architectures, especially convolutional neural networks (CNNs), can extract intrinsic features from raw image data without the requirement of more effort. Therefore CNNs have widely been adopted in CAD systems, resulting in a groundbreaking performance in biomedical applications, especially diagnostic pathology [4].

CNNs are similar to conventional neural networks in such a way that they are built on neurons with learnable weights and biases. Each neuron takes some inputs and then performs a transformation process. The whole network still represents a differentiable function that transforms a raw image data to a class score. Moreover, a loss function (e.g., softmax) is still apparent on the last (denoted as fully connected) layer and all the learning procedures for conventional neural networks are still performed. So, what is the difference between CNNs and conventional neural networks? The inputs of CNNs are images and so CNNs allow us to encode the characteristics of them into the architecture. The forward function can then be implemented more efficiently, thereby reducing the number of parameters in the network.

Thanks to the effectiveness of CNNs in CAD systems, the diagnosis process of breast cancer from histopathological images has aroused the interest of researchers. Especially with the releasement of the largest publicly available datasets, the studies have rapidly been increased to develop CNN-based automated systems to deal with these datasets. Spanhol et al. [5] used a variant of AlexNet [6] to form an automated breast cancer diagnosis method on histopathological images created based on a set of pixel patches generated by using random strategies and sliding window. The obtained accuracy was between 81% and 89%. In another work, Spanhol et. al. [7] extracted deep features using a pretrained CNN architecture (BVLC CaffeNet) and then built a classification model on the extracted deep features using a logistic regression classifier. According to the results, it was possible to obtain promising results through deep features as well as training a CNN variant from scratch. Bayramoglu et al. [8] introduced the following two CNN-based architectures: (1) single-task CNN variant was used to detect the malignancy and (2) multitask CNN variant was designed to detect both the malignancy and the magnification degree. The accuracy obtained by these CNN-based architectures was between 80% and 84%. Wei et al. [9] proposed a CNN-based breast cancer detection methodology which applied improved data augmentation and transfer learning strategies. The proposed methodology reached nearly 97% accuracy. Pratiher and Chattoraj [10] introduced an end-to-end deep ensemble methodology based on manifold learning L-ISOMAP and stacked sparse auto-encoder to detect breast cancer. Different from the previous studies, Nahid and Kong [11] did not only use the raw image data as an input but also utilized handcrafted features and frequency-domain information to carry out the training process of a CNN architecture. The overall accuracy obtained by this study was between 94% and 97%. In another study, Nahid et al. [12] introduced a combined version of CNN and long short-term memory to extract meaningful features and then individually applied softmax layer and support vector machines as classifiers. The reported accuracy was between 90% and 91%. Bardou et al. [13] introduced a comparative study of deep features and handcrafted features. According to the results, the classification of deep features was superior to that of handcrafted features. In summary, the aforementioned studies wrapped around CNNs showed a significant performance for breast cancer diagnosis from histopathological images. However, the training process of a CNN architecture requires high memory and computation resources. Moreover, its performance may deteriorate due to local stagnation and overfitting problems. To cover this issue, pretrained CNN-based architectures became an alternative to the conventional CNNs trained from scratch. For instance, Shallu and Mehra [14] compared the performance of a fine-tuned VGG16 with that of a CNN trained from scratch through a logistic regression classifier. According to the results, VGG16 obtained better diagnosis performance. Xiang et al. [15] investigated the performance of a fine-tuned InceptionV3 to diagnose breast cancer from histopathological datasets. It can therefore be indicated that the studies for diagnosis of breast cancer have not come to an end yet.

In this chapter, we aim to utilize CNN-based architectures to reflect which deep learning architecture performs well among a variety of deep learning architectures. To achieve this goal, we use CNNs trained from scratch and pretrained CNN-based architectures on the invasive ductal carcinoma (IDC) dataset. The aim of this chapter is to give comprehensive understanding of deep CNN models to researchers/experts who are interested in breast cancer diagnosis, and researchers who would like to have an insight of several methods in this field.

The remainder of the chapter is as follows. In Section 13.2, we present the dataset used in experiments and then briefly describe the methods with implementation codes. In Section 13.3, we present the results with discussions. Finally, we conclude the chapter with future trends.

13.2 Materials and methods

13.2.1 Dataset

We choose the IDC histopathological image dataset [16] to carry out experiments for breast cancer diagnosis. IDC is regarded as the most common type among all breast cancers. Pathologists typically grade whole mount samples on the basis of the IDC present in the region. As a

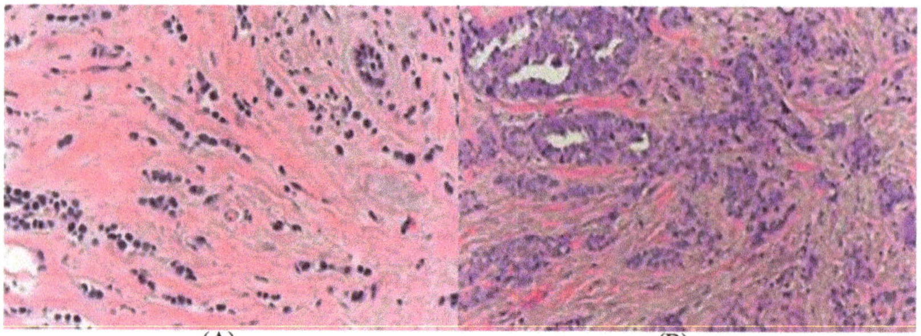

(A) (B)

FIGURE 13.1 Microscopic patterns of invasive (A) lobular carcinoma and (B) ductal carcinoma.

result, identifying the exact regions of IDC inside of a given mount slide is one of the common preprocessing steps for automatic aggressiveness grading. Fig. 13.1 shows histopathological image examples of malignant breast tumor.

The original dataset involves 162 whole mount slide images scanned at 40×. In total, 277,524 patches with size 50×50 (198,738 IDC negative and 78,786 IDC positive) were obtained from the dataset. To alleviate the class-imbalance problem, we choose 78,786 image samples from the category of IDC negative. Accordingly, the total number of image samples is 157,572. After addressing the class-imbalance problem, the splitting process is started. We split the data into 75% train and 25% test set. Also, we separate 25% of the train set as the validation set. Finally, we convert the target variable of each set into a matrix of binary values to use with the loss function "categorical_crossentropy." The splitting procedure is shown in the following section.

13.2.2 Methods

In this section, we consider the methods we incorporated to accomplish the task of breast cancer diagnosis. We implement the corresponding methods by using Python and TensorFlow. The evaluation metrics are calculated through the scikit-learn package. The number of epochs is set to 10 for all deep learning architectures and the random seed is set to 41.

13.2.2.1 Convolutional neural network-based diagnosis method

CNNs are specialized types of neural network architecture firstly designed for 2D image data. Not only working with images, CNNs can also work with 1D and 3D data. CNNs involve the following layers: convolutional layer, pooling layer, and fully connected layer. The design of a CNN architecture starts with the convolutional layer. Additional convolutional layers or

```
# Train/Validation/Test Splitting
X_train, X_test, y_train, y_test = train_test_split(X, y,
                         test_size=0.25, random_state=42)
X_train, X_valid, y_train, y_valid = train_test_split(X_train,
                     y_train, test_size=0.25, random_state=42)
y_train = to_categorical(y_train)
y_test = to_categorical(y_test)
y_valid = to_categorical(y_valid)
```

pooling layers can be added to the initial convolutional layer. The final layer is the fully connected layer. While convolutional layers learn local patterns from images by using 2D windows (also called filters), fully connected layers extract global patterns from the feature space of their input. Proportionally to the number of layers, the complexity of a CNN architecture increases. Therefore it is possible to identify larger portions of the image. To be specific, while small local patterns (e.g., color and edge) are extracted from images in earlier layers, latter layers detect larger patterns of the features (e.g., ear and eye).

The novelty of a CNN architecture is the convolutions with convolutional layer, the main components of which are input, a feature detector, and a feature map. Convolutions are carried out over a feature map. For a color image as an input, the feature map will be three dimensions: two of them are spatial axes (height and width) and the other one is the depth axis (also known as the channel axis). For an RGB image, the depth axis will be 3 since it has three channels: red, green, and blue. Another component of the convolutional layer, the feature detector (called a kernel or a filter) checks whether the feature is present in the image by moving across the receptive fields. The feature detector is a 2D array of weights representing the image part. Filter sizes can vary, but a traditional 3×3 matrix is typically the most common; this also determines how large a receptive field is. When a filter operation is applied to a region of an image, the dot product is calculated between the input and the filter. The calculated dot product is then used to build an output array. This process is repeated by shifting the filter by a stride until the filter has crossed the entire image. The final output array built by a number of dot products between the input and the filter is known as a feature map or a convolved feature. Each output value does not have a direct connection to each pixel from the input image, that is, it only connects to the corresponding region where the filter is carried out [17]. Accordingly, convolutional and pooling layers are referred to as partially connected layers.

Pooling layers also called downsampling reduce the number of parameters in the input by sweeping a filter over the entire input. Different from the convolutional layer, the filter applied in pooling layers does not have any weights. To be specific, the filter processes the receptive field values to generate the output array by applying an aggregation function. There are two main pooling types: max pooling and average pooling. While max pooling generates the output array by assigning the pixel with the maximum value within the receptive field, average pooling generates the output array by calculating the average value within the receptive field. Max pooling is more general than average pooling. After a convolution operator, a transformation (e.g., ReLU) is applied to the output array (feature map) to introduce nonlinearity to the model.

As convolutional and pooling layers are partially connected, the pixel value of an input image does not have a direct relationship with the class labels. To cover this issue, fully connected layers use the output array (feature map) obtained from partially connected layers to carry out a classification task. While partially connected layers usually use ReLU as an activation function, fully connected layers generally use a softmax function to generate probabilities within the range of 0 and 1 representing the predicted class labels of the input image data.

Thanks to the success in a variety of fields, CNNs have also been applied for breast cancer diagnosis. How a CNN is implemented to diagnose breast cancer from histopathological images is shown in Fig. 13.2. Firstly, the preprocessing stage is carried out to prepare the

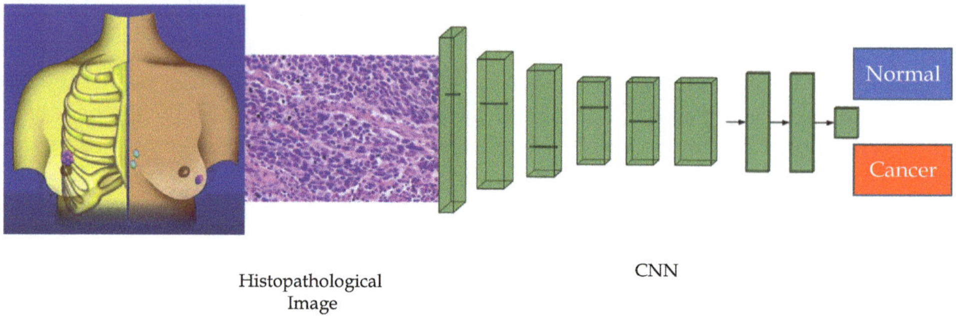

FIGURE 13.2 The general methodology of CNN architecture for breast cancer diagnosis from histopathological images. *CNN*, Convolutional neural network.

histopathological image dataset for the training stage. In this stage, we should first deal with data origin problems such as noise, artefacts, inconsistency, and class-imbalance. Then, we split the preprocessed data into the train and the test image sets. In the training stage, a CNN model is built on the train image set. Finally, the performance of the trained model is evaluated on the test image set to verify the diagnosis performance of the model on the unseen dataset. A sample CNN model with 6 layers is shown as below.

```
import tensorflow as tf
from tensorflow import keras
model = Sequential()
model.add(Conv2D(32, (3,3), activation = 'relu', input_shape = (50,50,3)))
model.add(Conv2D(32, (3,3), activation = 'relu'))
model.add(MaxPooling2D(pool_size = (2,2)))
model.add(Dropout(0.3))

model.add(Conv2D(64, (3,3), activation ='relu'))
model.add(Conv2D(64, (3,3), activation ='relu'))
model.add(MaxPooling2D(pool_size = (2,2)))
model.add(Dropout(0.3))

model.add(Conv2D(128, (3,3), activation ='relu'))
model.add(Conv2D(128, (3,3), activation ='relu'))
model.add(MaxPooling2D(pool_size = (2,2)))
model.add(Dropout(dropout_conv))

model.add(Flatten())
model.add(Dense(256, activation = "relu"))
model.add(Dropout(0.3))
model.add(Dense(2, activation = "softmax"))

model.summary()
```

13.2.2.2 *Pretrained convolutional neural network-based diagnosis method*

A number of pretrained CNN architectures have been introduced due to the motivation of addressing new datasets, such as MNIST, CIFAR-10, and competitions such as ImageNet. Some of these architectures are given as follows.

1. *VGG* [18], which is treated as one of the successful CNN architectures, was first proposed to win the ImageNet competition in 2014. VGG does not have a complex structure, that is, VGG16 applies convolutional layers with 3 × 3 sized filters and a stride 1 and uses same padding and max pooling with 2 × 2 sized filters and stride 2. The convolution and max pooling layers are arranged consistently over the whole architecture. At the end of partially connected layers, two fully connected layers are leveraged. The number 16 in VGG16 represents that the architecture has 16 layers. A fixed-sized RGB image is used to train the model, and the only preprocessing that takes place at the training stage is subtracting the mean RGB values computed for each pixel on the training set. Proportionally to the network depth, the training process of VGG slows down and the parameter size becomes quite large. VGG is easy to explainable and works properly for classical classification problems. However, the large number of parameters causes high computational cost. A sample VGG16 model with 7 CNN layers is shown as below.

```
# VGG16 Model
import tensorflow as tf
from tensorflow import keras
base_model=tf.keras.applications.VGG19(input_shape=(50,50,3),
                        include_top=False,weights="imagenet")
model=Sequential()
model.add(base_model)
model.add(Dropout(0.5))
model.add(Flatten())
model.add(BatchNormalization())
model.add(Dense(2048,kernel_initializer='he_uniform'))
model.add(BatchNormalization())
model.add(Activation('relu'))
model.add(Dropout(0.5))
model.add(Dense(1024,kernel_initializer='he_uniform'))
model.add(BatchNormalization())
model.add(Activation('relu'))
model.add(Dropout(0.5))
model.add(Dense(1024,kernel_initializer='he_uniform'))
model.add(BatchNormalization())
model.add(Activation('relu'))
```

```
model.add(Dropout(0.5))
model.add(Dense(1024,kernel_initializer='he_uniform'))
model.add(BatchNormalization())
model.add(Activation('relu'))
model.add(Dropout(0.5))
model.add(Dense(1024,kernel_initializer='he_uniform'))
model.add(BatchNormalization())
model.add(Activation('relu'))
model.add(Dropout(0.5))
model.add(Dense(1024,kernel_initializer='he_uniform'))
model.add(BatchNormalization())
model.add(Activation('relu'))
model.add(Dropout(0.5))
model.add(Dense(1024,kernel_initializer='he_uniform'))
model.add(BatchNormalization())
model.add(Activation('relu'))
model.add(Dropout(0.5))
model.add(Dense(2,activation='softmax'))
model.summary()
```

2. *ResNet* [19] is an exotic type of architectures that uses a network of microarchitecture modules unlike conventional sequential pretrained architectures such as AlexNet and VGG. Microarchitecture represents the building blocks used to build the entire network. ResNet has a much deeper architecture than VGG but has a smaller number of actual weighting parameters. This is because ResNet leverages global average pooling rather than fully connected layers. ResNet addresses vanishing gradients, accelerates the training speed, provides higher accuracy in classification problems, and detects redundant extracted features. On the other hand, ResNet has an increased complex architecture. Moreover, skip connections between layers cause extra dimensionality. The ResNet50 model is generated as follows.

```
#Import ResNet from Keras
from keras.applications.resnet50 import ResNet50
base_model=ResNet50(input_shape=(50,50,3),include_top=False,weights="imagenet")
```

3. *DenseNet* [20] connects each layer to every other layer. For L layers, the total number of direct connections is $L(L+1)/2$. Each layer takes the feature maps as inputs generated by the preceding layers. In other words, the input of a layer in a DenseNet architecture is the concatenation of feature maps of previous layers. The architecture in DenseNet is divided into DenseBlocks, where the dimensionality of a feature map remains constant, but the number of filters changes between them. Convolution and pooling are the first layers in DenseNet. Then, there is a dense block followed by a transition layer, and finally a dense block followed by a classification layer. DenseNet addresses vanishing gradients, enhances feature reuse, and reduces the

parameter size. The DenseNet169 model is generated as follows.

```
#Import DenseNet from Keras
from keras.applications.DenseNet169 import DenseNet169
base_model=DenseNet169(input_shape=(50,50,3),include_top=False,weights="imagenet")
```

4. *MobileNet* [21] applies the same convolution as CNNs to filter images, but its method differs from that of CNN in such a way that it performs depth convolution and point convolution different from the conventional convolution as done by CNNs. Accordingly, the efficiency of CNN is increased and so it is possible to integrate MobileNet in the mobile systems. In other words, it is possible to obtain a better response in a short time due to the time efficiency. The MobileNet model is generated as follows.

```
#Import MobileNet from Keras
from keras.applications.mobilenet import MobileNet
base_model= MobileNet(input_shape=(50,50,3),include_top=False,weights="imagenet")
```

13.3 Results and discussions

How a pretrained CNN architecture is implemented to diagnose breast cancer from histopathological images is shown in Fig. 13.3. As described in Section 13.2.2.1, the preprocessing stage is carried out and then the training stage is performed on the train data. Finally, the performance of the pretrained model is verified on the test data.

13.3.1 Experimental setup

To make the established CNN-based model ready for training, the following three cases are considered as part of the compilation stage:

1. An optimizer: It is a function or an algorithm that is used to update the weights and learning rate of a deep learning model based on the loss function value. Accordingly, it helps to reduce the overall loss and increase the accuracy. There are variety of optimizers such as gradient descent, stochastic gradient descent, AdaGrad, root mean square (RMS Prob), and Adam. As a deep learning model generally consists of millions of parameters, choosing the best weights for it can be a daunting task. Therefore it is necessary to choose a suitable optimizer for your application.
2. A loss function: The function is used to quantify how good or bad the model performs. Loss functions are mainly investigated in two groups: regression loss (e.g., mean squared error, mean squared logarithmic error, and mean absolute error) and classification loss (e.g., binary cross entropy, hinge, and categorical cross entropy). While regression loss is used with continuous target variables, classification loss is used with discrete target variables.
3. Metrics: Metrics are different from loss functions. While loss functions are used to update the model parameters through an optimizer, metrics are used to measure or visualize the performance of model during training and testing. Metrics can be

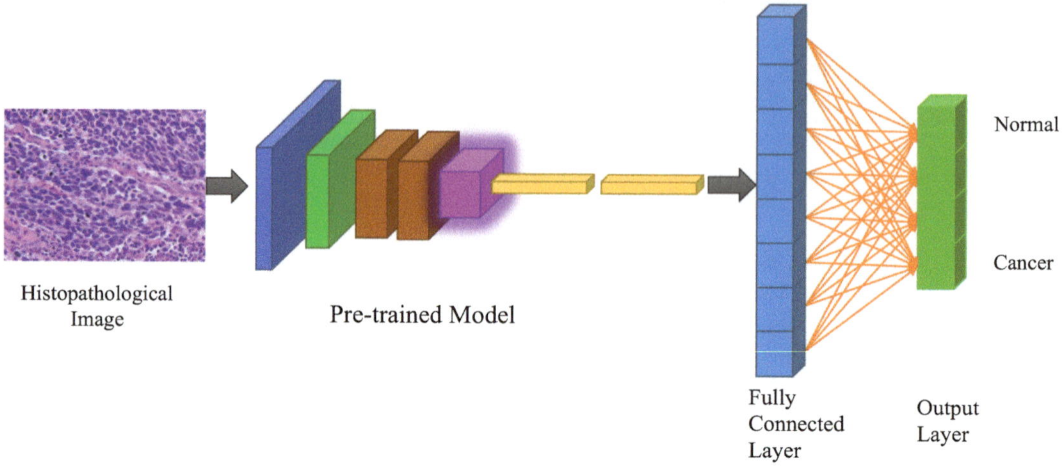

FIGURE 13.3 The general methodology of pretrained architecture for breast cancer diagnosis.

divided into the following groups: regression metrics (e.g., mean absolute error, mean squared error, and root mean squared error) and classification metrics (e.g., accuracy, confusion matrix, F1-score, and receiver operating characteristic curve [ROC]).

In this study, the compilation stage is implemented as follows.

```
def f1_score(y_true, y_pred): #taken from old keras source code
    true_positives = K.sum(K.round(K.clip(y_true * y_pred, 0, 1)))
    possible_positives = K.sum(K.round(K.clip(y_true, 0, 1)))
    predicted_positives = K.sum(K.round(K.clip(y_pred, 0, 1)))
    precision = true_positives / (predicted_positives + K.epsilon())
    recall = true_positives / (possible_positives + K.epsilon())
    f1_val = 2*(precision*recall)/(precision+recall+K.epsilon())
    return f1_val

METRICS = [
      tf.keras.metrics.BinaryAccuracy(name='accuracy'),
      tf.keras.metrics.Precision(name='precision'),
      tf.keras.metrics.Recall(name='recall'),
      tf.keras.metrics.AUC(name='auc'),
        f1_score,
]


lrd = ReduceLROnPlateau(monitor = 'val_loss',patience = 5,verbose = 1,factor = 0.75,
                                                                        min_lr = 1e-10)
mcp = ModelCheckpoint('model.h5')
es = EarlyStopping(verbose=1, patience=5)

model.compile(optimizer='Adam', loss='binary_crossentropy',metrics=METRICS)
history=model.fit(X_train, y_train,validation_data=(X_valid, y_valid),verbose = 1,
                                                epochs  = 20,callbacks=[lrd,mcp,es])

model.evaluate (X_test, y_test, verbose=1)
```

13.3.2 Experimental results

The results of a variety of CNN and pretrained architectures are presented in Table 13.1 in terms of the accuracy, F1-score, and ROC. According to the results, CNNs with 2 and 3 layers cannot perform as well as other CNNs models having more layers. Moreover, it is observed that increasing the layer size from 4 to 9 in CNNs does not significantly enhance the diagnosis performance, that is, CNNs with layers between 4 and 9 have similar performance in detecting breast cancer. It can therefore be stated that CNNs established with more layers do not always guarantee better classification performance. When considering pretrained CNN architectures, it is observed from Table 13.1 that pretrained architectures can achieve similar or mostly better diagnosis performance than conventional CNNs due to their well-designed network model. However, the F1-score obtained by ResNet50, VGG19, ResNet101, and DenseNet121 is lower than 0.9, that is, they cannot perform well compared to other pretrained architectures. It should also be notified that a combined VGG19 architecture with 7 layers obtains similar performance with conventional VGG19 despite additional modules. This means the network architecture even with deeper structure does not perform well if not well-prepared and well-designed. The F1-score more than 0.9 is obtained by VGG16 and MobileNet, and these pretrained architectures are followed by MobileNetV2 and DenseNet121. It can therefore be suggested that deep learning architectures, especially pretrained architectures, can be successfully applied to diagnose breast cancer from histopathological images.

TABLE 13.1 Results of deep learning architectures.

Classifiers	Train accuracy	Val accuracy	Test accuracy	F1-score	ROC area
CNN 2 Layer	0.8455	0.85	0.846	0.8455	0.8
CNN 3 Layer	0.846	0.85	0.846	0.846	0.85
CNN 4 Layer	0.8741	0.881	0.8808	0.8741	0.9429
CNN 5 Layer	0.8737	0.8767	0.8779	0.8737	0.9726
CNN 6 Layer	0.88	0.88	0.88	0.8826	0.95
CNN 7 Layer	0.8757	0.8784	0.8785	0.8784	0.9433
CNN 8 Layer	0.87	0.86	0.8692	0.87	0.93
CNN 9 Layer	0.88	0.88	0.87	0.88	0.95
ResNet50	0.8731	0.8837	0.8792	0.8731	0.94
VGG16	0.934	0.9211	0.922	0.934	0.9786
VGG19	0.87	0.85	84.5	0.87	0.93
MobileNet	0.94	0.91	0.91	0.94	0.95
DenseNet169	0.89	0.86	0.85	0.89	0.93
DenseNet121	0.91	0.91	0.9	0.91	0.95
MobileNetV2	0.92	0.92	0.91	0.92	0.97
ResNet101	0.88	0.86	0.85	0.87	0.95

13.4 Conclusion

In this chapter, we leveraged deep learning architectures to carry out a comparative study on detecting breast cancer from histopathological images. According to the results, we obtained 92.2% test accuracy and 0.934 F1-score with VGG16 in the IDC dataset. It can therefore be revealed that deep learning architectures, especially pretrained models, are really good alternatives for the diagnosis process of the breast cancer. We would also want to notify that increasing the deepness of an architecture does not verify better diagnosis performance if not well-designed and well-prepared. For the future, we would like to work upon transfer learning to improve the effectiveness of the diagnosis process. To achieve this, we will first design a deep architecture by integrating well-designed modules to each other. Moreover, we will propose a diagnosis methodology based on a feature selection method to select the most appropriate features from the feature set extracted by deep architectures for classification.

References

[1] S. Boumaraf, et al., A new transfer learning based approach to magnification dependent and independent classification of breast cancer in histopathological images, Biomed. Signal. Process. Control. 63 (102192) (2021).
[2] R. Rashmi, K. Prasad, C.B.K. Udupa, Breast histopathological image analysis using image processing techniques for diagnostic purposes: a methodological review, J. Med. Syst. 46 (1) (2021) 7.
[3] X. Zhou, et al., A comprehensive review for breast histopathology image analysis using classical and deep neural networks, IEEE Access. 8 (2020) 90931–90956.
[4] F.A. Zeiser, et al., Breast cancer intelligent analysis of histopathological data: a systematic review, Appl. Soft Comput. 113 (107886) (2021).
[5] F.A. Spanhol, et al., Breast cancer histopathological image classification using Convolutional Neural Networks, in: 2016 International Joint Conference on Neural Networks (IJCNN), 2016.
[6] A. Krizhevsky, I. Sutskever, G.E. Hinton, ImageNet classification with deep convolutional neural networks, Commun. ACM 60 (6) (2017) 84–90.
[7] F.A. Spanhol, et al., Deep features for breast cancer histopathological image classification, in: 2017 IEEE International Conference on Systems, Man, and Cybernetics (SMC), 2017.
[8] Bayramoglu, N., J. Kannala, J. Heikkilä, Deep learning for magnification independent breast cancer histopathology image classification, in: 2016 23rd International Conference on Pattern Recognition (ICPR), 2016.
[9] B. Wei, et al., Deep learning model based breast cancer histopathological image classification, in: 2017 IEEE 2nd International Conference on Cloud Computing and Big Data Analysis (ICCCBDA), 2017.
[10] S. Pratiher, S. Chattoraj, Manifold learning & stacked sparse autoencoder for robust breast cancer classification from histopathological images. arXiv, 2018, 1806.06876.
[11] A.-A. Nahid, Y. Kong, Histopathological breast-image classification using local and frequency domains by convolutional neural network, Information 9 (1) (2018).
[12] A.-A. Nahid, M.A. Mehrabi, Y. Kong, Histopathological breast cancer image classification by deep neural network techniques guided by local clustering, BioMed. Res. Int. (2018) 2362108. 2018.
[13] D. Bardou, K. Zhang, S.M. Ahmad, Classification of breast cancer based on histology images using convolutional neural networks, IEEE Access. 6 (2018) 24680–24693.
[14] Shallu, R. Mehra, Breast cancer histology images classification: training from scratch or transfer learning? ICT Express 4 (4) (2018) 247–254.
[15] Z. Xiang, et al., Breast cancer diagnosis from histopathological image based on deep learning, in: 2019 Chinese Control and Decision Conference (CCDC), 2019.
[16] A. Janowczyk, A. Madabhushi, Deep learning for digital pathology image analysis: a comprehensive tutorial with selected use cases, J. Pathol. Inform. 7 (1) (2016) 29. ISSN 2153–3539, https://doi.org/10.4103/2153-3539.186902.
[17] IBM Cloud Education, Convolutional neural networks https://www.ibm.com/cloud/learn/convolutional-neural-networks, 2020.
[18] K. Simonyan, A. Zisserman, Very deep convolutional networks for large-scale image recognition, in: 3rd International Conference on Learning Representations (ICLR2015), 2015.
[19] K. He, et al., Deep residual learning for image recognition, in: 2016 IEEE Conference on Computer Vision and Pattern Recognition (CVPR), 2016.
[20] G. Huang, et al., Densely connected convolutional networks, in: IEEE Conference on Computer Vision and Pattern Recognition (CVPR2017), 2017.
[21] A.G. Howard, et al., MobileNets: efficient convolutional neural networks for mobile vision applications, CoRR (2017). abs/1704.04861.

CHAPTER

14

Artificial intelligence based Alzheimer's disease detection using deep feature extraction

Manav Nitin Kapadnis[1], *Abhijit Bhattacharyya*[2] *and* *Abdulhamit Subasi*[3,4]

[1]Department of Electrical Engineering, Indian Institute of Technology Kharagpur, Kharagpur, West Bengal, India [2]Department of Electronics and Communication Engineering, National Institute of Technology Hamirpur, Hamirpur, Himachal Pradesh, India [3]Institute of Biomedicine, Faculty of Medicine, University of Turku, Turku, Finland [4]Department of Computer Science, College of Engineering, Effat University, Jeddah, Saudi Arabia

OUTLINE

14.1 Introduction

Alzheimer's disease (AD) is considered as one of the progressive fatal neurodegenerative disorders that affect elderly population and results in decline of cognitive abilities. The latest statistical survey concludes that majority of the dementia patients (60%–80%) suffer from AD and the number will increase as the population become elderly. It is predicted that by the year 2050,

Applications of Artificial Intelligence in Medical Imaging
DOI: https://doi.org/10.1016/B978-0-443-18450-5.00007-4

worldwide the total number of AD patients can reach up to 152 million [1] and 1 out of 85 persons will be affected by AD [2]. In contrast to other brain disorders, AD has significantly high mortality rate and is rising each year.

AD has multiple stages such as predementia, early, middle, and advanced stages. In the predementia stage, the common symptoms include mild cognitive impairment (MCI) with forgetfulness that mimics the natural aging process. In the early AD stage, person suffers from impairment of executive functions, learning, and memory, often leading to language difficulty. Patient feels more speech difficulty, reading and writing skills are largely impeded in the middle stage. In the advanced stage, AD patients exhibit apathy and fail to perform even simple tasks independently. Eventually, the patients become immobilized and death occurs [3,4].

The early diagnosis of this disease can reduce the surging pervasiveness and high mortality. Till date, AD has no curable treatment and only after early identification of the disorder, the disease progression can be slowed with cognition enhancing drugs, physical exercises, and proper lifestyle management. The diagnosis of AD is commonly carried out based on patient's illness history and also using physiological and neurological features. Patient's medical history can be received from relatives and assessing patient's behavior [5].

In recent times, for the diagnosis of AD, clinicians generally rely on patient's brain image analysis, as current imaging systems provide wide range of information about the subject's health condition. Commonly used imaging techniques are computed tomography (CT), positron emission tomography-CT (PET-CT), and magnetic resonance imaging (MRI). A CT image can be used for diagnosing dementia by inspecting the different brain region sizes that includes frontal lobe, temporal lobe, and hippocampus. PET-CT provides information about different types of brain functions such as (1) fluorodeoxyglucose (FDG) PET-CT are generally used for measuring glucose levels in the brain, (2) amyloid PET-CT provide insight about beta-amyloid protein level, (3) tau PET-CT are utilized for tau (protein responsible for formation of neurofibrillary tangles in the nerve cells) detection [6,7]. For example, higher level of beta-amyloid protein can be confirmed by a positive amyloid PET-CT scan that can assist to confirm AD. Other conditions such as head trauma, stroke, and tumors contributing to dementia can also be disregarded using the aforementioned diagnostic methods [8]. The MRI scan measures the volume alteration at characteristic positions to analyze AD which can provide up to 87% analytical accuracy [9,10].

More often the appraisal is performed on temporoparietal cortical atrophy and mesial temporal lobe atrophy (i.e., entorhinal cortex and hippocampus). Direct estimation is quantified by measuring the volume loss of hippocampal or parahippocampal tissue, whereas indirect estimation is based on the parahippocampal fissure's magnification.

Recently, the accessibility of advanced noninvasive imaging methods has gained significant attention of the researchers developing reliable and precise tools for brain condition monitoring. In fact, many computer-assisted systems for AD diagnosis and severity analysis have been proposed and implemented in literature [11–13]. In general, the primary steps involved in a computer-aided AD diagnosis system are listed as follows: (1) brain imagery collection using a recommended and standard imaging technique, (2) enhancement of the brain images with a suitable image enhancement tool, (3) extraction of automatic/handcrafted discriminatory features from the enhanced images, (4) selection of dominant features using feature selection techniques, (5) building a classification model for brain image categorization, and (6) validation of the built system using new test images.

In this work, we aim to build AI-driven AD diagnosis method using brain MRI images. More specifically, we propose different transfer learning (TL) models [pretrained deep neural networks (DNNs)] for automatic learning and extraction of discriminatory features for AD diagnosis. The privilege of using DNNs is that features are optimally tuned for discriminating different classes of images. Convolutional neural networks (CNNs) are very effective in extracting spatial patterns from the input images. Finally, different machine learning (ML) models are explored to classify the deep features for AD diagnosis. A detailed comparative study of the performance of different deep feature extraction models has been presented in detail in order to choose the best combination of deep feature-ML model for the AD detection.

14.2 Background/literature review

In the area of medical imaging the AD diagnosis plays a critical role in maintaining the cognitive capability of an individual. In fact, the persons suffering from AD face serious problems that impact thinking, memory, and other living activities. The human intellectual activities are highly affected in later stage of their lives due to implacable destruction of the nerve cells over time [14]. In the United States only, the number of persons suffering from AD has approximately reached to five million, most of them are aged between 80 and 90 years [15]. Early detection of AD is carried out using various therapies and advanced diagnostic techniques. The MRI is a potent tool for investigation of clinical signs of AD in the brain. However, MRI requires manual inspection of the workflow and thus acquisition procedure of MRI is sluggish [16].

Recently, several efficient AD diagnosis techniques have been proposed using image processing and ML techniques, which can become alternative to the manual systems. Patil et al. [17] proposed an effective image processing system for AD detection using MRI images. They computed atrophy with wavelet, K-means, watershed, and other customized algorithms. Their obtained results can be useful for the early diagnosis of AD. Kaur and Kaur [18] employed different image enhancement techniques for AD detection. They used corrected red and green ingredients and found the best AD detection approach by computing sensitivity and specificity measures. Zhao et al. [19] used through gradient echo plural contrast imaging (GEPCI) with MRI scan for the detection of AD. The GEPCI enhances the resolution of the affected brain area in the MRI and can effectively identify the damaged brain tissues caused by AD.

Sankari and Adeli [20] used probabilistic neural network (PNN) for AD diagnosis using MRI scans. In the first stage, they computed total brain volume and atrophy rate. Then several features, namely, shape, correlation, and contrast were extracted and given as input for classification. Their study reveals that PNN performs better than the KNN and support vector machine (SVM) with respect to sensitivity, specificity, and accuracy. Plant et al. [21] proposed an MRI image-based novel pipeline for the AD diagnosis. The classifiers, namely, Bayes statistics, SVM, and voting feature interval (VFI) were combined for AD pattern matching in the imagery. They obtained the MRI from thirty-two AD patients, which were used for feature extraction purpose. The authors achieved significant prediction accuracy after selecting the discriminatory features. Zhang et al. [22] introduced a novel ML framework for MRI-based AD diagnosis with significant performance. In the first stage, they performed skull stripping to remove the extracted regions, and thereafter features were computed based on stationary wavelet entropy. The computed features were fed to a neural network with single hidden layer, and weights and biases were optimized via particle swarm optimization (PSO).

Wang et al. [23] presented single slice (SS) method based on wavelet entropy and perceptron learning for AD detection using MRI scans. Several features were extracted, including wavelet transform-based SS features, wavelet entropy, and wavelet orientation. The authors employed multilayer perceptron with biogeography-based optimization algorithm for the classification of the extracted features that outperformed the existing methods. Zhang et al. [24] introduced an AD detection framework using MRI scans based on undersampling method. Principal component analysis along with singular value decomposition methods were used for feature computation and discriminatory feature selection. Further, decision tree (DT) and SVM classifiers were utilized for achieving significant detection performance.

Zhang and Wang [25] computed displacement field (DF) in MRI for detecting abnormalities present in normal brain for AD detection. The discrete wavelet transform (DWT)-based features were computed and feature dimensionality was reduced using PCA. The final set of selected features were categorized using three different classifiers, namely, SVM, twin SVM (TSVM), and generalized eigenvalue proximal SVM. The authors concluded that DF is useful in AD diagnosis when MRI scans were utilized. El-Dahshan et al. [26] described an MRI-based AD classification method using a hybrid approach. Their framework includes three basic steps that are as follows: feature extraction, dimensionality reduction, and classification. They extracted DWT-based MRI features and then feature reduction was carried by irrelevant points using PCA. Finally, two classification algorithms, namely, ANN and KNN were employed for the classification of normal and AD MRI scans. Wang et al. [27] introduced a novel AD classification algorithm using the features of Zernike moment (ZM), followed by a linear regression (LR) classifier. The ZM was used to extract features with lengths between 10 and 256 from each MRI image. The computed features were classified LR that achieved an accuracy of 97.51%, outperforming the existing methods [28]. In another work, Wang et al. [29] presented a 3D DF estimation-based method for discrimination of AD and healthy subjects. They extracted features using 3D DF method and selected statistically significant features using Bhattacharyya distance, Welch's *t*-test (WTT), and Student's *t*-test. Finally, selected features were classified using SVM and TSVM classifiers.

Zhang et al. [30] presented a computer-aided diagnostic (CAD) system for AD detection with MRI images. The authors used maximum interclass variance for selecting key slices of 3D MRI. Afterward, for each slice set eigenbrain was generated. Then, significant eigenbrains were selected using WTT and fed to SVM classifier with different kernels. Further, the prediction accuracy was notably improved using PSO algorithm. Zhang et al. [31] introduced a CAD system to detect AD from MRI scans. From each MRI scan wavelet entropy and Hu moment invariant features were extracted. Then extracted features were classified using computation of generalized eigenvalues with SVM. Hett et al. [32] presented multi-textural (MTL) pipeline for feature computation from MRI scans. The AD structural information was estimated via MTL approach. Further, adaptive fusion method was applied for fusing texture grading features computed using 3D Gabor filter. Their work achieved significant performance improvement over existing biomarker methods. Gao et al. [33] employed deep learning for obtaining early-stage AD information and classification. They fused 2D and 3D CNN that provided significant performance with a softmax layer.

Ayadi et al. [34] described a hybrid method for extracting features from brain MRI scans and proposed a classification system. Initially, DWT-based features were extracted from the test images, and further, Bag-of-Words method adopted for grouping key image features. Finally, several ML techniques, such as random forest (RF), AdaBoost, KNN, and SVM were employed for classification. The recent work of Acharya et al. [13] uses MRI scans and employs different quantitative techniques, including filtering, feature extraction, feature selection using Student's *t*-test, and KNN classifier for the diagnosis of AD. The shearlet transform-based feature computation method achieved superior performance over the alternative methods.

14.3 Artificial intelligence models

14.3.1 Deep feature extraction techniques

Pretrained neural networks from freely available platforms such as ImageNet are also available for usage in different applications and datasets off-the-shelf. Except for the fully linked layer, practically all of the pretrained weights in the CNN are used without change. The fully connected layer's weights are derived from the dataset at hand. Because the labels of the class in a certain context may differ from those of ImageNet, the last layer's preparation is critical. The weights in the preceding layer, on the other hand, are useful because they learn different types of shapes in the images and may be used for nearly any type of categorization. Furthermore, the last layer's activation function can also be applied for clustering. It is worth noting that the utilization of pretrained CNNs is so popular that training is hardly begun from the scratch [35].

Deep learning methods need lots of examples or a big data. We can use TL since the number of samples in our dataset is lower than modern data standards, that is, the number of MRI scans for the classification model is less. Taking characteristics acquired on one topic and applying them to a new, related issue is known as TL [36]. By integrating knowledge from the source task, export learning aims to improve learning in the target task. The source task must be learned on a bigger dataset than the target task, and the source and target tasks must be similar in nature before utilizing TL. There are three main approaches that might help students learn more effectively through transfer. The first is the first feasible output in the target task using only the transferred information, as contrasted to an uninformed person's initial performance before any further learning. Second, the time it takes to fully comprehend the objective mission, notwithstanding the information provided, in comparison to the time it takes to comprehend it from scratch. Third, the maximum output that can be achieved in the target role in comparison to the maximum output that can be achieved without transition.

VGG16, VGG19, DenseNet169, DenseNet121, ResNet101, InceptionV3, InceptionResNetV2, MobileNet, and MobileNetV2 are some of the models we used. The VGG network design was introduced by Simonyan and Zisserman [37], which is recognized by its simplicity, utilizing only three universal layers constructed on top of each other in increasing intricacy. Max pooling is a high-volume capacity method. ResNet is a form of "exotic architecture" built on micro-architecture modules, as opposed to traditional sequential network designs such as AlexNet, VGG, and OverFeat. The term "micro-architecture" refers to the set of "building pieces" used to put the network together. The ResNet design, initially proposed by He et al. [38], has shown that extraordinary deep networks may be outfitted using conventional stochastic gradient descent and residual modules. The purpose of the inception module is to operate as a "multi-level feature extractor" by computing one, three, and five brackets all inside the same network module. The filter outputs are then stacked along the channel dimension before being transferred to the following network layer. The initial implementation of this idea was Google LeNet, but subsequent implementations were simply known as Iteration vN, where N is the Google version number. François Chollet, the Keras library's inventor and major maintainer, offered the name Xception. The Xception architecture is an expansion of the standard that replaces the Inception architecture.

One of the challenges that image analysts encounters is that labeled training data may not be suitable for a particular purpose. Consider the following scenario: you have a set of images that must be combined to get a single image. Although retrieval applications do not use labels, semantic compatibility between features is critical. In other cases, you might want to classify a dataset using a specific collection of markings that are not accessible in large numbers and are not the part of a larger tool such as ImageNet. This causes issues since neural networks require a lot of training data to build from start. However, the most important thing to remember about image data is that the characteristics retrieved from an image are just that: features from a specific dataset are particularly reusable across data sources [35].

The following code snippet gives us a template for replicating our model architecture for feature extraction using a TL model, (importing of libraries step is skipped):

```
input_tensor = Input(shape=(224, 224, 3))
base_model = VGG16(include_top=False, weights='imagenet', input_shape=(224,224,3))

x = base_model.output
x =  Dropout(0.5)(x)
x = Flatten()(x)
x = BatchNormalization()(x)
x = Dense(1024,kernel_initializer='he_uniform')(x)
x = BatchNormalization()(x)
x = Activation('relu')(x)
x =  Dropout(0.5)(x)
x = Dense(1024,kernel_initializer='he_uniform')(x)
x = BatchNormalization()(x)
x = Activation('relu')(x)
x =  Dropout(0.5)(x)
x = Dense(1024,kernel_initializer='he_uniform')(x)
x = BatchNormalization()(x)
x = Activation('relu')(x)
x =  Dropout(0.5)(x)


predictions = Dense(4, activation='softmax')(x)


model_feat = Model(inputs=base_model.input,outputs=predictions)


train_features = model_feat.predict(x_train)
test_features = model_feat.predict(x_test)
```

In the above code snippet, in order to use a different pretrained CNN model, you just have to replace the VGG16 with the choice of model you wish to use.

14.3.2 Classification techniques

14.3.2.1 *Artificial neural network*

An artificial neural network (ANN) with a single hidden layer has limited capability, but by adding more hidden layers, the ANN can learn more intricate data. This is the idea underlying DNNs, which start with raw input networks and learn increasingly intricate data by integrating values from previous layers. The goal of DNN is to learn the degree of features by upgrading concepts with the least amount of human interaction possible [39–41]. Medical image analysis and computer-aided decision support systems have inspired a lot of interest in deep learning. The detection of illness has been greatly improved because to recent breakthroughs in AI. To examine various datasets, a variety of deep learning methods are applied. It's critical to understand the intrinsic brain mechanisms that may be exploited by various scanning technologies to diagnose AD. DNNs are an AI algorithm inspired by the way human brain functions and is structured. To solve artificial intelligence (AI) difficulties, a variety of DNN topologies were developed. The DNN's operation is divided into two phases. To adjust the network parameters during training, the first phase uses training data, which is a proportion of the original data collected. Following training, test data is used to determine if the model is properly trained to detect previously unseen instances [41–43].

The following code snippet gives an in-depth detail about the model architecture used for DNN:

```
model = Sequential()
model.add(Flatten(input_shape=(224,224,3)))
model.add(Dense(units=4, activation='relu'))
model.add(Dense(units=8, activation='relu'))
model.add(Dense(units=16, activation='relu'))
model.add(Dense(units=4))
model.add(Activation('softmax'))

def f1_score(y_true, y_pred): #taken from old keras source code
    true_positives = K.sum(K.round(K.clip(y_true * y_pred, 0, 1)))
    possible_positives = K.sum(K.round(K.clip(y_true, 0, 1)))
    predicted_positives = K.sum(K.round(K.clip(y_pred, 0, 1)))
    precision = true_positives / (predicted_positives + K.epsilon())
    recall = true_positives / (possible_positives + K.epsilon())
    f1_val = 2*(precision*recall)/(precision+recall+K.epsilon())
    return f1_val
```

```
METRICS = [
        tf.keras.metrics.BinaryAccuracy(name='accuracy'),
        tf.keras.metrics.Precision(name='precision'),
        tf.keras.metrics.Recall(name='recall'),
        tf.keras.metrics.AUC(name='auc'),
          f1_score,
]

def exponential_decay(lr0, s):
    def exponential_decay_fn(epoch):
        return lr0 * 0.1 **(epoch / s)
    return exponential_decay_fn

exponential_decay_fn = exponential_decay(0.01, 5)

lr_scheduler = tf.keras.callbacks.LearningRateScheduler(exponential_decay_fn)

model.compile(optimizer= tf.keras.optimizers.Adam(
    learning_rate = 0.001, beta_1 = 0.9, beta_2 = 0.999, epsilon = 1e-
07, amsgrad=False, name='Adam'),loss='categorical_crossentropy',metrics=METRICS)

history = model.fit(train_features, y_1, epochs=20,batch_size=32,validation_split=0.2)
```

14.3.2.2 K-nearest neighbor

KNN, also known as k-nearest neighbor, is a supervised ML technique for classification and regression problems. One of the easiest algorithms to learn is K-closest neighbor. KNN is nonparametric, that is, it is not a normal distribution. It does not make any assumptions about the underlying data. The K-closest neighbor method is sometimes known as a lazy algorithm since it does not learn during the training phase but does learn during the testing phase. It is a distance-based method. A lot of studies in the past have worked on different modifications of this KNN algorithm for AD. Aruchamy et al. [44] modified the KNN algorithm along with methods such as image processing feature extraction, and feature reduction and achieved a significant increase in the classification performance. Furthermore, Dinu and Ganesan [45] presented a feature selection utilizing the *t*-test technique for combined regression and classification using an instance-based KNN classifier for AD detection.

The following code snippet gives a deep detail about the model architecture used for KNN:

```
# Getting Train and validation accuracy
x_train_acc, x_val_acc, y_train_acc, y_val_acc = train_test_split(train_features,
      y_train,test_size = 0.2,
      stratify = y_train,
      shuffle = True,
      random_state = 42)
X_test,y_test=test_features,y_test


Model = KNeighborsClassifier()
Model_trained = model.fit(X_train, y_train)
y_pred_train= Model_trained.predict(X_train)
y_pred_val = Model_trained.predict(X_val)
y_pred_test = Model_trained.predict(X_test)
```

14.3.2.3 *Support vector machine*

The SVM is a supervised ML approach for classifying and predicting data. It is, however, mainly used to tackle categorization problems. Each data item is represented as a point in *n*-dimensional space (where *n* is the number of features), with the value of each feature being the SVM algorithm's position value. Then, to complete classification, we find the hyperplane that clearly divides the two categories. Rabeh et al. [46] implemented SVM for AD detection along with feature extraction from image such as Gaussian blur, edge detection features, among others. They achieved impressive results for early Alzheimer's detection on MRI images. Alam et al. [47] suggested an innovative method for identifying AD from healthy controls by employing dual-tree complex wavelet transforms, main coefficients from MRI transaxial slices, linear discriminant analysis, and TSVM.

The following code snippet gives a deep detail about the model architecture used for SVM:

```
# Getting Train and validation accuracy
x_train_acc, x_val_acc, y_train_acc, y_val_acc = train_test_split(train_features,
                        y_train,test_size = 0.2,
                        stratify = y_train,
                        shuffle = True,
                        random_state = 42)
X_test,y_test=test_features,y_test


Model = SVC()
Model_trained = model.fit(X_train, y_train)
y_pred_train= Model_trained.predict(X_train)
y_pred_val = Model_trained.predict(X_val)
y_pred_test = Model_trained.predict(X_test)
```

14.3.2.4 *Random forest*

RF is a supervised ML technique for classification and regression problems. It employs the majority vote for classification and the average for regression to build DTs from several examples. The capacity of the RF algorithm to handle datasets with both continuous and categorical variables, as in regression and classification, is one of its most valuable characteristics. It outperforms others when it comes to categorization challenges. Recent studies [48–50] implemented RF for AD detection using various image formats and achieved excellent results due to the robustness of RF. Ali et al. [51] developed a new variation in RF algorithm for optimum feature extraction which in turn proposed a new method for feature extraction along with a better version of RF.

The following code snippet gives a deep detail about the model architecture used for RF classifier:

```
# Getting Train and validation accuracy
x_train_acc, x_val_acc, y_train_acc, y_val_acc = train_test_split(train_features,
                        y_train,test_size = 0.2,
                        stratify = y_train,
                        shuffle = True,
                        random_state = 42)
X_test,y_test=test_features,y_test


Model = RandomForestClassifier()
Model_trained = model.fit(X_train, y_train)
y_pred_train= Model_trained.predict(X_train)
y_pred_val = Model_trained.predict(X_val)
y_pred_test = Model_trained.predict(X_test)
```

14.3.2.5 *AdaBoost*

AdaBoost, also known as Adaptive Boosting, is a ML technique used as an ensemble approach. The most common method used with AdaBoost is DTs with one level, or DTs with only one split. These trees are also known as decision stumps. This method generates a model by giving all of the data points the same weight. It therefore gives improperly classified points a larger weight. All points with higher weights are given more weight in the following model. It will continue to train models until it receives a smaller error. Recent studies such as Refs. [52–54] show the effectiveness of AdaBoost approach in Alzheimer's disease classification.

The following code snippet gives a deep detail about the model architecture used for AdaBoost classifier:

```
# Getting Train and validation accuracy
x_train_acc, x_val_acc, y_train_acc, y_val_acc = train_test_split(train_features,
                                y_train,test_size = 0.2,
                                stratify = y_train,
                                shuffle = True,
                                random_state = 42)
X_test,y_test=test_features,y_test


Model = AdaBoostClassifier()
Model_trained = model.fit(X_train, y_train)
y_pred_train= Model_trained.predict(X_train)
y_pred_val = Model_trained.predict(X_val)
y_pred_test = Model_trained.predict(X_test)
```

14.3.2.6 XGBoost

XGBoost is an ensemble learning approach. It is not always possible to rely just on the results of a single ML model. Ensemble learning is a method for systematically integrating the prediction skills of several learners. As a consequence, a single model has been created that incorporates the results of many models. The ensemble's foundation learners, or models, may come from the same learning algorithm or from different learning algorithms. Ensemble learners are often used in two different ways: bagging and boosting. Despite the fact that these two procedures may be used to a wide range of statistical models, DTs are the most widely employed. Recent studies such as Refs. [55,56] show the effectiveness of XGBoost approach in Alzheimer's disease classification.

The following code snippet gives a deep detail about the model architecture used for XGBoost classifier:

```
# Getting Train and validation accuracy
x_train_acc, x_val_acc, y_train_acc, y_val_acc = train_test_split(train_features,
                        y_train,test_size = 0.2,
                        stratify = y_train,
                        shuffle = True,
                        random_state = 42)
X_test,y_test=test_features,y_test


Model = XGBClassifier()
Model_trained = model.fit(X_train, y_train)
y_pred_train= Model_trained.predict(X_train)
y_pred_val = Model_trained.predict(X_val)
y_pred_test = Model_trained.predict(X_test)
```

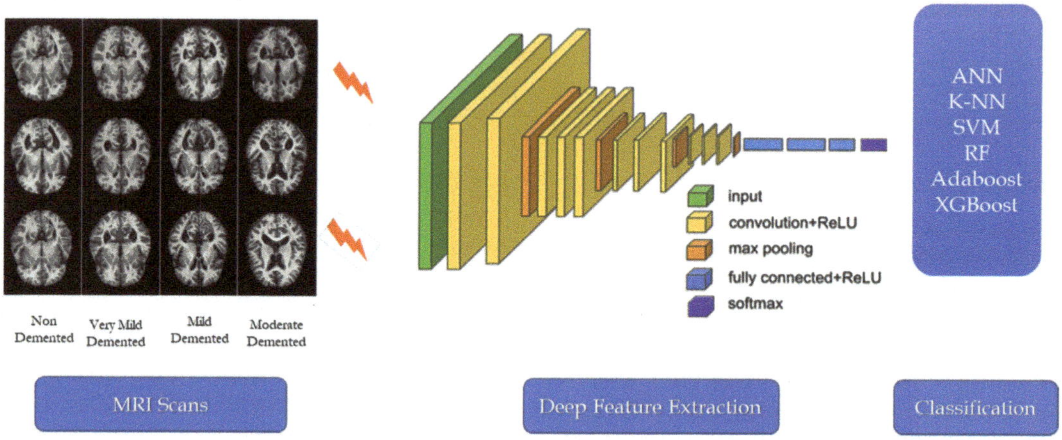

FIGURE 14.1 General framework for Alzheimer's disease detection utilizing artificial intelligence methods.

14.4 Alzheimer's disease detection using artificial intelligence

The general framework for AD detection using AI methods is presented in Fig. 14.1.

14.4.1 Experimental data

We utilize a publicly available MRI dataset that is taken from the Kaggle. The data samples are shown in Fig. 14.2. The dataset

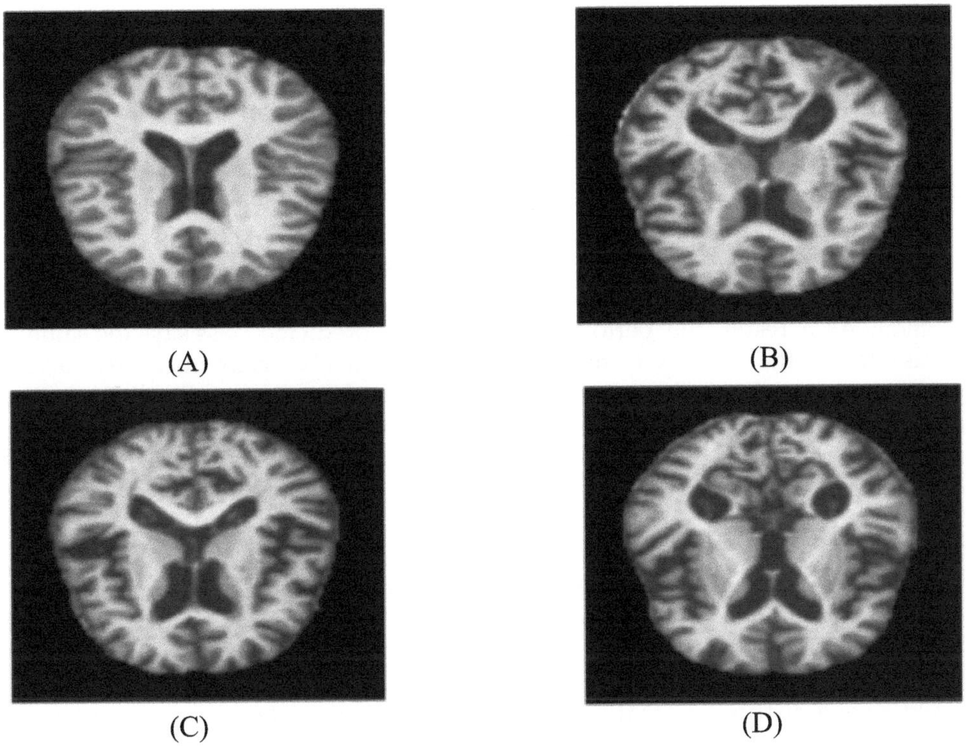

FIGURE 14.2 MRI scans for different subjects: (A) nondemented, (B) mild demented, (C) very mild demented, and (D) moderate demented. *MRI*, Magnetic resonance imaging.

TABLE 14.1 Class distribution of magnetic resonance imaging dataset.

	Nondemented	Very mild demented	Mild demented	Moderate demented	Total
Train	2048	1434	574	41	4097
Val	512	358	143	11	1024
Test	640	448	179	12	1279

includes total of 5480 images. We divided this dataset into training, validation, and test sets. The training and test sets are provided by the creator of datasets themselves. The training set is further split into training and validation set, 80% and 20%, respectively. The training, validation, and testing set contains around 4097, 1024, and 1279 images, respectively. Table 14.1 gives a description of class distribution of the four classes in the each of the three datasets.

14.4.2 Performance evaluation measures

The performance measurements for the training set and the test set cannot be presumed to be identical. Although the training set contains the majority of the instances, the test set is intended to be more realistic. A lack of data is a significant challenge for ConvNets. In this example, the methodologies utilized to evaluate the performance measurements are still debatable. In the case of the training set, the classifier is fine-tuned to produce the best performance measurements. The most essential thing to remember when training is that no instances from the test set must be included in the classifier construction. As a result, the performance measurements from the test set may be expected to be similar to the performance measures from the control set. A classifier assigns a categorization to an image. It is believed to be successful if the category matches the specified category. If there is a discrepancy, it is presumed to be a mistake. The majority of performance metrics are dependent on a classifier's error rate. A large number of examples are included in the training set. To have the best performance measurements, the validation dataset should have a distribution that is nearly identical to the test set. The validation set should be used to fine-tune parameters, while the test set should be used to determine the final values of performance measures [57].

A confusion matrix is a table, which is often employed to illustrate the performance of a classification model on a set of test data for which the true values are known. All the metrics we have used to evaluate the performance of the model can be assessed using the confusion matrix (see Fig. 14.3).

The observations that are accurately predicted and hence represented in green are true positives (TP) and true negatives (TN). We wish to reduce false positives (FP) and false negatives (FN), thus they are presented in red.

TP are correctly predicted positive values, demonstrating that the value of the actual class and the value of the projected class are both yes. For instance, if the actual class value shows that this patient survived, and the anticipated class also suggests the same. TN are correctly predicted negative values, implying that the value of the actual class is zero and the value of the predicted class is zero as well. For instance, if the real class states the patient did not survive and the predicted class says the same.

FP and FN occur once your actual class is different from the predicted class. FP are when the actual class is not the same as the predicted class. For instance, if the actual class implies that this patient did not survive, but the forecast class implies that this patient will. FN are situations in which the real class is yes, but the expected class is no. For instance, if the patient's actual class value indicates that he or she survived, while the predicted class value indicates that the person would die.

Accuracy is the most straightforward performance metric, because it is simply the ratio of correctly predicted observations to total observations.

$$\text{Accuracy} = (TP + TN)/(TP + FP + FN + TN)$$

Precision is the ratio of correctly predicted positive observations to the total predicted positive observations.

$$\text{Precision} = TP/TP + FP$$

Recall is the ratio of correctly predicted positive observations to all observations in actual class.

$$\text{Recall} = TP/TP + FN$$

F1-score is the weighted average of precision and recall. Hence, this score takes both FP and FN into account. Naturally, it is not as simple

	Predicted class		
Actual Class		Class = Yes	Class = No
	Class = Yes	True Positive	False Negative
	Class = No	False Positive	True Negative

FIGURE 14.3 Example of a confusion matrix.

to understand as accuracy, but F1 is generally more effective than accuracy, particularly if you have an uneven class distribution.

$$\text{F1} - \text{score} = 2 \times (\text{Recall} \times \text{Precision}) / (\text{Recall} + \text{Precision})$$

The in-built metric class of scikit learn library is used in order to implement all these functions. The code snippet for the same is given below:

```
train_accuracy = np.round(accuracy_score(y_train, y_pred_train),4)*100
train_precision = np.round(precision_score(y_train, y_pred_train,average='weighted'),4)
train_recall = np.round(recall_score(y_train, y_pred_train, average='weighted'),4)
train_F1 = np.round(f1_score(y_train, y_pred_train, average='weighted'),4)
train_confusion_matrix = confusion_matrix(y_train,y_pred_train)

val_accuracy = np.round(accuracy_score(y_val, y_pred_val),4)*100
val_precision = np.round(precision_score(y_val, y_pred_val, average='weighted'),4)
val_recall = np.round(recall_score(y_val, y_pred_val, average='weighted'),4)
val_F1 = np.round(f1_score(y_val, y_pred_val, average='weighted'),4)
val_confusion_matrix = confusion_matrix(y_val,y_pred_val)

test_accuracy = np.round(accuracy_score(y_test, y_pred_test),4)*100
test_precision = np.round(precision_score(y_test, y_pred_test, average='weighted'),4)
test_recall = np.round(recall_score(y_test, y_pred_test, average='weighted'),4)
test_F1 = np.round(f1_score(y_test, y_pred_test, average='weighted'),4)
test_confusion_matrix = confusion_matrix(y_test,y_pred_test)


print()
print('------------------------ Train Set Metrics---------------------')
print()
print("accuracy : {}%".format(train_accuracy))
print("F1_score : {}".format(train_F1))
print("Recall : {}".format(train_recall))
print("Precision : {}".format(train_precision))
print("Confusion Matrix :\n {}".format(train_confusion_matrix))
```

```
print()
print('----------------------- Validation Set Metrics----------------')
print()
print("accuracy : {}%".format(val_accuracy))
print("F1_score : {}".format(val_F1))
print("Recall : {}".format(val_recall))
print("Precision : {}".format(val_precision))
print("Confusion Matrix :\n {}".format(val_confusion_matrix))

print()
print('----------------------- Test Set Metrics----------------------')
print()
print("accuracy : {}%".format(test_accuracy))
print("F1_score : {}".format(test_F1))
print("Recall : {}".format(test_recall))
print("Precision : {}".format(test_precision))
print("Confusion Matrix : {}".format(test_confusion_matrix))
```

14.4.3 Experimental results

We performed feature extraction with different pretrained models and fed them to ML models such as ANN, KNN, SVM, RF, AdaBoost, and XGBoost classifier.

VGG16

Model	Training accuracy	Validation accuracy	Test accuracy	F1-score	Recall	Precision
ANN	74.44	75.05	75.02	0.0164	0.0086	0.5238
KNN	60.35	42.19	42.73	0.4099	0.4273	0.4036
SVM	50	50	50	0.3333	0.5	0.25
Random Forest	100	43.65	46.02	0.4321	0.4602	0.4219
AdaBoost	49.58	49.51	49.45	0.3866	0.4945	0.3925
XGBoost	91.89	47.56	46.64	0.4257	0.4664	0.4201

RF and XGBoost have higher train accuracy as compared to others, but at the same time, these classifiers seem to be overfitting. ANN on the other hand shows greater test accuracy than all of them, which in turn is deemed to be good for AD classification.

VGG19

Model	Training accuracy	Validation accuracy	Test accuracy	F1-score	Recall	Precision
ANN	75.82	76.61	77.56	0.4869	0.4292	0.5677
KNN	63.21	45.8	44.69	0.4273	0.4469	0.4246
SVM	51.86	51.56	51.09	0.3758	0.5109	0.4307
Random Forest	100	49.51	46.09	0.431	0.4609	0.4183
AdaBoost	50.56	51.37	50	0.3952	0.5	0.4534
XGBoost	89.97	49.41	48.52	0.4509	0.4852	0.4457

Here too like in VGG16 we find out that RF and XGBoost have a very high training accuracy since they beat all the models in train accuracy but at the same time these classifiers can be seen overfitting to a great extent. Here unlike previous model all classifiers are overfitting to a certain extent.

ResNet 50

Model	Training accuracy	Validation accuracy	Test accuracy	F1-score	Recall	Precision
ANN	75.82	76.61	77.56	0.4869	0.4292	0.5677
KNN	63.21	45.8	44.69	0.4273	0.4469	0.4246
SVM	51.8599	51.559	51.09	0.3758	0.5109	0.4307
Random Forest	100	49.51	46.0899	0.431	0.4609	0.4183
AdaBoost	50.56	51.37	50	0.3952	0.5	0.4534
XGBoost	89.97	49.41	48.52	0.4509	0.4852	0.4457

RF and XGBoost beat all others like usual in the training accuracy but are seen to be overfitting again. Here ANN can be considered as a good classifier with less overfitting.

ResNet101

Model	Training accuracy	Validation accuracy	Test accuracy	F1-score	Recall	Precision
ANN	76.38	77.49	76.11	0.2971	0.2048	0.561
KNN	64.62	46.88	48.59	0.4658	0.4859	0.4603
SVM	50	50	50	0.333	0.5	0.25
Random Forest	100	48.73	48.67	0.4664	0.4867	0.4598
AdaBoost	51.51	50.88	48.91	0.4424	0.4891	0.4437
XGBoost	90.23	49.32	48.05	0.4573	0.4805	0.4508

RF and XGBoost have a very high train accuracy as they beat all classifiers once again. None of the classifiers seem to give good test results.

MobileNetV2

Model	Training accuracy	Validation accuracy	Test accuracy	F1-score	Recall	Precision
ANN	75.53	77.05	74.78	0.374	0.3041	0.493
KNN	61.1299	41.41	42.97	0.407	0.4297	0.4003
SVM	50	50	50	0.3333	0.5	0.25
Random Forest	100	41.99	44.14	0.4135	0.4414	0.4001
AdaBoost	49.02	47.36	48.75	0.3889	0.4875	0.3839
XGBoost	90.11	43.36	46.72	0.4297	0.4672	0.4285

All classifiers are overfitting in this case, except ANN, which shows consistent accuracy of around ~75%.

MobileNet

Model	Training accuracy	Validation accuracy	Test accuracy	F1-score	Recall	Precision
ANN	75.67	76.05	75.06	0.1112	0.0641	0.5093
KNN	60.72	42.97	42.89	0.4028	0.4289	0.3957
SVM	50	50	50	0.3333	0.5	0.25
Random Forest	100	44.04	44.61	0.4098	0.4461	0.3963
AdaBoost	49.27	48.93	49.22	0.3872	0.4922	0.4242
XGBoost	89.33	46.29	47.27	0.4317	0.4727	0.4297

All of the classifiers are overfitting in this case, except ANN, which shows consistent accuracy of around ~75%. Here ANN can be considered a good classifier for AD classification.

InceptionV3

Model	Training accuracy	Validation accuracy	Test accuracy	F1-score	Recall	Precision
ANN	74.76	75.07	75.04	0.003	0.0016	1
KNN	60.77	41.21	42.03	0.3978	0.4203	0.3882
SVM	50	50	50	0.3333	0.5	0.25
Random Forest	100	45.61	42.5	0.3987	0.425	0.3859
AdaBoost	49.41	49.41	49.06	0.397	0.4906	0.4176
XGBoost	91.55	45.51	43.83	0.4073	0.4383	0.3998

Here, just like the previous cases, the classifiers are overfitting, apart from ANN, which shows consistent accuracy of around ~75%. Here ANN can be considered a good classifier for AD classification InceptionV3.

InceptionResNetV2

Model	Training accuracy	Validation accuracy	Test accuracy	F1-score	Recall	Precision
ANN	75	75	75	0.6229	0.4938	0.8935
KNN	62.23	46.39	44.379	0.4257	0.4438	0.4226
SVM	50.83	50.88	50.16	0.3621	0.5016	0.3894
Random Forest	100	46.88	45.39	0.4278	0.4539	0.4151
AdaBoost	47.49	44.73	43.44	0.3927	0.4344	0.3613
XGBoost	91.039	47.85	45.78	0.4259	0.4578	0.4117

In this case, XGBoost and RF achieve good accuracy on training set, but both of them perform poorly on test sets. ANN is a better classifier, as we can see from previous results as well as from this.

DenseNet169

Model	Training accuracy	Validation accuracy	Test accuracy	F1-score	Recall	Precision
ANN	75.11	76.2	75.02	0.5004	0.5004	0.5004
KNN	60.99	41.21	44.3	0.4178	0.443	0.4123
SVM	50	50	50	0.3333	0.5	0.25
Random Forest	100	42.58	44.92	0.4123	0.4492	0.401
AdaBoost	50.1499	49.02	48.83	0.3871	0.4883	0.4168
XGBoost	88.4	43.55	44.769	0.4147	0.4477	0.4071

Here, just like the previous cases, the classifiers are overfitting, apart from ANN, which shows consistent accuracy of around ~75%. Here ANN can be considered a good classifier for AD classification with DenseNet169.

DenseNet121

Model	Training accuracy	Validation accuracy	Test accuracy	F1-score	Recall	Precision
ANN	75	75	75	0.6609	0.5023	0.9853
KNN	62.5	42.77	42.97	0.4111	0.4297	0.4025
SVM	50.0199	50.29	50.16	0.3398	0.5016	0.4258
Random Forest	100	46.19	44.84	0.415	0.4484	0.4076
AdaBoost	50.27	49.8	50.08	0.3686	0.5008	0.3889
XGBoost	91.63	46.19	45.39	0.4142	0.4539	0.4021

Here, just like the previous cases, the classifiers are overfitting, apart from ANN, which shows consistent accuracy of around ~75%. Here ANN can be considered a good classifier for AD classification with DenseNet121 as well.

Xception

Model	Training accuracy	Validation accuracy	Test accuracy	F1-score	Recall	Precision
ANN	77.94	78.05	76.06	0.6667	0.3542	0.5317
KNN	61.839	45.019	44.92	0.4254	0.4492	0.42
SVM	50.149	50.29	50.16	0.339	0.5016	0.368
Random Forest	100	46.68	47.81	0.4494	0.4781	0.43485
AdaBoost	51.339	51.559	51.41	0.4318	0.5141	0.4886
XGBoost	91.02	49.32	47.89	0.4392	0.4789	0.4311

With Xception deep feature extraction, ANN achieved good results as well.

14.5 Discussion

AI methods have been studied in the neuroimaging research field due to their advantages over typical diagnostic procedures that use mass-univariate statistical methodologies. In contrast, AI models can differentiate effects at the subject level, whereas mass-univariate statistical approaches employ changes at the category level. DNN is being more widely employed in neuroimaging, resulting in technical improvements in computer vision by outperforming other state-of-the-art detection approaches [58,59]. DNN differs from typical AI approaches in that it learns raw data features without relying on extraction and selection characteristics. DNN also utilizes a nonlinear transformation hierarchy that is best suited for recognizing distributed, complex, and dynamic patterns [41,42].

The existing evidence suggests that DNN might be effective for the AD diagnosis, and that the binary classification accuracy for discrimination between control participants and patients was good (above 95%) [60–66]. As previously stated, the absence of suitable ways to combat DNN classifier overfitting might explain these large findings. DNN is considered as a dynamic technique that employs a large range of hyperparameters and integrates multiple forms inside the same model [41,42]. CNNs are a type of ANN that is activated by the visual cortex of the human brain. CNNs have shattered computer vision records in a variety of competitions, indicating a very promising method [58]. In terms of detecting AD, CNNs have achieved the most impressive results. Local networking and weight sharing are two key aspects of CNNs, which result in a drastically decreased number of weights and computationally viable network. The CNNs are not only utilized for discrimination between AD and MCI patients, but the study has also had a high level of effectiveness so far. Structural MRI [60,64], CT imaging [67], and resting-state fMRI [65] were all used to achieve high efficiency. Hosseini-Asl et al. [64] took a different strategy, using a pretrained CNN model to the Alzheimer's dementia dataset, which they then altered and used a different test dataset. The results for HC versus AD, HC versus MCI, AD versus MCI, and HC versus AD versus MCI were all very positive. These trials, together with CNN-based models reported significant performance in other research domains, demonstrate CNNs as effective method for detecting AD [42]. Furthermore, these encouraging findings show how AI could link real-world therapeutic experience and neuroimaging data [41].

It could be interesting to map the consecutive layers weights back to original neuroimage to determine the regions that have the most influence on the AD diagnosis due to a range of nonlinearities [63]. This information, on the other hand, might be beneficial in clinical neuroimaging that is used to diagnose problems. Importantly, a predictor with high performance can categorize people using irrelevant variables rather than therapeutically important data. Another issue is that an accurate model without understanding of underlying neurofunctional or neuroanatomical variations may be insignificant for medical purposes. To solve this problem, deconvolution and input modification techniques might be applied. For visualization, the input adjustment approach is used, which comprises methodological input modification, the activation functions of artificial neurons to the next layers, and the computation of the network's output alterations. The occlusion method [68], which includes the section of the input image that impacts the output category's likelihood, is an example of this approach. The deconvolution approach, on the other hand, aims to analyze the impact of input characteristics to the output. Convolution also entails specifying an efficient activation function in the output neuron, followed by assessment of each neuron's influence in the appropriate layers [69] and DeconvNet [41,68]. Furthermore, in comparison to other deep learning models, TL approaches are quite successful at detecting AD.

14.6 Conclusion

In image-based diagnostics and disease diagnosis, AI technologies are becoming increasingly successful. To reach their full potential, several technological and pragmatic solutions are required. This chapter uses MRI to demonstrate the application of AI approaches in the detection of AD. In addition, AI technologies for the identification of AD that result in serious health problems were examined. When it comes to comparing AI algorithms, TL models have outperformed other deep learning models for detecting AD. As a result, numerous AI approaches for detecting AD are investigated. It is reported that TL approaches have obtained the best results in detecting AD.

AI has shown to be effective in diagnosing AD and has the potential to make substantial advancements in neurological disorders. However, various improvements are required to fully comprehend AI's capabilities in the AD diagnosis. Because AI methods are sophisticated, a dataset with significantly larger samples is required at first, rather than moderate or small sample sizes. To accomplish so, multicenter partnerships are necessary, in which data is collected across sites using the same recording circumstances and scanning methodologies. Then, by combining diverse AI techniques, it is feasible to create significant AI improvements. Finally, the application of AI to imagine consistent scores might be applied for prospective medical employment research [42]. Finally, the ability of AI systems to learn complex features through nonlinear transformations might lead to promising findings in the detection of AD. Although considerable roadblocks remain, the findings presented here give a fundamental example of AI algorithms' potential relevance in the development of future AD prognostic and diagnostic markers [41].

In this chapter, we showed some findings from brain MRI scans utilizing several AI models for identifying AD, which is the fifth largest cause of mortality among Americans aged 65 and over [70]. Initially, we experimented with feature extraction by using various different TL models such as VGG16 and ResNet 50 among others. We further tried supervised ML models on these extracted features as comparison of their performance on direct images and extracted features. We finally conclude that feature extraction along with gradient boosting methods such as XGBoost and AdaBoost classifier gives promising results for AD detection.

References

[1] F. Richard, J. Zeisel, K. Bennett, Design, dignity, dementia: dementia-related design and the built environment, World Alzheimer Rep. 2020 (2020).
[2] R. Brookmeyer, E. Johnson, K. Ziegler-Graham, H.M. Arrighi, Forecasting the global burden of Alzheimer's disease, Alzheimers Dement. 3 (3) (2007) 186–191. Available from: https://doi.org/10.1016/j.jalz.2007.04.381.
[3] G.A. Malik, N.P. Robertson, Treatments in Alzheimer's disease, J. Neurol. 264 (2) (Feb. 2017) 416–418. Available from: https://doi.org/10.1007/s00415-017-8395-1.
[4] The Lancet, The three stages of Alzheimer's disease, Lancet 377 (9776) (2011) 1465. Available from: https://doi.org/10.1016/S0140-6736(11)60582-5.
[5] J. Neugroschl, S. Wang, Alzheimer's disease: diagnosis and treatment across the spectrum of disease severity, Mt. Sinai J. Med. J. Transl. Pers. Med. 78 (4) (2011) 596–612. Available from: https://doi.org/10.1002/msj.20279.
[6] E. Merlo Pich, et al., Imaging as a biomarker in drug discovery for Alzheimer's disease: is MRI a suitable technology? Alzheimers Res. Ther. 6 (4) (2014) 51. Available from: https://doi.org/10.1186/alzrt276.
[7] M.N. Braskie, A.W. Toga, P.M. Thompson, Recent advances in imaging Alzheimer's disease, J. Alzheimers Dis. 33 (s1) (2012) S313–S327. Available from: https://doi.org/10.3233/JAD-2012-129016.
[8] R. Ossenkoppele, et al., Prevalence of amyloid PET positivity in dementia syndromes: a *meta*-analysis, JAMA 313 (19) (2015) 1939. Available from: https://doi.org/10.1001/jama.2015.4669.
[9] J.F. Norfray, J.M. Provenzale, Alzheimer's disease: neuropathologic findings and recent advances in imaging, Am. J. Roentgenol. 182 (1) (2004) 3–13. Available from: https://doi.org/10.2214/ajr.182.1.1820003.
[10] L.J. Whalley, Spatial distribution and secular trends in the epidemiology of Alzheimer's disease, Neuroimaging Clin. N. Am. 22 (1) (2012) 1–10. Available from: https://doi.org/10.1016/j.nic.2011.11.002.

[11] M. Tanveer, A.H. Rashid, M.A. Ganaie, M. Reza, I. Razzak, K.-L. Hua, Classification of Alzheimer's disease using ensemble of deep neural networks trained through transfer learning, IEEE J. Biomed. Health Inf. 26 (4) (2022) 1453–1463. Available from: https://doi.org/10.1109/JBHI.2021.3083274.

[12] S. Dwivedi, T. Goel, M. Tanveer, R. Murugan, R. Sharma, Multi-modal fusion based deep learning network for effective diagnosis of Alzheimers disease, IEEE Multimed. (2022) 1. Available from: https://doi.org/10.1109/MMUL.2022.3156471.

[13] U.R. Acharya, et al., Automated detection of Alzheimer's disease using brain MRI images – a study with various feature extraction techniques, J. Med. Syst. 43 (9) (2019) 302. Available from: https://doi.org/10.1007/s10916-019-1428-9.

[14] N.A. Mathew, R.S. Vivek, P.R. Anurenjan, Early Diagnosis of Alzheimer's Disease from MRI Images Using PNN, in: 2018 International CET Conference on Control, Communication, and Computing (IC4), Thiruvananthapuram, Jul. 2018, pp. 161–164. Available from: https://doi.org/10.1109/CETIC4.2018.8530910.

[15] R. Varatharajan, G. Manogaran, M.K. Priyan, R. Sundarasekar, Wearable sensor devices for early detection of Alzheimer disease using dynamic time warping algorithm, Clust. Comput. 21 (1) (2018) 681–690. Available from: https://doi.org/10.1007/s10586-017-0977-2.

[16] B.C. Dickerson, et al., Alzheimer-signature MRI biomarker predicts AD dementia in cognitively normal adults, Neurology 76 (16) (2011) 1395–1402. Available from: https://doi.org/10.1212/WNL.0b013e3182166e96.

[17] C. Patil et al., Using image processing on MRI scans, in: 2015 IEEE International Conference on Signal Processing, Informatics, Communication and Energy Systems (SPICES), Kozhikode, India, Feb. 2015, pp. 1–5. Available from: https://doi.org/10.1109/SPICES.2015.7091517.

[18] A. Kaur, P. Kaur, A comparative study of various exudate segmentation techniques for diagnosis of diabetic retinopathy, Int. J. Curr. Eng. Technol. 46 (2016) 142–146.

[19] Y. Zhao, et al., In vivo detection of microstructural correlates of brain pathology in preclinical and early Alzheimer disease with magnetic resonance imaging, NeuroImage 148 (2017) 296–304. Available from: https://doi.org/10.1016/j.neuroimage.2016.12.026.

[20] Z. Sankari, H. Adeli, Probabilistic neural networks for diagnosis of Alzheimer's disease using conventional and wavelet coherence, J. Neurosci. Methods 197 (1) (2011) 165–170. Available from: https://doi.org/10.1016/j.jneumeth.2011.01.027.

[21] C. Plant, et al., Automated detection of brain atrophy patterns based on MRI for the prediction of Alzheimer's disease, NeuroImage 50 (1) (2010) 162–174. Available from: https://doi.org/10.1016/j.neuroimage.2009.11.046.

[22] Y. Zhang, et al., Multivariate approach for Alzheimer's disease detection using stationary wavelet entropy and predator-prey particle swarm optimization, J. Alzheimers Dis. 65 (3) (2018) 855–869. Available from: https://doi.org/10.3233/JAD-170069.

[23] S.-H. Wang, et al., Single slice based detection for Alzheimer's disease via wavelet entropy and multilayer perceptron trained by biogeography-based optimization, Multimed. Tools Appl. 77 (9) (2018) 10393–10417. Available from: https://doi.org/10.1007/s11042-016-4222-4.

[24] Y. Zhang, S. Wang, Z. Dong, Classification of Alzheimer disease based on structural magnetic resonance imaging by kernel support vector machine decision tree, Prog. Electromagn. Res. 144 (2014) 171–184. Available from: https://doi.org/10.2528/PIER13121310.

[25] Y. Zhang, S. Wang, Detection of Alzheimer's disease by displacement field and machine learning, PeerJ 3 (2015) e1251. Available from: https://doi.org/10.7717/peerj.1251.

[26] E.-S.A. El-Dahshan, T. Hosny, A.-B.M. Salem, Hybrid intelligent techniques for MRI brain images classification, Digit. Signal. Process. 20 (2) (2010) 433–441. Available from: https://doi.org/10.1016/j.dsp.2009.07.002.

[27] S.-H. Wang, et al., Alzheimer's disease detection by pseudo Zernike moment and linear regression classification, CNS Neurol. Disord. - Drug. Targets 16 (1) (2017) 11–15. Available from: https://doi.org/10.2174/1871527315666161111123024.

[28] H.T. Gorji, J. Haddadnia, A novel method for early diagnosis of Alzheimer's disease based on pseudo Zernike moment from structural MRI, Neuroscience 305 (2015) 361–371. Available from: https://doi.org/10.1016/j.neuroscience.2015.08.013.

[29] S. Wang, Y. Zhang, G. Liu, P. Phillips, T.-F. Yuan, Detection of Alzheimer's disease by three-dimensional displacement field estimation in structural magnetic resonance imaging, J. Alzheimers Dis. 50 (1) (2015) 233–248. Available from: https://doi.org/10.3233/JAD-150848.

[30] Y. Zhang, et al., Detection of subjects and brain regions related to Alzheimer's disease using 3D MRI scans based on eigenbrain and machine learning, Front. Comput. Neurosci. 9 (2015). Available from: https://doi.org/10.3389/fncom.2015.00066.

[31] Y. Zhang, S. Wang, P. Sun, P. Phillips, Pathological brain detection based on wavelet entropy and Hu moment invariants, Biomed. Mater. Eng. 26 (s1) (2015) S1283–S1290. Available from: https://doi.org/10.3233/BME-151426.

[32] K. Hett, V.-T. Ta, J.V. Manjón, P. Coupé, Adaptive fusion of texture-based grading for Alzheimer's disease classification, Comput. Med. Imaging Graph. 70 (2018) 8–16. Available from: https://doi.org/10.1016/j.compmedimag.2018.08.002.

[33] X.W. Gao, R. Hui, Z. Tian, Classification of CT brain images based on deep learning networks, Comput. Methods Prog. Biomed. 138 (2017) 49–56. Available from: https://doi.org/10.1016/j.cmpb.2016.10.007.

[34] W. Ayadi, W. Elhamzi, I. Charfi, M. Atri, A hybrid feature extraction approach for brain MRI classification based on bag-of-words, Biomed. Signal. Process. Control. 48 (2019) 144–152. Available from: https://doi.org/10.1016/j.bspc.2018.10.010.

[35] C.C. Aggarwal, Neural Networks and Deep Learning: A Textbook, Springer, Cham, Switzerland, 2018. Available from: https://link.springer.com/book/10.1007/978-3-319-94463-0.

[36] S.J. Pan, Q. Yang, A survey on transfer learning, IEEE Trans. Knowl. Data Eng. 22 (10) (Oct. 2010) 1345–1359. Available from: https://doi.org/10.1109/TKDE.2009.191.

[37] K. Simonyan, A. Zisserman, Very deep convolutional networks for large-scale image recognition. arXiv preprint arXiv:1409.1556, https://arxiv.org/abs/1409.1556, Sep. 2014.

[38] K. He, X. Zhang, S. Ren, J. Sun, Deep residual learning for image recognition, ArXiv151203385 Cs, http://arxiv.org/abs/1512.03385, Dec. 2015 (accessed 01.03.22).

[39] Y. Bengio, Learning deep architectures for AI, Found. Trends® Mach. Learn. 2 (1) (2009) 1–127. Available from: https://doi.org/10.1561/2200000006.

[40] E. Alpaydin, Introduction to Machine Learning, Third edition, The MIT Press, Cambridge, Massachusetts, 2014.

[41] A. Subasi, Use of artificial intelligence in Alzheimer's disease detection, Artificial Intelligence in Precision Health, Elsevier, 2020, pp. 257–278. Available from: https://doi.org/10.1016/B978-0-12-817133-2.00011-2.

[42] S. Vieira, W.H.L. Pinaya, A. Mechelli, Using deep learning to investigate the neuroimaging correlates of psychiatric and neurological disorders: methods and applications, Neurosci. Biobehav. Rev. 74 (2017) 58–75. Available from: https://doi.org/10.1016/j.neubiorev.2017.01.002.

[43] M. Deepika Nair, M.S. Sinta, M. Vidya, A study on various deep learning algorithms to diagnose Alzheimer's disease, in: D. Pandian, X. Fernando, Z. Baig, F. Shi, (Eds.), Proc. International Conference on ISMAC in Computational Vision and Bio-Engineering 2018 (ISMAC-CVB), vol. 30, Cham: Springer International Publishing, 2019, pp. 1705–1710. Available from: https://doi.org/10.1007/978-3-030-00665-5_157.

[44] S. Aruchamy, V. Mounya, A. Verma, Alzheimer's disease classification in brain MRI using modified kNN algorithm, in: 2020 IEEE International Symposium on Sustainable Energy, Signal Processing and Cyber Security (iSSSC), Gunupur Odisha, India, Dec. 2020, pp. 1–6. Available from: https://doi.org/10.1109/iSSSC50941.2020.9358867.

[45] A.J. Dinu, R. Ganesan, Early detection of Alzheimer's disease using predictive k-NN instance based approach and T-test method, Int. J. Adv. Trends Comput. Sci. Eng. 8 (1.4) (Sep. 2019) 29–37. Available from: https://doi.org/10.30534/ijatcse/2019/0581.42019.

[46] A.B. Rabeh, F. Benzarti, H. Amiri, Diagnosis of Alzheimer diseases in early step using SVM (support vector machine), in: 2016 13th International Conference on Computer Graphics, Imaging and Visualization (CGiV), Beni Mellal, Morocco, Mar. 2016, pp. 364–367. Available from: https://doi.org/10.1109/CGiV.2016.76.

[47] S. Alam, G.-R. Kwon, J.-I. Kim, C.-S. Park, Twin SVM-based classification of Alzheimer's disease using complex dual-tree wavelet principal coefficients and LDA, J. Healthc. Eng. 2017 (2017) 8750506. Available from: https://doi.org/10.1155/2017/8750506.

[48] P.J. Moore, T.J. Lyons, J. Gallacher, Alzheimer's Disease Neuroimaging Initiative, Random forest prediction of Alzheimer's disease using pairwise selection from time series data, PLoS One 14 (2) (2019) e0211558. Available from: https://doi.org/10.1371/journal.pone.0211558.

[49] A. Sarica, A. Cerasa, A. Quattrone, Random Forest algorithm for the classification of neuroimaging data in Alzheimer's disease: a systematic review, Front. Aging Neurosci. 9 (329) (2017). Available from: https://doi.org/10.3389/fnagi.2017.00329.

[50] A.V. Lebedev, et al., Random Forest ensembles for detection and prediction of Alzheimer's disease with a good between-cohort robustness, NeuroImage Clin. 6 (2014) 115–125. Available from: https://doi.org/10.1016/j.nicl.2014.08.023.

[51] M.S. Ali, Md. K. Islam, J. Haque, A.A. Das, D.S. Duranta, M.A. Islam, Alzheimer's disease detection using m-Random Forest algorithm with optimum features extraction, in: 2021 1st International Conference on Artificial Intelligence and Data Analytics (CAIDA), Riyadh, Saudi Arabia, Apr. 2021, pp. 1–6. Available from: https://doi.org/10.1109/CAIDA51941.2021.9425212.

[52] A. Savio, M. García-Sebastián, M. Graña, J. Villanúa, Results of an Adaboost approach on Alzheimer's disease detection on MRI, in: J. Mira, J.M. Ferrández, J.R. Álvarez, F. de la Paz, F.J. Toledo, (Eds.). Bioinspired applications in artificial and natural computation, vol. 5602, Berlin, Heidelberg: Springer Berlin Heidelberg, 2009, pp. 114–123. Available from: https://doi.org/10.1007/978-3-642-02267-8_13.

[53] J.H. Morra, Z. Tu, L.G. Apostolova, A.E. Green, A.W. Toga, P.M. Thompson, Comparison of AdaBoost and

support vector machines for detecting Alzheimer's disease through automated hippocampal segmentation, IEEE Trans. Med. Imaging 29 (1) (2010) 30–43. Available from: https://doi.org/10.1109/TMI.2009.2021941.

[54] G. Battineni, et al., Improved Alzheimer's disease detection by MRI using multimodal machine learning algorithms, Diagnostics 11 (11) (2021) 2103. Available from: https://doi.org/10.3390/diagnostics11112103.

[55] L. Akter, Ferdib-Al-Islam, Dementia identification for diagnosing Alzheimer's disease using XGBoost algorithm, in: 2021 International Conference on Information and Communication Technology for Sustainable Development (ICICT4SD), Dhaka, Bangladesh, Feb. 2021, pp. 205–209. Available from: https://doi.org/10.1109/ICICT4SD50815.2021.9396777.

[56] Y. Shmulev, M. Belyaev, Predicting conversion of mild cognitive impairments to Alzheimer's disease and exploring impact of neuroimaging, ArXiv180711228 Cs Stat, http://arxiv.org/abs/1807.11228, Jul. 2018 (accessed 06.03.22).

[57] I.H. Witten, E. Frank, Data Mining: Practical Machine Learning Tools and Techniques with Java Implementations, Morgan Kaufmann, San Francisco, California, 2000.

[58] A. Krizhevsky, I. Sutskever, G.E. Hinton, ImageNet classification with deep convolutional neural networks, Commun. ACM 60 (6) (May 2017) 84–90. Available from: https://doi.org/10.1145/3065386.

[59] Q.V. Le et al., Building high-level features using large scale unsupervised learning, ArXiv11126209 Cs, http://arxiv.org/abs/1112.6209, Jul. 2012 (accessed 06.03.22).

[60] A. Payan, G. Montana, Predicting Alzheimer's disease: a neuroimaging study with 3D convolutional neural networks, ArXiv150202506 Cs Stat. [Online]. Available: http://arxiv.org/abs/1502.02506, Feb. 2015 (accessed 09.01.21).

[61] H.-I. Suk, D. Shen, Deep learning-based feature representation for AD/MCI classification, in: C. Salinesi, M. C. Norrie, Ó. Pastor (Eds.), Advanced Information Systems Engineering, vol. 7908, Springer Berlin Heidelberg, Berlin, Heidelberg, 2013, pp. 583–590. Available from: 10.1007/978-3-642-40763-5_72.

[62] H.-I. Suk, S.-W. Lee, D. Shen, Hierarchical feature representation and multimodal fusion with deep learning for AD/MCI diagnosis, NeuroImage 101 (2014) 569–582. Available from: https://doi.org/10.1016/j.neuroimage.2014.06.077.

[63] H.-I. Suk, S.-W. Lee, D. Shen, The Alzheimer's disease neuroimaging initiative, Latent feature representation with stacked auto-encoder for AD/MCI diagnosis, Brain Struct. Funct. 220 (2) (Mar. 2015) 841–859. Available from: https://doi.org/10.1007/s00429-013-0687-3.

[64] E. Hosseini-Asl, G. Gimel'farb, A. El-Baz, Alzheimer's Disease diagnostics by a deeply supervised adaptable 3D convolutional network, ArXiv160700556 Cs Q-Bio Stat, http://arxiv.org/abs/1607.00556, Jul. 2016 (accessed 09.01.21).

[65] S. Sarraf, G. Tofighi, Classification of Alzheimer's disease using fMRI data and deep learning convolutional neural networks, ArXiv160308631 Cs,. [Online]. Available: http://arxiv.org/abs/1603.08631, Mar. 2016, (accessed 06.03.22).

[66] H.-I. Suk, S.-W. Lee, D. Shen, The Alzheimer's disease neuroimaging initiative, Deep sparse multi-task learning for feature selection in Alzheimer's disease diagnosis, Brain Struct. Funct. 221 (5) (Jun. 2016) 2569–2587. Available from: https://doi.org/10.1007/s00429-015-1059-y.

[67] X.W. Gao, R. Hui, A deep learning based approach to classification of CT brain images, in: 2016 SAI Computing Conference (SAI), London, United Kingdom, Jul. 2016, pp. 28–31. Available from: https://doi.org/10.1109/SAI.2016.7555958.

[68] M.D. Zeiler, R. Fergus, Visualizing and understanding convolutional networks, ArXiv13112901 Cs, http://arxiv.org/abs/1311.2901, Nov. 2013 (accessed 06.03.22).

[69] J.T. Springenberg, A. Dosovitskiy, T. Brox, M. Riedmiller, Striving for simplicity: the all Convolutional Net, ArXiv14126806 Cs, http://arxiv.org/abs/1412.6806, Apr. 2015, (accessed 06.03.22).

[70] W. Thies, L. Bleiler, 2012 Alzheimer's disease facts and figures Alzheimer's Association, Alzheimers Dement. 8 (2) (2012) 131–168. Available from: https://doi.org/10.1016/j.jalz.2012.02.001.

Index

Note: Page numbers followed by "*f*" and "*t*" refer to figures and tables respectively.

D

FSC
www.fsc.org
MIX
Papier | Fördert
gute Waldnutzung
FSC® C083411

Kolibri 360 GmbH
Ferdinand-Jühlke-Straße 7
99095 Erfurt, Deutschland
produktsicherheit@kolibri360.de